"A rigorous, in-depth analysis of successful ...
has long been an under-researched, often o...
study. Yet the critical importance collabo...
requires urgent attention. In this groundbreaking volume, Halewood, ...
and Louafi have assembled an international group of leading scholars and practitioners to systematically guide us through the complex terrain of crop genetic resources and agricultural biodiversity as global commons. The thoroughness of the analysis along with the lessons learned from practical applications will serve as indispensable tools for students of all types of global-commons resources."

—*Charlotte Hess, co-editor with Elinor Ostrom of 'Understanding Knowledge as a Commons' and Associate Dean for Research, Collections, and Scholarly Communication for Syracuse University Library*

"The preservation and enhancement of agrobiodiversity is of huge importance in a world that shall witness more disruptive climate shocks in the future, and in which food-deficit regions shall be increasingly dependent on food-surplus regions. This volume makes a strong case for governing plant genetic resources in ways that promote the evolution and conservation of agrobiodiversity, and to ensure that they are available to be used by all regions to adapt better to a changing environment. Yet, it is more than just another book about the governance of natural resources by the best experts in the field: it is also an indispensable tool to understand the future of agriculture in a world of dwindling resources and biodiversity loss."

—*Olivier de Schutter, United Nations Special Rapporteur on the Right to Food*

"These meticulous studies of the FAO Treaty on plant genetic resources for food and agriculture are of major theoretical and empirical importance to scholars and practitioners seeking a workable, transnational regime to govern all genetic resources for research and benefit sharing under the evolving international legal framework."

—*Jerome H. Reichman, Bunyan S. Womble Professor of Law, Duke Law School, USA*

CROP GENETIC RESOURCES AS A GLOBAL COMMONS

Farmers have engaged in collective systems of conservation and innovation – improving crops and sharing their reproductive materials – since the earliest plant domestications. Relatively open flows of plant germplasm attended the early spread of agriculture; they continued in the wake of (and were driven by) imperialism, colonization, emigration, trade, development assistance and climate change. As crops have moved around the world, and agricultural innovation and production systems have expanded, so too has the scope and coverage of pools of shared plant genetic resources that support those systems. The range of actors involved in their conservation and use has also increased dramatically. This book addresses how the collective pooling and management of shared plant genetic resources for food and agriculture can be supported through laws regulating access to genetic resources and the sharing of benefits arising from their use. Since the most important recent development in the field has been the creation of the multilateral system of access and benefit-sharing under the International Treaty on Plant Genetic Resources for Food and Agriculture, many of the chapters in this book will focus on the architecture and functioning of that system. The book analyzes tensions that are threatening to undermine the potential of access and benefit-sharing laws to support the collective pooling of plant genetic resources, and identifies opportunities to address those tensions in ways that could increase the scope, utility and sustainability of the global crop commons.

Michael Halewood is a Senior Scientist and Head of the Policy Research and Support Unit at Bioversity International, Rome, Italy.

Isabel López Noriega is a legal specialist in the Policy Research and Support Unit of Bioversity International in Rome.

Selim Louafi is based at the Centre de Coopération Internationale en Recherche Agronomique pour le Devéloppement (CIRAD), Montpellier, France.

Issues in Agricultural Biodiversity
Series editors: Michael Halewood and Danny Hunter, Bioversity International

This series of books is published by Earthscan in association with Bioversity International. The aim of the series is to review the current state of knowledge in topical issues associated with agricultural biodiversity, to identify gaps in our knowledge base, to synthesize lessons learned and to propose future research and development actions. The overall objective is to increase the sustainable use of biodiversity in improving people's well-being and food and nutrition security. The series' scope is all aspects of agricultural biodiversity, ranging from conservation biology of genetic resources through social sciences to policy and legal aspects. It also covers the fields of research, education, communication and coordination, information management and knowledge sharing.

Published titles:

Crop Wild Relatives
A manual of *in situ* conservation
Edited by Danny Hunter and Vernon Heywood

The Economics of Managing Crop Diversity On-farm
Case studies from the Genetic Resources Policy Initiative
Edited by Edilegnaw Wale, Adam Drucker and Kerstin Zander

Plant Genetic Resources and Food Security
Stakeholder perspectives on the International Treaty on Plant Genetic Resources for Food and Agriculture
Edited by Christine Frison, Francisco López and José T. Esquinas

Crop Genetic Resources as a Global Commons
Challenges in international law and governance
Edited by Michael Halewood, Isabel López Noriega and Selim Louafi

Forthcoming titles:

Community Biodiversity Management
Promoting resilience and the conservation of plant genetic resources
Edited by Walter S. de Boef, Abishkar Subedi, Nivaldo Peroni and Marja Thijssen

Farmers' Crop Varieties and Farmers' Rights
Challenges in taxonomy and law
Edited by Michael Halewood

Diversifying Food and Diets
Using agricultural biodiversity to improve nutrition and health
Edited by Jessica Fanzo, Danny Hunter, Teresa Borelli and Federico Mattei

CROP GENETIC RESOURCES AS A GLOBAL COMMONS

Challenges in international law and governance

Edited by
Michael Halewood, Isabel López Noriega
and Selim Louafi

First published 2013
by Routledge
2 Park Square, Milton Park, Abingdon, Oxon, OX14 4RN

Simultaneously published in the USA and Canada
by Routledge
711 Third Avenue, New York, NY 10017

Routledge is an imprint of the Taylor & Francis Group, an informa business

© 2013 Bioversity International

All rights reserved. No part of this book may be reprinted or reproduced or utilised in any form or by any electronic, mechanical, or other means, now known or hereafter invented, including photocopying and recording, or in any information storage or retrieval system, without permission in writing from the publishers.

Trademark notice: Product or corporate names may be trademarks or registered trademarks, and are used only for identification and explanation without intent to infringe.

British Library Cataloguing in Publication Data
A catalogue record for this book is available from the British Library

Library of Congress Cataloging in Publication Data
Crop genetic resources as a global commons : challenges in international law and governance / edited by Michael Halewood, Isabel López Noriega and Selim Louafi. -- 1st ed.
p. cm.
Includes bibliographical references and index.
1. Germplasm resources, Plant--Law and legislation. 2. Crops--Germplasm resources. 3. Plant diversity conservation--Law and legislation. 4. International Treaty on Plant Genetic Resources for Food and Agriculture (2001)
I. Halewood, Michael, 1963- II. López Noriega, Isabel. III. Louafi, Sélim.
K3876.C76 2012
343.07'61523--dc23
2012012769

ISBN13: 978-1-84407-892-9 (hbk)
ISBN13: 978-0-84407-893-6 (pbk)
ISBN13: 978-1-84977-681-3 (ebk)

Typeset in Baskerville
by Taylor & Francis Books

Printed and bound by CPI Group (UK) Ltd, Croydon, CR0 4YY

For my parents, the first farmers in my life, Deane and Bill
Michael Halewood

For my inspiring educators and colleagues: Enrique, Alejandro, Michael, Gerald, Devra, Toby, Pablo
Isabel López Noriega

For all my former colleagues at the International Treaty Secretariat
Selim Louafi

CONTENTS

List of tables — xii
List of figures — xiii
List of contributors — xiv
Acknowledgements — xvii

1 The global crop commons and access and benefit-sharing laws: examining the limits of international policy support for the collective pooling and management of plant genetic resources — 1
Michael Halewood, Isabel López Noriega and Selim Louafi

PART I
Setting the scene: countries' interdependence on plant genetic resources for food and agriculture and the imperative of international cooperation — 37

2 Demonstrating interdependence on plant genetic resources for food and agriculture — 39
Marleni Ramirez, Rodomiro Ortiz, Suketoshi Taba, Leocadio Sebastián, Eduardo Peralta, David E. Williams, Andreas W. Ebert and Anne Vézina

3 Flows of crop germplasm resources into and out of China — 62
Fuyou Wang

4 Crop and forage genetic resources: international interdependence in the face of climate change — 78
Julian Ramirez-Villegas, Andy Jarvis, Sam Fujisaka, Jean Hanson and Christoph Leibing

5 Changing rates of acquisition of plant genetic resources by international gene banks: setting the scene to monitor an impact of the International Treaty 99
Michael Halewood, Raj Sood, Ruaraidh Sackville Hamilton, Ahmed Amri, Ines Van den Houwe, Nicolas Roux, Dominique Dumet, Jean Hanson, Hari D. Upadhyaya, Alexandra Jorge and David Tay

PART II
The history and design of the International Treaty's multilateral system of access and benefit-sharing 133

6 A brief history of the negotiations on the International Treaty on Plant Genetic Resources for Food and Agriculture 135
José Esquinas-Alcázar, Angela Hilmi and Isabel López Noriega

7 The design and mechanics of the multilateral system of access and benefit sharing 150
Daniele Manzella

8 Protecting the interests of the multilateral system under the Standard Material Transfer Agreement: the third party beneficiary 164
Gerald Moore

9 Plant genetic resources under the management and control of the contracting parties and in the public domain: how rich is the ITPGRFA's multilateral system? 177
Carlos M. Correa

10 Efforts to get the multilateral system up and running: a review of activities coordinated by the Treaty's Secretariat 187
Selim Louafi and Shakeel Bhatti

PART III
Critical reflections 197

11 Assessment of progress to make the multilateral system functional: incentives and challenges at the country level 199
Isabel López Noriega, Peterson Wambugu and Alejandro Mejías

12 From negotiations to implementation: global review of achievements, bottlenecks and opportunities for the Treaty in general and for the multilateral system in particular 226
Godfrey Mwila

13 The multilateral system of access and benefit sharing: could it have been constructed another way? 243
 Clive Stannard

14 The moving scope of Annex 1: the list of crops covered under the multilateral system 265
 Bert Visser

15 Building a global information system in support of the International Treaty on Plant Genetic Resources for Food and Agriculture 283
 Caroline Ker, Selim Louafi and Myriam Sanou

16 Collective action challenges in the implementation of the multilateral system of the International Treaty: what roles for the CGIAR centres? 310
 Selim Louafi

17 International and regional cooperation in the implementation of the International Treaty on Plant Genetic Resources for Food and Agriculture 329
 Gerald Moore

18 The evolving global system of conservation and use of plant genetic resources for food and agriculture: what is it and where does the Treaty fit in? 344
 Toby Hodgkin, Nicole Demers and Emile Frison

19 Institutionalizing global genetic resource commons for food and agriculture 368
 Tom Dedeurwaerdere

Index 392

TABLES

1.1	Analytical framework – excludability and rivalry	32
3.1	Average yields of rice, wheat and maize in China compared to yields of experimental crops, 1995–97 and 2005–7	63
3.2	Ratio of International Potato Center parentage in released and adopted potato varieties, 1997 and 2007	70
3.3	Flow of germplasm out of/into China	72
4.1	Likely changes in interdependence for 17 crops globally for 2050	87
5.1	Distributions and acquisitions by all CGIAR centres' gene banks	102
5.2	CGIAR centres' acquisitions, 2005–10	114
5.3	Materials sent to international collections hosted by CGIAR centres, 2007–2010 (inclusive), with permission to re-distribute using the SMTA	115
5.A1	Gene bank acquisitions: CIAT, CIMMYT and CIP, 1980–2004	121
5.A2	Gene bank acquisitions: ILRI, WARDA and ICRAF, 1980–2004	122
5.A3	Gene bank acquisitions: IPGRI, ICRISAT and IITA, 1980–2004	122
5.A4	Gene bank acquisitions: ICARDA and IRRI, 1980–2004	123
5.A5	Holdings that the centres may distribute using the Standard Material Transfer Agreement	127
15.1	Potential users and beneficiaries of the global information system	289
18.1	Multinational elements of the current global system	350
18.2	Some aspects of PGRFA conservation and use where international elements can play a significant role beyond that of the national programmes	356
19.1	Biophysical, institutional and community characteristics of the genetic resource commons	376

FIGURES

4.1	Overview of changes in interdependence.	88
4.2	Changes in Euclidean distances pooled by crop for all countries having one or more hectares on each crop versus the average Euclidean distance between current climates and future climates of each country pooled by crop.	89
5.1	Total accessions acquired by all CGIAR centres.	103
5.2	Acquisitions by the International Rice Collection hosted by the IRRI, 1994–2009.	113
11.1	Number of accessions held by international gene banks and research centres.	205
11.2	Location of PGRFA identified as included in the multilateral system, 2011.	206
11.3	Portion of Annex 1 accessions that have been identified as included in the multilateral system, out of the total of accessions held worldwide.	207
18.1	System-level description of conservation of PGRFA in terms of benefits and outcomes.	348
19.1	Genetic biodiversity of the Matsoni Yoghurt in Georgia.	378
19.2	Example of a collective breeding process with collective ownership.	385

CONTRIBUTORS

Ahmed Amri, International Centre for Agricultural Research in the Dry Areas (ICARDA), Aleppo, Syria

Shakeel Bhatti, Secretary of the International Treaty on Plant Genetic Resources for Food and Agriculture, Rome, Italy

Carlos M. Correa, University of Buenos Aires, Buenos Aires, Argentina

Tom Dedeurwaerdere, University of Louvain, Louvain, Belgium

Nicole Demers, Bioversity International, Rome, Italy

Dominique Dumet, International Institute of Tropical Agriculture (IITA), Ibadan, Nigeria

Andreas W. Ebert, The World Vegetable Centre (AVRDC), Taiwan

José Esquinas Alcázar, Chair on Studies on Hunger and Poverty of the University of Cordoba, Spain

Emile Frison, Bioversity International, Rome, Italy

Sam Fujisaka, International Center for Tropical Agriculture (CIAT), Cali, Colombia

Michael Halewood, Bioversity International, Rome, Italy

Jean Hanson, International Livestock Research Institute (ILRI), Addis Ababa, Ethiopia

Angela Hilmi, Universidad Politécnica de Valencia

Toby Hodgkin, Bioversity International, Rome, Italy

Contributors

Andy Jarvis, International Center for Tropical Agriculture (CIAT), Cali, Colombia; CGIAR Research Program on Climate Change, Agriculture and Food Security (CCAFS), CIAT, Cali, Colombia

Alexandra Jorge, International Livestock Research Institute (ILRI), Nairobi, Kenya

Caroline Ker, Research Center Information, Law and Society, University of Namur, Namur, Belgium

Christoph Leibing, International Center for Tropical Agriculture (CIAT), Cali, Colombia

Isabel López Noriega, Bioversity International, Rome, Italy

Selim Louafi, Centre International de Recherche Agronomique pour le Développement (Cirad), UMR AGAP, Montpellier, France

Daniele Manzella, United Nations Food and Agriculture Organization, Rome, Italy

Alejandro Mejías, (graduate student) New York University School of Law

Gerald Moore, Bioversity International, Rome, Italy

Godfrey Mwila, Global Crop Diversity Trust, Rome, Italy

Rodomiro Ortiz, International Center for the Improvement of Wheat and Maize (CIMMYT), Mexico

Eduardo Peralta, Autonomous National Institute for Agricultural Research (INIAP), Quito, Ecuador

Julian Ramirez-Villegas, International Center for Tropical Agriculture (CIAT), Cali, Colombia; CGIAR Research Program on Climate Change, Agriculture and Food Security (CCAFS), CIAT, Cali, Colombia; School of Earth and Environment, University of Leeds, Leeds, UK

Marleni Ramirez, Bioversity International, Cali, Colombia

Nicolas Roux, Bioversity International, Montpellier, France

Ruaraidh Sackville Hamilton, International Rice Research Institute (IRRI), Los Baños, Philippines

Myriam Sanou, Research Center Information, Law and Society, University of Namur, Namur, Belgium

Leocadio Sebastian, Bioversity International, Serdang, Malaysia

Raj Sood, Bioversity International, Rome, Italy

Clive Stannard, University of Leicester

Suketoshi Taba, International Center for the Improvement of Wheat and Maize (CIMMYT), Mexico

David Tay, Centro Internacional de la Papa/International Potato Centre (CIP), Lima, Peru

Hari D. Upadhyaya, International Crops Research Institute for the Semi-Arid Tropics (ICRISAT), Patancheru, India

Ines Van den Houwe, Bioversity International, Louvain, Belgium

Anne Vézina, Bioversity International, Montpellier, France

Bert Visser, Centre for Genetic Resources, Wageningen, The Netherlands

Peterson Wambugu, Kenyan Agriculture Research Institute, Nairobi, Kenya

Fuyou Wang, Chinese Academy of Tropical Agricultural Sciences, Haikou, China

David E. Williams, CGIAR System-wide Genetic Resources Programme, Rome, Italy

ACKNOWLEDGEMENTS

We would like to recognize the contributions the external reviewers of chapters in this book. Double and triple thanks to those of you who reviewed two chapters or more. Elizabeth Arnaud, Ana Berretta, Tom Dedeurwaerdere, Adam Drucker, Luigi Guarino, Jean Hanson, Evanson Kamau, Ruth Meinzen-Dick, Gerald Moore, Javad Mozafari, Ruaraidh Sackville Hamilton, Maria José Sampaio, Lim Eng Siang, Mary Taylor, Bert Visser.

1

THE GLOBAL CROP COMMONS AND ACCESS AND BENEFIT-SHARING LAWS

Examining the limits of international policy support for the collective pooling and management of plant genetic resources

Michael Halewood, Isabel López Noriega and Selim Louafi

This book addresses how the collective pooling and management of plant genetic resources for food and agriculture (PGRFA) can be supported through international law. The collective pooling of plant genetic resources is not new – it has always been a vital form of support for agricultural development, crop improvement and agricultural research. What is new is that such pooling has become the object of regulation, including international laws controlling access to genetic resources and the sharing of benefits arising from their use.

One of the objectives of this book is to evaluate the current state of access and benefit-sharing laws and how they influence internationally coordinated efforts to pool and manage genetic resources. Since the most important recent development in the field has been the creation of the multilateral system of access and benefit sharing (multilateral system) under the International Treaty on Plant Genetic Resources for Food and Agriculture (ITPGRFA), many of the chapters in this book will focus on the architecture and functioning of this system.[1] Another objective is to understand some of the tensions that are presently threatening to frustrate the potential of access and benefit-sharing laws – again, with a particular focus on the Treaty's multilateral system – to support the collective pooling and management of these resources. A third objective is to identify opportunities to address these tensions in proactive and constructive ways.

In this introduction, we provide a quick overview of plant genetic resource-pooling practices at farmer, community, research network and international levels. We also provide a snapshot of how access and benefit-sharing laws are (and are not) supporting those practices. We present an analysis of two issues related to the design of the multilateral system that are giving rise to tensions that are threatening to undermine the system's potential to support the collective pooling and management of PGRFA. We conclude with an overview of the chapters that follow.

Collective pooling of PGRFA

Farmers have been engaged in collective systems of conservation and innovation – openly sharing planting materials and conserving them through use – since the earliest crop domestications. Relatively open flows of plant germplasm attended the spread of agriculture and have subsequently followed (or been driven by) imperialism, colonization, emigration, trade, development assistance and climate change. As crops have moved around the world, and as the scope of agricultural innovation and production systems have shifted and expanded, so too has the scope and coverage of pools of shared plant genetic resources that support those systems.

Farmers' ancestral practices in germplasm conservation and innovation can still be found in small farming systems and in remote areas. These practices rely, to a great extent, on the exchange of planting materials between farmers. Although farmers prefer to select and save seed from their own harvest, seed and harvest loss, experimentation and the establishment of new households make them acquire seed from other farmers or from the market (Almekinders et al., 1994; Badstue et al., 2006; Hodgkin et al., 2007; McGuire, 2007). Farmer-to-farmer seed exchange often forms dynamic networks where, by virtue of social relations, interdependence and reciprocity rules, farmers have access to other farmers' seed stock. In this way, a decentralized and changing pool of germplasm is created.

A number of studies show that, in some seed networks, certain farmers play a predominant role in germplasm conservation, generation and supply. Analysis of exchange patterns in Yucatan (Mexico), Rajasthan (India) and East Java (Indonesia) show that some farmers or even villages are specialized in seed production and supply for certain crops. They are known for reliably and regularly producing and selling seed (Hodgkin et al., 2007; Linnemann and Siemonsma, 1987; Weltzein and Vom Brocke, 2001). In Kaski and Bara, Nepal, community members consider some farmers to be 'nodal farmers' since they maintain a relatively larger amount of crop diversity than other members of the community. These farmers look for new cultivars, conserve them and experiment with them. They thereby play an important role in the flow of genetic material as they give seed to other farmers (Subedi et al., 2003).

In addition to informal exchange networks, farmers generate and participate in organized germplasm exchange systems, such as seed fairs and community seed banks. Seed fairs have been in existence for centuries in some parts of the world. In the Andes, for example, people from different communities often congregate in religious festivities, normally at the end of the harvest, and sell, buy and exchange plant genetic resources and related knowledge (Tapia and Rosa, 1993). Typically, in a response to crisis situations such as war, long droughts and dramatic loss of local diversity, community seed banks have represented a much more explicit form of common pooling of plant genetic resources at the local level. Local germplasm and associated knowledge are collected from within the community, markets and neighbouring villages, stored in seed banks, multiplied, and then distributed to farmers whenever they need it (Shrestha et al., 2008). Often, these efforts are supported by international and national development agencies.

Germplasm exchange patterns expanded over larger scales as cultivation spread to new areas (Kloppenburg and Kleinman, 1988). The early development of agriculture has involved slow extensions of domestication processes and the movements of crops and germplasm beyond their natural area. These movements experienced a dramatic increase with the first contacts between Europe and the Americas. A vast array of plants was introduced from the Americas to Europe and vice versa. This so-called 'Columbian exchange' (Crosby, 1972) also witnessed the establishment of botanic gardens to receive, nurture, classify and transship exotic plants (Brockway, 1988). The subsequent mercantilist expansion of plantation systems oriented toward export contributed to further movements of PGRFA across continents, mainly through the initiative of European governments and trading companies.

The advent of scientific breeding after the rediscovery of Mendel's laws in the early twentieth century (first developed in the 1880s) helped to guide the search for new material on the basis of known genetic characteristics (Byerlee and Dubin, 2010; Pistorius, 1997). Breeders' collections were first conscious efforts to pool genetic resources from all around the world for improvement purposes (Garrison Wilkes, 1988). Public sector breeders' practices of relatively open systems of germplasm exchange through informal networks has since become widespread and well documented (Byerlee and Dubin, 2010; SGRP, 2011).

The famous Russian scientist Nicolay Vavilov (2007), during the same period, identified the major centres of crop diversity and first collected an enormous number of plants not for breeding purposes but, rather, for genetic study. Pooling efforts very much remained of an *ad hoc* nature and were linked to short-term use (Frankel, 1988). The development and maintenance of these shared pools was often more a by-product of the way in which the genetic resources were used rather than an end in itself. Often the resources were *de facto* pooled as a function of the way they were used and openly available between farmers, breeders and research partners in both the formal and informal research and development sectors. When the pooled resources no longer served a purpose, the members of the pool would dissipate and the resource pool would fall into disuse. Since crop diversity is a function of the interaction of environment, plant reproduction systems and human intervention, this diversity can disappear in the absence of the latter.

After the Second World War, as a result of progress in genetics, the emergence of disease resistances in improved varieties and increased contact between scientists worldwide, there was increased interest and demand for a more systematic approach to exchanging PGRFA (Kloppenburg, 1988a). The Food and Agriculture Organization (FAO) decided in 1948 to establish a world clearing house to increase cooperation in plant exploration and the recording of living collections (Whyte and Julèn, 1963). This clearing house consisted in building a catalogue of available material around the world to foster its utilization in international breeding programmes. However, the FAO efforts were not restricted to information pooling – it also oversaw the physical delivering of samples, especially in relation to breeders in developing countries.

By the end of the 1950s, the FAO had distributed approximately 20,000 seed envelopes per year (Pistorius, 1997). However, until the late 1960s, the main part of the worldwide exchanges of genetic material was taking place within a network of so-called 'plant introduction stations', which were mainly centred in Western Europe, the United States, Australia, New Zealand and the Soviet Union. These stations established working collections for research purposes and were consequently not equipped to conserve material for a longer term (Kloppenburg, 1988a).

The establishment of the green revolution in the 1960s as a response to food insecurity increased the demand for access to genetic diversity for plant breeding. At the same time, the spread of monoculture practices increased the perception of the risks of genetic diversity erosion. As a result, the FAO, along with the International Biological Program, decided to hold the first technical conference on plant genetic resources in 1967 with the aim of establishing a consensus on the importance of biodiversity erosion and an agreement on collective conservation actions. This conference succeeded in putting PGRFA on the international agenda for the first time, but the participants could not agree on a common approach, so a panel was established to try to develop proposals for coordinated actions. Building upon the existence of some collections established under some important international breeding programmes funded by the Ford and Rockefeller foundations, the Consultative Group on International Agricultural Research (CGIAR) was created in 1971. It gathered existing and newly created agronomic research institutes around the world under this new umbrella hosted by the World Bank.

Three major governance features characterized this new international organization dedicated to PGRFA conservation. The first one relates to the fact that it was intentionally created outside the UN framework, guided by international agricultural research centres overseen by a consortium of donors under the umbrella of the CGIAR. The second feature relates to the choice to exclusively emphasize *ex situ* conservation (conserving material outside of its natural area – in cold storage facilities) over the virtual exclusion of *in situ* conservation (conserving PGRFA through their use on farms and subject to complex interactions between environment, genotype and farmer interventions). Finally, the third feature relates to the nature of the material collected. Fierce debates took place during the 1967 technical conference on this matter. Some individuals argued for a utilitarian approach, focusing only on the conservation of crops of major agronomic interest, while others argued for an ecosystem approach – which was closely aligned with *in situ* conservation – that was expanded in scope to conserve all kinds of materials, including those with no known value, from within specific areas. Consistent with the choice to emphasize *ex situ* conservation, the CGIAR centres dedicated themselves to a mono-specific approach – each centre being responsible for only a limited number of crops of major importance for food security.

This network constituted the first effort towards systematizing and formalizing the conservation and pooling of PGRFA at the international level. It was backed up technically by the newly created International Board on Plant Genetic Resources (IBPGR), which was in charge of coordinating collection missions, establishing common conservation criteria and description standards and establishing an international network of base collections. The IBPGR quickly became a rallying point for people concerned with genetic resources and contributed to the expanded collection, supporting complementary activities and a clearing house for information (Frankel, 1988). The CGIAR centres (and the IBPGR, in particular), in coordination with all of the countries that facilitated the new collecting missions, transformed the pre-existing systems of local and distributed common pools of PGRFA into a global pool or pools of PGRFA. This 'jump' from distributed and informally managed local pools to a formalized and globally networked pool contributed to PGRFA becoming a subject of global interest, international legal regulation and concern.

'Hyperownership' trends undermine PGRFA pooling

The internationalization of *ex situ* PGRFA collections took place in a context of rising tensions about intellectual property rights and perceived inequities concerning who bears the cost of conservation and who benefits (commercially) most from its use. Most biological diversity is located in the tropics, in developing countries, and is often in areas inhabited by the poorest, most marginalized people in those countries. The centres of diversity of most domesticated crops are located in developing country regions (Harlan, 1992). However, the technological capacity to exploit biological diversity for commercial gain resides principally in developed countries. The emergence of biotechnology has contributed to a further disproportionate concentration of technological capacity in the North. This development, coupled with the insistence of developing countries to include intellectual property protections in the Uruguay Round of trade negotiations under the General Agreement on Tariffs and Trade, has exacerbated international North–South political tensions.[2] Developing countries ultimately agreed to the Agreement on Trade-Related Aspects of Intellectual Property Rights (TRIPs Agreement) so as to make gains in other aspects of the negotiation, but they remained generally unhappy with the requirements to create protections for foreign-owned biologically based technologies.[3] Their discomfort has been heightened by the fact that Europe and the United States have, for decades, also exerted bilateral pressure on developing countries, outside the multilateral framework of the World Trade Organization, to provide intellectual property protection for their biologically based products that are stronger than those required by the TRIPs Agreement.

The development of the access and benefit-sharing rules under the 1993 Convention on Biological Diversity (CBD) were partly a response to this situation.[4] If developed countries were able to exercise restrictive control over advanced biologically based technologies using intellectual property rights, developing countries could exercise their sovereign rights to regulate and restrict access to the biological and genetic resources within their borders. The escalation in policies to control access and use of genetic resources has been referred to as the 'seed wars' (Aoki, 2008), and the situation has been referred to as a trend towards 'hyperownership' (Safrin, 2004). While the impact of these restrictive policies on commons-based pooling of PGRFA has not been fully documented – some of the chapters in this book attempt to provide such documentation – anecdotal evidence strongly suggests that they have had a negative impact on the willingness of a range of actors to make PGRFA widely available, with negative consequences for scientific research, and agricultural development (Jinna and

Jungcourt, 2009; UNEP, 2006) and for the conservation of PGRFA (Halewood, 2012; Ruiz and Vernooy, 2011).

The ITPGRFA was designed to strike a compromise position by establishing a set of rules and procedures to encourage member states to pool that portion of the PGRFA within their control that they feel comfortable sharing, and to direct a portion of financial benefits derived from their commercial use to commonly identified conservation and use-related priorities. The Treaty stands somewhere between the situation that prevailed before the onslaught of intellectual property and access and benefit-sharing laws, when all PGRFA were *de facto* considered public domain and free to use by all, and the race-to-the-bottom scenario that has ensued since the 1980s in terms of establishing forms of private control over PGRFA.

A snapshot of the current state of international legal support for PGRFA pooling

By far the most directly relevant international law affecting the collective pooling and management of plant genetic resources is the ITPGRFA. The Treaty was adopted in 2001 and came into force in 2004. It establishes the 'multilateral system of access and benefit sharing' for 64 specifically identified crops and forages. The creation of the multilateral system represents an extremely important breakthrough in the history of international policy development *vis-à-vis* plant genetic resources. The multilateral system was developed to support the virtual pooling,[5] and management of plant genetic resources, as inputs into plant breeding, agricultural research and conservation efforts.[6]

It is important to note that limited forms of PGRFA pooling are of course possible on a voluntary basis between the owners of proprietary technologies through mechanisms such as patent pools (van Overwalle, 2008). Like the Treaty's multilateral system, these pools are created to lower the high transaction costs and associated limitations on research and development that would otherwise result from a system of exclusive controls over inputs into research and development. Patent pools are generally limited to a narrow (compared to the multilateral system) range of advanced, privately owned technologies, controlled by private owners, with a relatively small number of private 'club' members. Patent pooling rules are designed primarily with the objective of increasing freedom to operate for members of the pool (and possibly, in some cases, for a defined group of beneficiaries, such as the resource poor in developing countries). The multilateral system established under the Treaty is concerned with another form of PGRFA pooling, that is, the pooling of shared PGRFA which are not generally subject to private ownership and controls, and that are to be made available on an international (potentially global) scale. The rules governing participation and membership are designed to increase members' contributions to the overall conservation and sustainable use of PGRFA as well as to promote equitable distribution of benefits associated with the use of the resources, notably not back to the providers of the resources used but, rather, to an international fund to support conservation and sustainable use generally.

According to some measures, there is considerable evidence of international enthusiasm for the ITPGRFA – for example, it has enjoyed one of the fastest rates of national ratification of any international environmental treaty, with a total of 127 member states to date. There is also evidence that it is having a positive impact on some countries' willingness to provide access to PGRFA and on the diversity of PGRFA that is being conserved and used overall. Examples of these positive effects are provided in a number of the chapters in this book. On the other hand, eight years have passed since it came into force, and the multilateral system is not operating near its full potential. Many of the actors one would have thought would be actively involved in the multilateral system, as conservers, users, providers and recipients of germplasm and information, are demonstrating reluctance to fully engage. It is evident that remedial measures are needed to invigorate the multilateral system and to create incentives for more active participation in it.

Beyond the ITPGRFA, there is virtually no concrete support within international access and benefit-sharing laws for the common pooling and management of genetic resources.[7] The CBD establishes basic principles that could be built upon to support the pooling of shared resources, but, so far, state parties to the CBD have not developed any such norms. Instead, emphasis on the implementation of the CBD at both the international level (that is, meetings of the Conferences of the Parties (COP)) and the national level concerns the creation of a legal means to close loopholes so that no one is able to get access to genetic resources without permission from the competent national authorities. The primary mechanism in this regard has been to legally compel access seekers to enter into process-heavy bilateral negotiations with competent national authorities on a case-by-case basis (Nijar et al., 2009). No infrastructure has been developed under the CBD to encourage providers to pool genetic resources and make them available as a part of the coordinated systems of conservation and use at the national level or regional level, let alone at the global level. In 2010, the CBD's COP adopted the Nagoya Protocol on Access and the Fair and Equitable Sharing of Benefits Arising from Their Utilization (Nagoya Protocol); it has not yet entered into force.[8] Its purpose is to deepen and make operational the CBD's commitments with respect to access and benefit sharing. Some clauses of the Nagoya Protocol reflect a growing appreciation among the parties that it is necessary to at least consider that it may be desirable, in the future, to develop, under specific conditions, some kind of pooling measures or specialized access and benefit-sharing arrangements for different sectors.[9] So there is potential under the Nagoya Protocol to begin to consider the various multilaterally inspired access and benefit-sharing arrangements to support pooling of genetic resources, especially in the food and agricultural sector. However, this consideration remains a relatively remote possibility.

The FAO's Commission on Genetic Resources for Food and Agriculture – the same body that hosted the negotiations of the ITPGRFA – has the specialist technical capacity to develop access and benefit-sharing norms supporting the pooling of genetic resources. In recent years, the commission has developed an interesting body of information papers dedicated to exploring the issue and has most recently decided to 'consider the need for, and modalities of, access and benefit-sharing arrangements for genetic resources for food and agriculture' and has established an inter-regional working group to engage in a preliminary analysis in the lead-up to the commission's next session in 2013 (CGRFA, 2011). However, so far, the commission has shied away from giving itself a norm-setting mandate.

Two design issues affecting actors' engagement in the multilateral system

This section addresses the question of why, to date, there has not been more vigorous engagement in the ITPGRFA's multilateral system by a number of key actors. In particular, we focus on two aspects related to the architectural design of the multilateral system that are contributing to this situation. We do not describe those aspects of the architecture of the multilateral system as design flaws. On the one hand, we are convinced that they are contributing to suboptimal performance of the multilateral system overall, while, on the other hand, we fully appreciate the range of competing pressures and visions that had to be accommodated during the negotiations of the Treaty. The Treaty and the multilateral system are remarkable achievements and high-water marks for how countries can work together under the United Nations to tackle complicated transnational conservation issues. As the following section highlights, there are understandable reasons (with which the authors are broadly sympathetic) for why some key actors have contributed to, and perpetuated, the design-related tensions that we focus on here. We are not suggesting that the multilateral system is fatally flawed. It is quite possible that the multilateral system as it currently stands will eventually 'pull in' a broader range of actors providing and receiving material and generating benefits to be shared. Such increased engagement could develop over time of its own accord, as more actors become aware of the benefits to be gained through the multilateral system. Meanwhile, increased engagement can be encouraged through the kinds of complementary, supportive efforts discussed later in this chapter. It is also possible that aspects of the multilateral system can be reformed to respond to some of the challenges we identify.

In this context we feel it is important to clearly state our opinion that access to genetic diversity in the multilateral system is by far the most important benefit for all member states. If this fact were more widely appreciated the design-related tensions we discuss below would lose much of their force. Raising awareness of the benefits that can accrue to each country (and to their companies, farmers, universities, etc) from participation in the multilateral system is a prerequisite for more sustained, proactive engagement. Meanwhile the issues we raise in the following sections need to be considered.

Design issue no. 1: insufficient policy reinforcement for the PGRFA commons – unclear boundaries and unenforced reciprocity

A condensed summary of the first challenge related to the design of the multilateral system is as follows. The multilateral system responds (mostly) to a vision of PGRFA as a commons. At the same time, however, the rules of engagement in the multilateral system – either by accident or by design – are loose enough that they allow actors inside and outside the system to behave in ways that undermine others' incentives to actively participate in the stewardship of the PGRFA commons. One set of actors that is culpable in this regard is made up of those who are unable or unwilling to shift away from the ideal of PGRFA as global public goods, and see them instead as goods that are managed and exploited by a limited range of actors on the basis of reciprocity. Other actors who are undermining the Treaty-supported PGRFA commons are free-riders who are able to benefit from the Treaty's multilateral system without contributing meaningfully to it. While their complicity with free-riders is inadvertent, the champions of global public goods end up supporting the free-riders by providing them with access to some of the world's most extensive collections of plant genetic resources without any regard for whether they have put anything into the multilateral system. This analysis suggests a fairly straightforward solution: reinforce the boundaries of the multilateral system and create, monitor and enforce a reciprocity requirement. Some of the necessary reforms are clearly realizable. However, other reforms may not be possible as a result of the fact that some key actors have competing, laudable, priorities (with fairly deep historical roots) that make some potential reforms to the multilateral system difficult and, in some cases, possibly not worthwhile. In the remainder of this section, we provide a more comprehensive account of this argument, defining and explaining the significance of the key concepts introduced in the preceding paragraph.

What is a commons and why do we say the Treaty's multilateral system responds to, and reinforces, a vision of PGRFA as a commons?

A commons is a resource or resources 'shared by a group of people that is subject to social dilemmas'. Commons are 'jointly used, managed by groups of varying sizes and interests'. Commons are not exclusively subject to either state or private/market controls. Instead, they are managed through the collective actions of interested parties who manage the resources in question through cooperative mechanisms (Hess and Ostrom, 2007). Elinor Ostrom has identified a number of design principles for successfully managed commons, including the importance of having defined geographical boundaries or membership; proportionality between rights to use common-pooled resources and contributions to their maintenance; resource users involved in creating and modifying the rules governing use; low transaction-cost rule review and modification; resource users (or people who answer to them) directly involved in monitoring compliance with the rules; graduated sanctions for rule violators; and low-cost conflict resolution mechanisms (Ostrom, 1990).

Until fairly recently, scholarship on 'traditional' commons focused on the management of natural resources that are both rivalrous (one person's use of the resource detracts from the ability of another to benefit from it) and non-excludable (it is impossible, or extremely expensive, to prevent others from using the resource). The geographical scope of traditional commons was limited, as was the number of actors involved in their management (Ostrom, 1990). More recently, 'new commons' literature has expanded to include consideration of human-made resources that are not necessarily rivalrous in nature and for which the geographic scope and membership of the commons is not limited. A prime example is understanding knowledge as a commons (Hess and Ostrom, 2007).

As the previous subsection attests, there is a long history of PGRFA being conserved, shared and managed through the coordinated actions of groups of people. The multilateral system provides administrative and policy support for the participation of a very large group of people – potentially limitless – in collective action conserving and using a large subset of PGRFA.[10] In the multilateral system, ownership of plant genetic resources is not clearly defined, nor does it need to be for the system to function efficiently. Nonetheless, members of the multilateral system have developed norms in support of the common stewardship of these resources, including shared rights to exploit those resources for specified purposes. Membership in the multilateral system is (in theory at least) limited to those who subscribe to these norms. Non-members are not entitled to access or use the pooled resources. Members (only) may collectively agree to alter the rules, by consensus.

The commons supported by the multilateral system is a 'new' commons in as much as PGRFA are at least partially man-made, being the product of complex interactions between plants' reproduction systems, the natural environment and human intervention. In the absence of continued use by humans (for examples, planting, seed selection, cross-breeding and regeneration of ex situ accessions) PGRFA diversity would cease to develop and much of it would disappear. (Halewood 2008; Schloen et al, 2011) Underuse of PGRFA, not overuse, is the primary driver behind its loss as a resource. This idiosyncrasy is one of the 'social dilemmas' that needs to be addressed through collective action in the stewardship of the PGRFA commons. PGRFA are spread out around the world, virtually pooled through the loose federation supported by the multilateral system. The multilateral system's membership is not subject to any practical limitations. Any state, and any number of actors within these states (and international organizations), can be members as long as they subscribe to 'the rules of the game'.

In general, genetic resources are also not rivalrous, or at least they are considerably less rival than the kinds of exhaustible natural resources that have been the focus of traditional commons literature.[11] Genetic resources are constituted by both physical and informational components. For the purposes of conservation, research and plant breeding, only small quantities of genetic material are required to be able to exploit the full informational component of the resource. The cost of producing or providing additional units of PGRFA for such purposes is relatively small, and supplies are relatively elastic, at least in the case of seeds in sustainably funded *ex situ* collections, or they can be collected or sourced from widely planted populations in farmers' fields. Clearly, PGRFA are more rivalrous in situations where supplies of the physical component (in which the information component is embedded) are limited, such as when a gene bank does not have funds to regenerate sufficient reproductive materials to provide samples or where the numbers of individuals of an *in situ* population are so small that the act of collecting a small number of plants creates the risk of extinction.

One relevant characteristic of PGRFA that distinguishes them from natural resources is that human intervention (for example, planting, seed selection, cross-breeding and regeneration of *ex situ* accessions) is a critical component of the development and conservation of crop and forage diversity (Halewood, 2008; Schloen et al., 2011). In the absence of continued use by humans, much of the existing plant genetic diversity would disappear. Underuse of PGRFA, not overuse, is the primary driver behind its loss as a resource. This idiosyncrasy is one of the 'social dilemmas' that needs to be addressed through collective action in the stewardship of the PGRFA commons.

Excluding others' uses of genetic resources is difficult and expensive. PGRFA share this quality with natural resources in the traditional commons. Indeed, PGRFA are even less excludable, given the ease with which a single seed, with all the genetic information embedded in it, can be moved from place to place. However, as Ravi Kanbur (2002) states, 'while rivalry can be characterized as a property given by a technology, excludability is man made.' And it is noteworthy that, as we highlighted earlier, national governments, private companies, university technology transfer departments, local communities and civil society organizations have demonstrated in the last 40 years a willingness to expend considerable financial resources to exclude (or at least try to exclude) others from using some subsets of PGRFA. They have done so through a variety of ways including technological means (for instance, creating hybrids or genetic use restriction technologies) and legal means (for instance, creating national and international intellectual property and access and benefit-sharing laws and investing in their administration and enforcement). Where these strategies have been successful, the PGRFA that are subject to these controls have been taken out of the potential scope of the PGRFA commons that is supported by the ITPGRFA. Indeed, the Treaty recognizes and 'works around' these situations where PGRFA are, or could be, subject to forms of restrictive controls (Halewood and Nnadozie, 2008). Article 11.2 of the Treaty explicitly states that only those PGRFA that 'are under the management and control of Contracting Parties and in the public domain' are automatically included in the multilateral system. All other PGRFA have to be voluntarily placed in the multilateral system, subject to whatever laws are in place in the countries concerned. As a result, a wide range of materials is not automatically included in the multilateral system, including those subject to intellectual property rights[12] and those that are under the management and control of individual farmers, companies, communities, provincial governments and so on. The end result is that most materials automatically included in the multilateral system are *ex situ* collections held by national governments. The Treaty also automatically includes those Annex 1 *in situ* PGRFA that are in the management and control of contracting parties and in the public domain. In practical terms, this appears to be limited to PGRFA on lands controlled by the national government that are not subject to competing claims of control by occupant farmers. Ultimately, in most countries, the practical operation of the Treaty's formula will result in a considerably higher proportion of *ex situ* material being automatically included in the multilateral system than *in situ* material.

What are global public goods? Why do we say that some key actors are unwilling or unable to shift away from the ideal of PGRFA as global public goods and why does that matter?

Economists define public goods as those that are non-rivalrous (one person's use of the good does not detract from another's use of the same resource) and non-excludable (it is available for everyone to use). Global public goods are non-rivalrous and non-excludable on a global scale (Morrissey et al., 2002). International public goods may be global, but the term leaves room for their being available for use only across a number of countries.

There are numerous organizations – some of the most important being the CGIAR centres that host international PGRFA collections and their financial supporters – that are deeply attached to the conceptualization of PGRFA as global public goods. The reasons for this attachment are both historic and practical. To understand them requires a rehearsal of the most salient shifts in the underlying legal status of PGRFA in international law and the efforts of the CGIAR centres to secure the legal status of the collections they host in particular.

Prior to 1983, international law was silent with respect to the conditions under which genetic resources could be accessed and used. Subject to a few notable exceptions, there were almost no national laws addressing the same issue. Plant genetic resources were, *de facto*, in the public domain, available to anyone for any purpose, without any restrictions, and without any concomitant obligations to participate in the stewardship of these resources. In 1983, the FAO's General Council adopted the non-binding International Undertaking on Plant Genetic Resources for Food and Agriculture (International Undertaking), which proclaimed the 'universally accepted principle that plant genetic resources are a heritage of mankind and consequently should be available without restriction'. The concept of the 'common interest of mankind' had previously appeared in the Treaty on Principles Governing the Activities of States in the Exploration and Use of Outer Space, including the Moon and Other Celestial Bodies,[13] and the idea of the 'common heritage of mankind' appeared in the Agreement Governing the Activities of States on the Moon and Other Celestial Bodies[14] and the Convention on the Law of the Sea.[15] In all three cases, the principle was associated with affirmation that the resources in question could not be subject to claims of national sovereignty or to other forms of appropriation (unless approved through an international mechanism), that all countries together shared a responsibility for conserving the resources for future generations, that the resources should be openly available for scientific research for peaceful purposes and that the results of that research should be shared. The 'heritage of mankind' in the International Undertaking appears to have been intended to have the same meaning.[16] On the basis of this concept, the International Undertaking sought to establish 'an internationally coordinated network of national, regional and international centres … under … FAO … to hold, for the benefit of the international community and on the principle of unrestricted exchange, base or active collections of the plant genetic resources of particular plant species'.

Meanwhile, the CGIAR centres were experiencing some insecurities with respect to the status of the collections for which they were responsible. In the early to mid-1980s, the management by the IBPGR of the network of base collections was subject to some high-profile criticism from some vocal civil society organizations and developing countries (Kloppenburg, 1988a; Mooney, 1983). Later, in the early 1990s, there were widely publicized rumours that the World Bank might appropriate the collections held by the centres, or that countries hosting CGIAR centres would seek to exercise exclusive controls over the collections, or that the CGIAR centres themselves would seek to do so (Halewood, 2010). In 1994, under pressure to end such controversies, the CGIAR centres signed agreements with the FAO, formally placing the crop and foraged collections within the International Undertaking's international network, affirming that they held the collections 'in trust for the benefit of the international community' and that the centres would not claim legal ownership over the germplasm or intellectual property rights with respect to the germplasm or its related information (SGRP, 2010). They agreed to pass on these conditions to whomever they distributed materials to from the collections. They also agreed that they would provide facilitated access, at no cost, to anyone requesting materials for the purpose of agricultural research. The agreements were approved in principle by the consortium of donors to the centres (that is, the CGIAR, chaired by the World Bank). They were greeted with enthusiasm by both the Commission on Plant Genetic Resources (CPGR) and the COP to the CBD (Halewood, 2010). The idea that the PGRFA collected from around the world, housed in international collections, should be treated as global public goods was firmly entrenched in the agreements between the centres and the FAO and condoned by both the United Nations Environment Programme and agricultural communities. In the years that followed, the centres developed internal policies and practices and information systems in line with their commitments to making 'in trust' materials available to anyone, anywhere in the world, who requested them.[17] These practices and policies became fully integrated into the self-conceptualization of the centres and in their approach to their mission. The 1994 in-trust agreements were meant to be a temporary measure until the ITPGRFA negotiations, which started in 1994, were completed. No one anticipated that those negotiations would take so long and that it would be 12 years before agreements could be signed between the CGIAR centres and the Governing Body of the Treaty.

In 2006, the CGIAR centres signed agreements with the Governing Body of the Treaty, placing the in-trust international crop and forage collections under the Treaty's framework. At that time, the CGIAR centres issued an interpretive statement to the effect that their agreements with the Governing Body did not prevent them from distributing materials from those collections to states that were not Treaty-member states. While non-parties are not *entitled* to receive materials from the multilateral system on request, there is nothing in the Treaty prohibiting member states from *voluntarily* providing materials that are in the multilateral system to those non-parties. The same is true for international institutions that sign agreements to place collections under the Treaty's framework. The CGIAR centres have continued to provide materials, indiscriminately, to non-parties since their agreements with the Governing Body came into force in January 2007. This is perfectly justifiable from the perspective of the centres, the agencies that fund them and the beneficiaries of their work – the CGIAR's mission is to promote food security in developing countries, and they have been maintaining the international collections, *qua* global public goods, as part of that effort for decades. In addition, very importantly, the international collections include materials that were originally collected from countries that are not presently members of the Treaty. As a result of these factors, the centres naturally resist the idea that a country should, by virtue of deciding not to join the Treaty or simply by being slow to ratify it, suffer discrimination in terms of getting less support for their agricultural research and development efforts. This is particularly significant when so far, at least 50 per cent of the material that is currently positively identified as being in the multilateral system (that is, listed on the Treaty webpages) is hosted by the CGIAR centres.

The problem with this situation – as far as the multilateral system is concerned – is that by making materials available to non-parties to the ITPGRFA, the CGIAR centres undermine some of the incentive that would otherwise have existed for countries to both ratify the Treaty and participate as providers of germplasm through the multilateral system. If countries can get access to the international crop and forage collections without joining the Treaty, they will be less motivated to ratify it because they enjoy the advantage of gaining access to pooled resources without incurring the 'cost' of having to place any of their own resources in it – a classic free-rider scenario. Furthermore, some member countries may be reluctant to deposit new diversity into the CGIAR centre-hosted collections – to date the most active 'pumps' of germplasm movement in the multilateral system – because they know that the CGIAR centres will make those materials available to non-members. Indeed, we know that these considerations have played a role in decisions made by competent national authorities in a few countries. In their understandable drive to continue treating PGRFA like global public goods, the CGIAR centres are undermining the incentives that exist (at least in principle) in the architecture of the Treaty – incentives based on the existence of a limited (or at least identified) community of users bound together by a reciprocal assumption of responsibilities for sharing and conserving resources.

In this context, it is also important to note that some developed country member states of the ITPGRFA – for example, the Netherlands and Germany – have also made multilateral system materials available to non-parties, using the same Standard Material Transfer Agreement (SMTA) that is used for all transfers of PGRFA in the multilateral system.[18] Prior to ratifying the Treaty, these countries often made materials available globally, and they, like the CGIAR centres, have decided to exploit the flexibility of the Treaty to continue this practice.

Meanwhile, the United States, which is not a party to the Treaty, is the second largest international supplier of PGRFA samples in the world (after the CGIAR centres). Through the National Genetic Resources Service (NGRS), the US Department of Agriculture (USDA) distributed, internationally, approximately 16,000 samples of ten crops per year throughout the 1990s, and another 45,000 samples domestically for each year (Fowler and Hodgkin, 2004). The NGRS does not use the SMTA when distributing material (unless it actually receives the material in question under a SMTA). Instead, it makes PGRFA available with minimal conditions – usually just a phytosanitary certificate – to most countries around the globe without discrimination (other than those countries included on the State Department's rolling list of enemy states). Just as the ability to get materials from the CGIAR centres without being a member of the Treaty undermines incentives to become a Treaty member (and make material available), so too does the ready availability of materials from the USDA's Agricultural Research Service (ARS). However, the availability of materials from the ARS also undermines incentives from actors hoping to receive materials from the multilateral system. There is some tentative evidence that recipients who want to avoid the benefit-sharing conditions of the multilateral system are seeking to obtain germplasm from the USDA's ARS instead of the CGIAR centres (Hammond, 2011).

Why are most Treaty state members less constrained than the CGIAR centres in shifting away from the concept of PGRFA as global public goods?

With the exception of the relatively small number of developed countries listed earlier, most ITPGRFA member states are clearly more comfortable departing from the concept of PGRFA as global public goods and embracing the concept of a PGRFA commons, based on membership-related benefits, collective management of shared resources and reciprocity. This difference between countries and the CGIAR centres can also be traced back to the negotiations in the CPGR in the 1980s and early 1990s concerning the development of the 'internationally coordinated network of ... collections of plant genetic resources' under the framework of the International Undertaking.

Originally, it was anticipated that national organizations hosting PGRFA collections, and/or the governments of the countries in which they were located, would sign agreements with the FAO – similar to those signed by the CGIAR centres in 1994 – in order to formalize participation in the internationally coordinated network. Negotiations for the development of such agreements started under the auspices of the CPGR in 1985 and continued until 1997 (when they were suspended pending the outcome of the negotiations of the ITPGRFA). Had such agreements been signed, they would have included commitments such as those made by the centres to treat PGRFA as global public goods. Such agreements might have had the effect of 'fixing' countries' conception and treatment of PGRFA as has been done in the case of the CGIAR centres. However, ultimately, no country ever signed any such agreements, and most countries' concept of the status of PGRFA shifted significantly in the years that followed, as reflected by the developments in the CPGR. The chronology of events culminating in the erosion of the 'heritage of mankind' in the International Undertaking will be familiar to many readers, but it is worth repeating the most salient details here. In 1983, a number of developed countries had abstained from the decision adopting the International Undertaking on the grounds that it failed to recognize plant variety protection rights. Pressures to compromise – to get the abstaining countries on board – led to the adoption of a resolution, by the FAO Council, in 1989, interpreting the International Undertaking's universally accepted principle of the heritage of mankind as being subject to the recognition of plant variety protection rights. Just two years later, in 1991, the FAO Council adopted a second resolution that recognized 'that nations have sovereign rights over their genetic resources'.[19] The heritage 'of mankind, while still on paper, had been effectively overridden.[20]

The end of PGRFA as common heritage, and the rise of national sovereignty, did not end efforts to develop international policies to support the establishment of internationally coordinated systems of conservation and use of PGRFA, including collective pooling and management aspects. Indeed, as argued elsewhere, the shift in the underlying understanding of the legal status of PGRFA appears to have contributed to countries' confidence to actually establish such a system in the form of the ITPGRFA (Halewood, forthcoming). Armed with the recognized right to control access to genetic resources in their jurisdictions, countries have demonstrated a willingness to formalize rules for commons-based pooling and use of PGRFA – but only with other countries that are willing to participate according to the same rules. This way of thinking about PGRFA, not as global public goods but, rather, as resources to be conserved and shared among a bounded community of users on the basis of reciprocity, is reflected in the basic infrastructure of the multilateral system, at least as far as member states are concerned.[21] Countries negotiating the Treaty were able to conceive of such a system on the basis of the dissolution of the concept of PGRFA as the heritage of mankind. They were also 'helped along' by the fact that the Treaty does not attempt to place all PGRFA in the multilateral system and allows for the exclusion of PGRFA subject to intellectual property rights, materials held by companies, individuals, communities and so on. The breakdown of the concept of all PGRFA as necessarily common heritage allowed countries to maintain restrictive controls over those PGRFA about which they feel possessive, while simultaneously allowing the rest to be included in the PGRFA commons supported by the Treaty's multilateral system. Developed countries have been most concerned about controlling access to materials that are the products of formal sector plant breeding or biotechnological manipulation, and they are more willing to treat landraces and wild relatives as openly available. Developing countries, on the other hand, have reacted to extensions of intellectual property rights over improved materials by stressing the importance of controlling the access to a wider set of biological and genetic resources, including farmers' varieties, wild relatives and so on.

How are free-riders able to benefit from the Treaty's multilateral system without contributing meaningfully to it? And why does this issue matter?

We have already highlighted the case of countries that are not members of the ITPGRFA, who are able to access materials under the Treaty from the CGIAR centres and some Treaty members without joining the Treaty and placing any materials in the multilateral system. It is also possible for Treaty members to engage in similar forms of free-riding. While state parties make a *de jure* commitment to make materials (that are under their management and control and in the public domain) available through the multilateral system, there is no requirement that they actually need to demonstrate that they have taken any active steps to do so before they are entitled to request and receive materials from others in the multilateral system. This is important because, in the absence of parties' publicly confirming which collections (or parts of collections) and which *in situ* materials within their countries are in the multilateral system, it is difficult (or impossible) for outsiders to know what is available from that country. Similarly, in the absence of minimum subsets of information about each accession (in collections) that is available, they are virtually useless. If all one knows about an accession is that it is rice, or wheat, or trifolium, it is unlikely to be requested. Each additional layer of passport, characterization, evaluation, climate-related and other forms of data adds to the potential utility of each accession. The Treaty anticipates the creation of a global information system through which such information could be shared.[22] However, the Governing Body has thus far not endorsed or created such a system. The absence of this anticipated global information system has certainly lessened the urgency that contracting parties might have otherwise felt to publish such information. Meanwhile, in 2010, the Treaty's Secretariat sent a notification to all Treaty focal points asking them to confirm what *ex situ* collections are included in the multilateral system.[23] (The request did not raise the more difficult question of identifying *in situ* materials in the multilateral system.) As of December 2011, only 17 per cent of the 126 member states have provided any such notifications. In any case, the Treaty does not create a legal requirement that members share information confirming what materials are in the multilateral system, or additional accession-level information, before they may receive materials through the multilateral system. So they may receive materials from the multilateral system without making any material practically available.

Free-riding is currently widespread among 'natural and legal persons'. Natural and legal persons located within member states are not required to place anything in the multilateral system as a precondition for their entitlement to request material through that system. Member states agree under the Treaty to encourage natural and legal persons to voluntarily make materials available through the multilateral system. So far, however, few natural and legal persons have actually voluntarily included new materials in the multilateral system, at least as far as one can tell from notifications to the Treaty's secretary.

Another form of possible free-riding by both member states and natural and legal persons has to do with the ability to obtain support from the benefit-sharing fund without having previously put any materials into the multilateral system. The Governing Body has developed rules to the effect that recipients of grants from the benefit-sharing fund must make PGRFA that have been developed (or regenerated, characterized or evaluated) available through the multilateral system (ITPGRFA, 2009). However, having made PGRFA available previously is not a *sine qua non* for applying for, or receiving, support.

These various forms of free-riding have arisen as a result of the fact that the Treaty does not include mechanisms to follow up on, and effectively enforce, the commons-informed logic that the Treaty otherwise largely embraces. This absence of effective mechanisms may reflect, *post-facto*, some naivety on the part of negotiators who designed the system regarding: (1) states' and natural and legal persons' self-interested behaviours (and their willingness to exploit free-riding opportunities) and (2) the difficulty that some participants under the Treaty's framework would have in breaking from their earlier conceptions of PGRFA as international public goods. The negotiators may have expected that, with the basic agreement of reciprocal facilitated access firmly embedded in the text of the Treaty, the relevant actors would 'follow suit' and adopt consistent, system-reinforcing policies and practices. One result of this free-riding is that many potential and actual participants are uncertain about each other's degree of commitment to the multilateral system, and they are unsure about whether the advantages of being in the system could not be enjoyed, at lower cost to themselves, by staying outside the system. Many potential and actual participants in the multilateral system exhibit relatively low levels of shared purpose and motivation.

Discussion of this issue would be incomplete without considering the provisions for monitoring and enforcing the Rules of Compliance recently adopted by the Governing Body in 2011.[24] In general, these rules do not reflect the low transaction cost, quick-response monitoring and enforcement principles that Elinor Ostrom derived from her observations of well-functioning traditional natural resource-based commons. Ostrom notes that in a successfully managed commons, the actual users of the concerned resource are the monitors of compliance with the rules governing use (Ostrom, 1990). The Rules of Compliance say that submissions regarding non-compliance to the compliance committee can only be made by contracting parties or the Governing Body itself.[25] In other words, the scientists, development workers and farmers who are the day-to-day users and conservers of resources in the multilateral system have no mechanisms to raise issues of non-compliance (other than through official government representatives to the compliance committee). While, on the one hand, this form of compliance mechanism is standard for international agreements – which are, after all, between states – it likely contributes to the fact that the complaints one often hears from researchers and breeders about germplasm not being made available are not being raised at the level of the Governing Body, where higher level political trade-offs, deals and the culture of consensus would be damaged by complaints by and against individual parties. The process for considering disputes under the rules is also slow. The Treaty's secretary has a month to forward complaints that have been received. The country that is the subject of a complaint has six months to respond. There is no time limit set yet on the deliberations of the committee.[26] Furthermore, the sanctions at the disposal of the Compliance Committee and the Governing Body are limited to requesting the contracting parties to develop an action plan to address the issue of non-compliance and request related progress reports. The Governing Body 'may take other actions it deems appropriate, including for capacity-building'. While these approaches to monitoring, compliance and enforcement may be typical of international environmental agreements – and for good reasons – they generally do not engage PGRFA users in ways that could contribute to the vitality of the multilateral system.

Before continuing, we want to make an earnest disclaimer about our use of the term 'free-riding'. The scenarios we are describing here are typical, in a generic sense, of what is called 'free-riding' in economics literature (Olson, 1971). For the sake of convenience, we have chosen to use this term in this chapter. However, we use the term cautiously because we fully appreciate that some countries that are not currently Treaty members have already contributed considerable germplasm to the international collections hosted by the CGIAR centres. The same is true of some Treaty member states that have not yet confirmed which materials will be included in the multilateral system. A number of such countries, from the 1970s to the 1990s – before the Treaty came into force – contributed thousands of accessions to the collections hosted by the CGIAR centres that are now available under the Treaty. So while they may be free-riding since the Treaty came into force, with the potential negative impacts on the system that we have described here, it is important to note that many such countries are, nonetheless, the source of significant amounts of material that were brought into the system by the CGIAR centres (and other gene banks).

What policy reforms could potentially ameliorate the situation? Are they feasible?

The answers to these questions are largely beyond the scope of this introductory chapter. Nonetheless, we will make a few observations that we hope will set the stage for some of the chapters that follow and highlight the issues that need to be addressed in the longer term. The proceeding analysis suggests that more clearly defining, monitoring and enforcing (1) the boundaries of the 'who is in and who is out' of the multilateral system and (2) the reciprocal obligations of all participants is essential to overcoming both free-riding and the knock-on disincentives of that free-riding for actors to place materials in the multilateral system. In short, the multilateral system could be invigorated if access to materials in the multilateral system was limited to Treaty member states, and natural and legal persons within those member states, who actively place materials in the system. Similarly, it is important to have put systems in place for regular monitoring by PGRFA users, for compliance with the rules and for appropriate sanctions that can be enforced relatively quickly and appropriately.

Can the boundaries of the PGRFA commons be delimited so that only Treaty member states can receive materials from the multilateral system? Our analysis has highlighted the fact that a number of actors in the multilateral system – the CGIAR centres and some European countries – are currently either unable or unwilling to stop making PGRFA available to non-member countries. Pursuant to their agreements with the Governing Body, the CGIAR centres are subject to the Governing Body's policy guidance. It is not entirely clear if it would be within the competence of the Governing Body to actually direct the centres to stop making materials available to non-parties, given that such a limitation is not included in the text of the Treaty. Equally importantly, it is clear, at least for the time being, that the Governing Body will not agree to provide such guidance, given that some member states of the Treaty have also decided to make multilateral system material available to non-parties using the SMTA and Governing Body decisions are usually made by consensus. In addition, some of the biggest financial donors to the CGIAR centres, and by extension to the international *ex situ* PGRFA collections hosted by the centres, are not yet members of the Treaty.

Countries that are currently making multilateral system materials available to non-parties have more latitude than the CGIAR centres to reverse their policy in this regard. For the time being, however, with very few exceptions (Altoveros et al., 2012) the issue has not received much critical attention, and the developed countries concerned are not under significant pressure to reconsider their policies in this regard. They are motivated, like the CGIAR centres, partly out of a sense of duty to users in poorer countries that have not, for whatever reason, ratified the ITPGRFA, and partly by the administrative efficacy of using the SMTA when possible. They also argue that providing materials to non-parties under the SMTA will make those non-parties more comfortable with the multilateral system and therefore more inclined to join the Treaty. This strategy may work. However, if in the years to come there is evidence that the strategy does not work – that is, that more countries do not join the Treaty – the Treaty parties concerned will feel more pressure to reconsider their policy.

Meanwhile, one developing country has recently 'taken matters into its own hands' and deposited materials with some of the CGIAR gene banks, stating that the centres concerned may redistribute this material using the SMTA but only to recipients in the Treaty member states. The country is clearly within its rights to make this request, although it is inconsistent with the CGIAR centres' policy (which they are within their rights to adopt), and it has caused them some discomfort in considering how to respond.

It is worth underscoring that while the rate of ratifications of the ITPGRFA has slowed down, it has not stopped. Some countries – including the United States – are currently working through domestic processes that could lead to ratification. Each new country joining the Treaty reduces the issue of free-riding non-parties. If Japan, China, Mexico, Russia and the United States joined the Treaty, it seems extremely likely that the remaining countries that are holding out would quickly follow suit, and the problem would be overcome. (An additional benefit of the United States joining would be that a significant alternative supply of PGRFA – from outside the multilateral system – would disappear.)

Is it possible to tighten up the multilateral system so that member states and natural and legal persons have to make materials available (or contribute in other ways to the maintenance of the commons) as a precondition for receiving materials? There is a mechanism built into the treaty to directly address this situation when it comes to natural and legal persons. Article 11 of the ITPGRFA states that the Governing Body will (1) assess the extent to which natural and legal persons are voluntarily placing materials in the multilateral system and (2) 'decide whether access shall continue to be facilitated to those natural and legal persons' that have not done so. The review should commence in 2013 at the fifth session of the Governing Body. Presumably, the Governing Body could decide to significantly tighten up rules regarding reciprocity, prohibiting natural and legal persons from obtaining materials from the multilateral system if they do not also contribute materials.

Meanwhile, some governments and international donors are already creating inducements for natural and legal persons to 'voluntarily' place materials in the multilateral system. For example, donors supporting PGRFA-related research and development can require recipients of their support to undertake to place PGRFA in the multilateral system and share related information as a condition of receiving financial support.

The situation is considerably more difficult when it comes to member states. There is no built-in Article 11-equivalent review with the objective of deciding whether member states should continue to enjoy facilitated access to PGRFA if they do not also provide it. The route forward therefore would involve case-by-case complaints through the compliance committee – a mechanism that is not developed to deal with free-riding complaints quickly or effectively. In this context, it would be useful if the additional rules to be developed for the compliance committee include provisions for more dynamic, responsive, user-information-driven processes.

Perhaps the most immediate, practical, non-coercive measure to help address this problem would be for the Governing Body to finally create the planned-for global information system. The global information system would provide a mechanism for countries to share information about the materials they are making available through the multilateral system, and eliminate an understandably attractive excuse for not doing so. It would also provide a basis for voluntary monitoring by interested organizations and parties concerning the steps others are taking to make materials and information available under the Treaty's framework.

Can mechanisms for more participatory monitoring and reporting be developed? Monitoring and reporting are areas where the strains associated with expanding the scope of commons from small and local to global are most apparent. International agreements that can provide legal and administrative support for 'nested' levels of networked activity in support of a global commons (as the Treaty does for the PGRFA commons) necessarily make states the bearers of rights and responsibilities. However, state-level representatives will rarely have first-hand experience seeking access to, or providing access to, PGRFA in the multilateral system. And as stated earlier, delegates to meetings of the Governing Body are subject to political incentives to *not* raise such issues. Researchers, breeders, development workers and farmers do have such experiences and are *de facto* monitors of the system by virtue of their day-to-day work. In the absence of disincentives possibly created by their own governments, they are much more likely to speak out freely and openly about abuses of the system. Again it would be useful if the Governing Body could open up more practical opportunities for users of the multilateral system to participate in monitoring and reporting abuses.

Similar criticisms can also be made of the potential sanctions provided for under the ITPGRFA and its compliance mechanisms, as they do not appear to be designed to address free-riding problems directly or quickly. It is beyond the scope of this introduction to explore possible ways to revise the compliance/enforcement measures or to identify complementary voluntarist efforts that concerned countries or organizations could engage in to address some of these challenges. For the time being, we will simply highlight the importance of addressing these issues in the future. Potential providers of genetic resources will be de-motivated to participate in the multilateral system if they have no faith that free-riders in the system will not be detected, reported and sanctioned.

Design issue no 2: the mandatory financial benefit-sharing provision: mixed incentives somewhere between multilateralism and bilateralism

The following section provides a condensed summary of the second design-related challenge. The multilateral system occupies a 'middle space' or compromise between fully multilateral and bilateral approaches to regulating access and benefit sharing. Not all of the component parts of the Treaty's multilateral system are entirely novel. Instead, some of them are drawn more or less unchanged from multilaterally and bilaterally conceived approaches. Not surprisingly, as a result, the multilateral system enjoys, and suffers from, a combination of the advantages and disadvantages of both bilateral and multilateral systems. In addition, because some of the components are drawn from different types of systems, in combination they give rise to some novel tensions that have the potential to dampen some actors' enthusiasm for participating in the multilateral system as conservers, users, providers and recipients of germplasm and information. Perhaps users will eventually overcome reservations about the extant benefit sharing formula. Alternatively, some rebalancing of the access and benefit-sharing conditions in the multilateral system – in particular, the mandatory financial benefit-sharing formula – could potentially create incentives for more active participation.

As stated above, we fully appreciate the history of the complex negotiations of benefit-sharing provisions under the Treaty, and we celebrate the formula included in the standard material transfer agreement as an outstanding achievement. Again, it may be that, as more actors become aware of the overriding benefits of facilitated access to plant genetic diversity through the multilateral system, they will be less concerned about aspects of the financial benefit-sharing formula. In the meantime, however, it is worthwhile examining options. In the remainder of this section, we provide a more comprehensive account of this argument, defining and explaining the significance of the key concepts introduced in the preceding paragraph.

Why do we say the Treaty's multilateral system occupies a middle space between bilateralism and multilateralism?

This question can best be answered by returning, once again, to the 1983 International Undertaking. The architects of the International Undertaking had in mind a truly multilateral system of access and benefit sharing – a system where all PGRFA – from any potential source – would be made available to anyone, without restrictions. Financial benefits derived through the commercialization of those resources would be shared through an international fund. The fund would be used to support the efforts of developing countries (and farmers in those countries) in their conservation and sustainable use-related activities. Support from the fund would not depend upon whether a resource from the country or farmer concerned was actually incorporated in the commercialized resource. And the payments into the fund would similarly not be tied to whether or not newly accessed materials were used in the development of commercialized products.[27]

As stated earlier, CBD member states, and the CBD's COP, have taken a bilateral approach to access and benefit sharing. Generally speaking, national laws implementing the CBD require that access and benefit-sharing terms are to be negotiated, as part of a single deal, between the providers and the recipients on a case-by-case basis. In some cases, competent national authorities handle the negotiations on behalf of the providers in their countries; in other cases, the authority must approve deals reached between the provider and the recipient. Access is not granted until there is agreement about the benefits to be shared. The quantum of benefits are meant to be linked in some way to the market value of the resource provided, although the precise amount is left to be negotiated between the supplier and the recipient. The invisible hand of the market is entrusted to play an important intermediary role in this regard. Benefits may be shared in many forms – up-front payments, royalties upon commercialization, technology transfer and training – but they are directed to the actual provider.

The ITPGRFA's multilateral system occupies a space somewhere between these two models. In the multilateral system, novel access and benefit-sharing agreements are not negotiated between transferring parties each time one seeks access to PGRFA. Instead, these terms are pre-agreed to and set out in the SMTA. As far as benefit sharing is concerned, the multilateral system is considered the supplier, and mandatorily shared financial benefits are directed to the multilateral system, not to the organizations, individuals or governments that actually provide the resource. In this way, both the negotiation of prior informed consent to provide access and the receipt of benefits are 'de-linked' from the countries, communities or legal individuals from whom those materials were originally collected (or who deposited them in collections through which they were made available). However, the requirement to share the monetary benefits under the multilateral system is not de-linked in the same way (Schloen et al., 2011). Mandatory financial benefit sharing is not triggered unless material accessed from the multilateral system is incorporated in actual new PGRFA products. And, then, the payments are directly linked to the sale of those particular products (in the form of a royalty payment of 1.1 per cent of gross sales minus 30 per cent to cover incidental costs).[28]

How does the multilateral system suffer from challenges associated with bilateralism?

A negative consequence of the Treaty's approach to mandatory financial benefit sharing is that it requires complex and relatively burdensome administrative and legal machinery to ensure that recipients of materials from the multilateral system share monetary benefits when they are supposed to. The requisite mechanisms include the obligation for all transferors to report each transfer they make to the Governing Body, the creation and maintenance of a storehouse of encrypted data concerning those transfers, maintaining a legal representative of the third party beneficiary interests of the multilateral system with the power to initiate legal proceedings where the benefit sharing has been breached and so on. Indeed, almost half of the enumerated clauses in the 12-page SMTA are devoted to establishing conditions, procedures and enforcement of mandatory financial benefit sharing. The Treaty's Governing Body has also had to dedicate a lot of time and financial resources to perfecting the rules and procedures that assist in keeping track of where materials are being sent, how the third party beneficiary can proceed and so on.[29] The resources expended to establish and maintain this system certainly far exceed the financial benefits that it has generated so far (there have not been any mandatory financial benefit-sharing payments yet) or are likely to be generated in the near (and possibly distant) future. The transaction costs are of course considerably lower than they would be for a fully bilateral system, but higher than they would be in the absence of the bilaterally-influenced approach to triggering benefit-sharing.

How does the multilateral system suffer from challenges associated with more broadly conceived multilateralism?

While tying benefit-sharing obligations to the actual uses of materials accessed from the multilateral system on a product-by-product basis is what one would expect to find in a bilaterally negotiated access agreement, not all aspects of the Treaty's benefit-sharing formula are bilaterally inspired. The 'triggers' (that is, any level of incorporation, commercialization and so on) for benefit sharing, and the amount that needs to be shared in each case, are fixed in the SMTA – they cannot be altered through agreement by the parties exchanging PGRFA. Such multilaterally inspired elements can work well if all of the intended users of that system are generally content with the established terms. However, as it turns out, a number of companies are not satisfied with the benefit-sharing conditions in the multilateral system, particularly those that would trigger the benefit sharing provisions by patenting new PGRFA products, or restricting their availability through technological means. The larger 'crop life' companies are avoiding, whenever possible, accessing materials through the multilateral system. As a result, the very users who should be generating contributions to the international benefit-sharing fund are not actually accessing materials from the system. On the other hand, so far, even those commercial users who have accessed materials from the multilateral system have not 'triggered' obligations to share financial benefits. The sole source of funds in the international benefit-sharing fund to date is from voluntary contributions from governments.

Another challenge associated with the Treaty's multilateralism concerns the complete de-linking of the countries, individuals or communities who provide PGRFA from the monetary benefits generated from the use of these materials. This approach to benefit sharing does not take advantage of a potential motivation for making material available through the multilateral system. Some would-be providers' reluctance to either (1) clarify what is already automatically in the multilateral system by virtue of their country's ratification of the Treaty or (2) voluntarily place additional materials in the system, could possibly be overcome if they knew they would receive a portion of the 1.1 per cent royalty fees that will be directed to the international benefit-sharing fund if those materials are commercially exploited.

How does the mixture of components from both kinds of access and benefit-sharing systems, married together under the Treaty, create novel challenges?

The asymmetry in the approach to access and benefit sharing in the Treaty's multilateral system (de-linked prior-informed consent, access and receipt of monetary benefits versus linked use and obligation to pay benefits) makes the system vulnerable to dissatisfaction on the part of providers of PGRFA, particularly in the case of developing countries. Having agreed to de-link the provision of access from the receipt of benefits, developing countries naturally tend to concentrate their attention (when it comes to monetary benefit sharing) on the state of the international benefit-sharing fund. Justifiably or not, the fund has become for many the most reliable indicator of the success or failure of the multilateral system. If the fund is not able to support a significant range of capacity building and sustainable use-related relevant activities in developing countries, these countries will likely be dissatisfied with the multilateral system as a whole. Past dissatisfaction with the amount of money available through the benefit sharing fund appears to have influenced some countries' sense of urgency with respect to (1) confirming what materials are automatically in the multilateral system, (2) voluntarily including additional materials and (3) encouraging natural and legal persons within their borders to do the same. In recent years, the levels of support to the fund have increased, and two rounds of projects have been supported from the fund; if levels of support continue to grow, the impact on countries' approach to the multilateral system should be positive. The fact that developing countries have also agreed to not de-link use in particular products from obligations to share benefits increases their negative reaction to the fact that commercial users are avoiding taking materials from the multilateral system and that shared monetary benefits are not flowing into the international fund.

Are there possibilities to reform the mandatory monetary benefit-sharing formula?

The forgoing analysis suggests that some reformulation of the mandatory financial benefit-sharing formula in the SMTA could create incentives for a broader range of actors to dynamically engage in the Treaty's multilateral system as conservers, users, providers and recipients of PGRFA.

Policy reform to address the current asymmetry of the mandatory financial benefit-sharing formula could go in either of two directions. One possibility would be to 're-link', in some way, the actual provision of PGRFA with sharing actual benefits associated with its commercial use. In this regard, it might be enough to simply redirect a proportion of the 1.1 per cent directed to the international fund to the original source of the material used, if that can be ascertained. The promise of such a payment could act as an incentive for would-be providers to place materials in the system in the future, especially when this material is of no known commercial value (as most PGRFA is). Such an approach would involve stepping back, at least partly, towards bilateralism, and would involve substantial additional costs in record keeping, tracking enforcement and so on.

Another possibility would be to 'de-link' the actual use of accessed materials from the obligation to share financial benefits associated with the commercial exploitation of PGRFA. De-linking would have the advantages that (1) the resource-intensive record-keeping system would not be necessary and (2) the incentives to seek (and stockpile) materials from outside the multilateral system would also be reduced as the obligation to share financial benefits would not be based on actual incorporation of materials accessed from the multilateral system in new products. Instead, the obligation to share benefits would be triggered by the act of commercializing any PGRFA, and the quantum would be calculated on some simple formula, such as the percentage of seed sales in the country concerned.[30] While de-linking in this way would not create a direct incentive for actors to place more materials in the multilateral system, it would potentially contribute, indirectly, to a greater level of confidence and comfort with the system on the part of those providers. Once actual incorporation was de-linked from an obligation to pay, then providers could potentially be more easily reconciled with the fact that provision of access is already de-linked from the right to receive a portion of financial benefits. This approach would involve extending the multilateral impulses of the Treaty's multilateral system further. It would also have the potential advantage of significantly lowering the costs of administering the system overall.

Article 13.2.d.ii of the ITPGRFA states that the Governing Body will assess 'whether the mandatory payment requirement in the [S]MTA shall apply also in cases where such commercialized products are available without restriction to others for further research and breeding'. Extending the requirement to pay when commercialized products are available without restriction would increase the amount of money directed to the international fund and should increase the level of satisfaction by would-be depositors. However, reforms as fundamental as those considered earlier concerning de-linking are beyond the scope of this scheduled review.

However, there is nothing to prevent key actors from voluntarily assuming practices that reflect the spirit of such policy reforms. The seed industry, for example, could establish non-binding guidelines for companies to voluntarily follow in order to make donations to the international benefit-sharing fund under the Treaty, based on the percentage of seed sales or some other metric. Once informed of such a voluntary practice, the Governing Body could express its appreciation and encourage companies to follow through. One would hope that, if these voluntarist efforts crystallized into consistent and predictable practices some time in the distant future, the

Governing Body would revise the benefit-sharing formula accordingly. One should not dismiss such a possibility too hastily. In the hallways outside formal negotiating sessions, one hears far more appreciation of the advantages of a much flatter system, with de-linked benefit sharing, than one would ever imagine based on official statements.

Before leaving this section, it is important to at least raise the possibility of renegotiating parts of the ITPGRFA or the SMTA if important policy reforms are beyond the scope of the scheduled reviews or remedial voluntary interventions by the key actors. For the time being, such drastic action does not appear worth contemplating. Renegotiating the Treaty or the SMTA would likely take years and millions of dollars. The full potential of the Treaty (as it is) has not yet been fully tested. The built-in reviews could contribute to significant invigoration of the multilateral system. And coordinated (imaginative) remedial voluntarism has hardly been explored. It is possible that by 'staying the course' that was set when the text of the Treaty was originally adopted and working on such reforms as are within the scope of the reviews created by the Treaty, the multilateral system will eventually become more firmly entrenched, more widely used and function in a very significant way to support the objectives of the Treaty.

The logic and structure of this book

Now that readers have been introduced to PGRFA pooling, the current state of access and benefit-sharing laws, and some of the challenges facing the implementation of the ITPGRFA, we want to move on to the rest of the book. The first part of the book – 'Countries' interdependence on PGRFA and the imperative of international cooperation' – is designed to deepen readers' appreciation of the extent to which countries rely on PGRFA that are (or were) located behind each others' borders. This part considers evidence that while countries' interdependence on PGRFA is increasing, their willingness to share those resources through internationally coordinated mechanisms generally is not.

In Chapter 2, entitled 'Demonstrating Interdependence on Plant Genetic Resources for Food and Agriculture', Marleni Ramirez et al. highlight how germplasm from many different countries was required to make advances in the recent improvement programmes of various crops.

Chapter 3, entitled 'Flows of Crop Germplasm Resources into and out of China', documents the high degree of China's dependence on germplasm from foreign sources as inputs into its agricultural research, plant breeding and development programmes as well as the valuable contributions China has made, in terms of germplasm, to other countries. Fuyou Wang's focus on China is particularly interesting, as China is one of a few countries – along with the United States, Mexico, Russia and Japan – that have considerable diversity of PGRFA and have not yet ratified the Treaty. The decision of such countries to ratify (or not) the Treaty has profound effects on the scope of the common pooled resources under the framework of the multilateral system. This chapter also highlights some of the very practical considerations that countries must make in terms of calculating the relative costs and benefits of participating in internationally coordinated systems for pooling and managing plant genetic resources.

Julian Ramirez-Villegas et al., in Chapter 4, entitled 'Crop and Forage Genetic Resources: International Interdependence in the Face of Climate Change', demonstrate that climate change is leading countries to be even more reliant on access to PGRFA from other countries, resulting in the concomitant importance of virtual common pooling of PGRFA under a favourable policy environment.

In Chapter 5 Halewood et al. highlight how the combination of high levels of political controversy and low levels of legal certainty (and satisfaction) concerning access and benefit-sharing rules have contributed to decreasing rates of new deposits of plant genetic resources to the international crop and forage collections hosted by the CGIAR centres. As the title of the chapter, 'Changing Rates of Acquisition of Plant Genetic Resources by International Gene Banks: Setting the Scene to Monitor an Impact of the International Treaty', suggests, it is hoped that the Treaty will eventually contribute to increased willingness of countries, companies, research organizations, farming communities and individuals to participate in systems of common pooling and use of PGRFA. The chapter reviews the evidence available to date concerning the Treatys impact in this regard.

The second part of the book focuses on the history, design and state of the implementation of the Treaty's multilateral system. After reading through the chapters in this part, readers will have a clear understanding about the structure and intended functioning of the multilateral system. They will also appreciate that implementation is moving ahead relatively slowly.

Chapter 6, by Jose Esquinas-Alcázar et al. – 'A Brief History of the Negotiations on the International Treaty on Plant Genetic Resources for Food and Agriculture' – provides a history of the international treatment of PGRFA from the first focused discussions of the need for coordinated conservation efforts in the FAO's Panel of Experts on Plant Exploration and Introductions in the 1960s, through the creation of the Commission on Plant Genetic Resources in 1983, through the seven-year negotiations of the Treaty, and finally to the subsequent negotiations of the SMTA.

Chapters 7 to 10 pick up where Chapter 6 ends, describing the architecture and practical day-to-day functioning of the multilateral system. In Chapter 7, entitled 'The Design and Mechanics of the Multilateral System of Access and Benefit Sharing', Daniele Manzella provides an overview of the most important components of the multilateral system. In Chapter 8, entitled 'Protecting the Interests of the Multilateral System under the Standard Material Transfer Agreement: The Third Party Beneficiary', Gerald Moore examines the evolution of the recognition of the third party beneficiary interests of the multilateral system. He reflects on how those interests are recognized in the right of a representative of those interests to bring legal actions, on behalf of the system, and how that helps to close an 'enforcement incentive problem' that exists in many multilateral environmental agreements.

Article 11.2 of the ITPGRFA states that the plant genetic resources that are 'in the management and control of the contracting parties and in the public domain' are automatically included in the multilateral system when a country ratifies the Treaty. However, those terms are not defined in the Treaty. In Chapter 9, entitled 'Plant Genetic Resources "under the Management and Control of the Contracting Parties and in the Public Domain": How Rich is the ITPGRFA's Multilateral System?' Carlos Correa suggests how these terms should be interpreted and the consequences for the range of materials that are, as a result, automatically included in the multilateral system.

In Chapter 10, entitled 'Efforts to Get the Multilateral System Up and Running: A Review of Activities Coordinated by the Treaty's Secretariat', Selim Louafi and Shakeel Bhatti provide a 'Secretariat's eye-view' of the range of ancillary procedures and mechanisms that have been (and are being) developed under the guidance of the Governing Body to support the implementation of the multilateral system. These include the establishment of the encrypted data store for information about all of the transfers in the multilateral system and the creation of an information management system that facilitates the access to, and exchange of, PGRFA within the multilateral system.

The third part of the book is dedicated to a critical analysis of the current state of implementation of the multilateral system and its position in the constellation of other mechanisms and systems dedicated to the conservation and sustainable use of genetic resources. After reading this section, readers will have a deepened appreciation of challenges to the fuller implementation of the multilateral system. Some of the chapters provide more detail and reflect different perspectives on issues highlighted in this introduction. Readers will also come to appreciate how the evolution of rules and practices related to the plant genetic resources commons are consistent with, and differ from, those that are evolving with respect to other commons.

In Chapter 11, entitled 'Assessment of Progress to Make the Multilateral System Functional: Incentives and Challenges at the Country Level,' Isabel López, Peterson Wambugu and Alejandro Mejías critically assess the progress that has (and has not) been made by countries to 'domesticate' the multilateral system. In so doing, they identify the range of incentives and disincentives that have informed countries' progress to date.

In Chapter 12, entitled 'From Negotiations to Implementation: Global Review of Achievements, Bottlenecks and Opportunities for the Treaty in General and the Multilateral System in Particular,' Godfrey Mwila also analyses factors contributing to countries' relatively slow rate of implementation of the multilateral system. He focuses on how a number of issues that were not resolved in the negotiations of the Treaty continue to affect many countries' willingness to fully engage with and proactively implement the multilateral system.

Clive Stannard, in Chapter 13, entitled 'The Multilateral System of Access and Benefit Sharing: Could It Have Been Constructed Another Way?' seeks to answer the question: could the multilateral system have been developed in some other ways that might have provided still more support for the common pooling and management of plant genetic resources? To answer this question, he looks back at some of the different components and characteristics of the multilateral system that were lost along the path of compromises that characterize international negotiations. Stannard underscores that it is possible to imagine a far simpler, more efficient, more horizontally constructed system of access and benefit sharing. Ultimately, he concludes, however, that the structure and function of the current multilateral system are the result of political compromises and that despite some of its shortcomings the multilateral system may be the best system that it is possible to create, given the relatively narrow common denominator in terms of political interests of countries involved in developing such systems.

Bert Visser, in Chapter 14 entitled 'The Moving Scope of Annex 1: The List of Crops Covered under the Multilateral System', examines the history of the negotiation of the list of crops and forages included in the multilateral system for insights as to whether it may be possible to expand the list in the future.

In Chapter 15, entitled 'Building a Global Information System in Support of the International Treaty on Plant Genetic Resources for Food and Agriculture', Caroline Ker et al. consider opportunities to develop a web-based plant genetic resources knowledge commons in ways that usefully link to, and support, the multilateral system. Such a system should include and be built on existing efforts to develop a federated 'global information system' to provide accession-level information about PGRFA in *ex situ* collections around the world. However, the authors contend that the genetic resources information commons should go beyond accession-level information to foster linkages with, and support for, activities of conservation and sustainable use of PGRFA. In the end, the authors imagine a 'one-stop' information commons where interested actors can find the most up-to-date and comprehensive information concerning all aspects of conservation and the sustainable use of PGRFA, farmers' rights, accession level information (including evaluation data), policies and laws affecting PGRFA.

Chapter 16, entitled 'Collective Action Challenges in the Implementation of the Multilateral System of the International Treaty: What Roles for the CGIAR Centres?' by Selim Louafi, analyses the role that the CGIAR can play in supporting the implementation of the Treaty's multilateral system and highlights how their mere existence and their experience in conserving, using and distributing PGRFA puts them in a situation where they can play a facilitating role as honest brokers.

Chapter 17, entitled 'International and Regional Cooperation in the Implementation of the International Treaty on Plant Genetic Resources for Food and Agriculture', is dedicated to analysing ways in which countries can organize themselves on a subregional and regional basis to take advantage of efficiencies in implementing the multilateral system. Gerald Moore considers a number of different possibilities, including the model developed by the South Pacific island states, which have opted to designate the regional gene bank, hosted by the Secretariat of the South Pacific, to act as their 'agent' in terms of receiving requests and supplying germplasm. The author also considers other initiatives under the Treaty's framework – for example, developing regional conservation plans and so on – where countries have coalesced on a regional basis to provide a broader contest for his analysis of the regionalized implementation of the multilateral system.

Toby Hodgkin et al. take a broad look, in Chapter 18, entitled 'The Evolving Global System of Conservation and Use of Plant Genetic Resources for Food and

Agriculture: What Is It and Where Does the Treaty Fit In?' at the status of the 'global system of conservation and use of PGRFA', which was first described in the International Undertaking in 1983 and which has been promoted at international levels, by the FAO and the Commission on Plant Genetic Resources ever since. The 'global system' has evolved over time, as new instruments, tools and practices have been adopted. In more recent years, the global system has expanded beyond the FAO's exclusive mandate, as the CBD has adopted programmes of work concerning agro-biodiversity and created standards for access and benefit sharing that affect some PGRFA (that is, those not in the multilateral system or held by an international organization that has an agreement with the Governing Body). The Treaty is by far the most important new instrument considered to be part of the 'global system'. Hodgkin et al. analyse the Treaty's position and function within the 'global system'.

In the final chapter, entitled 'Institutionalizing Global Genetic Resource Commons for Food and Agriculture,' Tom Dedeurwaerdere compares the coalescence of practices and rules concerning PGRFA as common pooled resources with ongoing developments in other 'new commons'.

Notes

1 International Treaty on Plant Genetic Resources for Food and Agriculture, 29 June 2004, www.planttreaty.org/texts_en.htm (last accessed 5 June 2011) [ITPGRFA].
2 General Agreement on Tariffs and Trade, 55 U.N.T.S. 194. During the 1979 FAO Council, Mexico raised questions about the legal status of the international collections managed by the Consultative Group of International Agricultural Research (CGIAR) centres, arguing that most of the material stored in these collections is collected in developing countries but mainly used by developed countries for their own needs, and when this material is returned it is under the form of high yielding varieties paid for by farmers at high prices while not always adapted to their specific needs (Pistorius, 1997).
3 Agreement on Trade Related Aspects of Intellectual Property Rights, Annex 1C of the Marrakech Agreement Establishing the World Trade Organization, 15 April 1994, 33 I.L.M. 15 (1994).
4 Convention on Biological Diversity, 31 I.L.M. 818 (1992).
5 The plant genetic resources for food and agriculture (PGRFA) in the multilateral system are not physically pooled in a single place. The pool is 'virtual' in the sense that the materials in the multilateral system are 'in' the pool and made part of a network of collections and *in situ* material subject to commonly agreed rules and administrative procedures and, eventually, using a common global information system.
6 The following PGRFA are not included within the multilateral system: all crops and forages not included in Annex 1 of the Treaty (the 64 crops and forages); all PGRFA of Annex 1 crops that are not either automatically included because they are 'in the management and control' of national governments 'and in the public domain' or voluntarily included; all genetic resources of all agricultural plants when they are used for non-food/non-feed purposes (for example, used for the production of biofuels, industrial oils and textiles). 'PGRFA under Development' as defined in the Standard Material Transfer Agreement (SMTA) – that is, material that is still being improved and that incorporates PGRFA accessed from the multilateral system – represent a grey area. PGRFA under development are only to be made available at the discretion of the developer. However, when they are made available, they must be transferred using the SMTA that is used for all transfers of materials in the multilateral system. However, the exception to the

exception is that providers of PGRFA under development may add terms and conditions to those contained in the SMTA, including restrictions on use, subsequent distribution and so on. So while PGRFA under development is arguably in the multilateral system, it is not treated like all other materials in that system.

7 Nagoya Protocol on Access and Benefit Sharing, 29 October 2012, www.cbd.int/abs/text/ (last accessed 5 June 2011).

8 See articles concerning transboundary cooperation for resources shared by different parties (Article 11) and the establishment of a global multilateral benefit-sharing mechanism for 'genetic resources that occur in transboundary situations or for which it is not possible to grant or obtain prior informed consent' (Article 10).

9 It may be useful to attempt to characterize a patent pool, as discussed earlier in Section B. Patented PGRFA are private goods. According to the analytical framework commonly referred to in the economics literature (a version of which is reproduced here), they are highly rivalrous and easy to exclude (or, if not easy, at least patent law is premised on the idea that exclusions can be efficiently enforced). Patent pools, when viewed from the inside, from the perspective of members, are low in rivalry (the rationale for the creation of the pool being that the good's value to the owner is not disproportionately subtracted from if used by the other members of the pool). They are highly rivalrous, however, when one's focus shifts to the potential impacts on value of the patented resources by unauthorized users outside the patent pool. Like the individual patents, patent pools are also highly excludable (or at least, again, patent law is premised on the notion that excludablity can be effectively enforced). Patent pools can therefore be described as either private goods (if emphasis is on the individual patents) or club goods (if emphasis is on the collected pool of patents and members). While the literature is not clear on this point, we feel that such pooling of privately owned patents would not constitute a commons, *per se*, and is not included in the focus of this book.

TABLE 1.1 Analytical framework – excludability and rivalry

		Rivalry	
		low	high
Excludibility	difficult	Public goods	Common pool resources
	easy	Toll or club goods	Private goods

Source: Adapted from Hess and Ostrom (2007).

10 For a list of PGRFA that are not included in the multilateral system, see note 6 in this chapter.

11 Rivalrous/non-rivalrous, excludable/non-excludable are not absolute categories. Most goods fall on a continuum somewhere between the possible extremes. Interestingly, Inge Kaul (2010) describes a 'global gene pool to promote biodiversity preservation' as rival goods that are deliberately kept public.

12 In this context, it is worth highlighting again the contrast between the commons-based pooling of a shared PGRFA as generally supported by the multilateral system and the pooling of private and proprietary resources in the private clubs that are sometimes created. Privately owned or proprietary PGRFA are, *ab initio*, excluded from the potential commons-based pool of shared resources included in the multilateral system. It is up to the owner of these technologies to decide, on a case-by-case basis, to enter into patent pools with other private owners.

13 Treaty on Principles Governing the Activities of States in the Exploration and Use of Outer Space, including the Moon and Other Celestial Bodies, 4 I.L.M. 386 (1967).

14 Agreement Governing the Activities of States on the Moon and Other Celestial Bodies, (1979) 18 I.L.M. 1434.

15 United Nations Convention on the Law of the Sea, 21 I.L.M. 1261 (1982).

16 International Undertaking on Plant Genetic Resources, 1983, www.fao.org/waicent/faoinfo/agricult/cgrfa/IU.htm (last accessed 5 June 2011) [International Undertaking].
17 See, for examples, the System-wide Information Network on Genetic Resources (SINGER), a globally accessible database providing information on the sources and transfers of all in-trust materials held in the international collections (see http://singer.cgiar.org/ (last accessed 5 June 2011)), the guidelines and published by the centres concerning the designation of in-trust materials pursuant to the 1994 agreements with FAO, the material transfer agreement developed for all transfers pursuant to the 1994 agreements (SGRP, 2010) and period reports to the Commission on Genetic Resources for Food and Agriculture and the Governing Body of the Treaty concerning materials held in the international collections and data on transfers (SGRP, 2011).
18 Standard Material Transfer Agreement, 16 June 2006, ftp://ftp.fao.org/ag/agp/planttreaty/agreements/smta/SMTAe.pdf (last accessed 5 June 2011) [SMTA].
19 International Undertaking, *supra* note 17, Resolution 3/91.
20 A number of factors contributed to the 1991 resolution. Many developing countries have continued to be uncomfortable with the compromise resolution over intellectual property rights in 1989, especially in light of the fact that developed countries were continuing to resist establishing financial mechanisms to support developing countries' implementation of International Undertaking priorities. In addition, the Commission on Plant Genetic Resources was presented with increasing evidence that both developed and developing countries were *de facto* exerting forms of sovereign control over plant genetic resources on their soil (Halewood, 2012).
21 It is true that the pre-existence of the international collections of PGRFA hosted by the CGIAR centres and the established practice of providing globally facilitated access to those collections made it more or less inevitable that the Treaty would embrace a multilateral approach to regulating access to those collections and sharing benefits associated with their use. However, member states have gone well beyond developing rules for the CGIAR centre-hosted collections and have made themselves participants in the multilateral system. As noted earlier, the multilateral system automatically includes PGRFA of Annex 1 crops and forages that are 'in the management and control' of Contracting Parties, 'and in the public domain'.
22 ITPGRFA, *supra* note 1, Article 17.
23 Information Relating to the Multilateral System of Access and Benefit-Sharing, Notification PL (40/31) GB4 MLS.
24 *Procedures and Operational Mechanisms to Promote Compliance and Address Issues of Non-Compliance*, Resolution 2/2011, www.itpgrfa.net/International/sites/default/files/R2_2011_en.pdf (last accessed 5 June 2011).
25 Similarly, Article 22 of the ITPGRFA, *supra* note 1, entitled, 'Dispute Resolution' includes standard clauses for international agreements concerning processes whereby disputes concerning 'interpretation or application of the Treaty' shall be subject to, variously, negotiation or meditation between the state parties concerned, and ultimately, potentially to be submitted to the international court of justice.
26 It is anticipated that additional rules concerning compliance will be submitted to the Governing Body at its fifth session in 2013.
27 There was no fixed decision in the International Undertaking (or made by the commission) concerning what the source of these contributions would be and what they would be based upon. Those are issues that were never resolved before the International Undertaking was eclipsed by the Treaty. Options included payments based on seed sales and/or voluntary contributions from governments.
28 Recipients of materials under the SMTA have the option to choose an alternative mandatory financial benefit-sharing scheme that does not turn on actual incorporation of materials in commercialized products. Recipients can choose to receive an unlimited number of samples of a particular crop or forage germplasm from the multilateral system over the course of ten years and, in return, pay 0.5 per cent of their gross sales of seeds of that crop over the same period to the multilateral system. However, very few recipients

have chosen this option to date. The difference in rates of payment (1.1 versus 0.5 per cent of gross sales) would have to be much higher for recipients to choose a benefit-sharing formula that applied to a wider range of their sold products.
29 The requirement to tie benefit sharing to actual uses of materials accessed from the multilateral system on a product-by-product basis came at the insistence of developed countries, presumably to reflect the preferred approach of their commercializing constituents (who also, presumably, did not like the idea of a flat tax on all seed sales as provided by the benefit-sharing mechanism).
30 This is precisely the approach adopted by the government of Norway, which has announced its intention to make an annual contribution to the benefit-sharing fund equal to 0.1 per cent of the value of all seeds sold in the country (ITPGRFA, 2008).

References

Almekinders, C.J.M., N.P. Louwaars and G.H. de Brujin (1994) 'Local Seed Systems and Their Importance for and Improved Seed Supply in Developing Countries', *Euphytica*, 78: 207–16.

Altoveros, N.C., T.H. Borromeo, N.A. Catibog, H. de Chavez, M.H.F. Dayo and M.L.H. Villavicencio (2012) 'Challenges and Opportunities for the Philippines to Implement the Multilateral System of Access and Benefit Sharing of the International Treaty on Plant Genetic Resources for Food and Agriculture', in I. López Noriega, M. Halewood and I. Lapeña (eds), *Incentives and Disincentives for Countries to Participate in the Multilateral System of Access and Benefit-Sharing of the International Treaty on Plant Genetic Resources for Food and Agriculture*. System-wide Genetic Resources Programme and Bioversity International, Rome.

Aoki, K. (2008) *Seed Wars*, Carolina Academic Press, Durham, NC.

Badstue, L.B., M.R. Bellon, J. Berthaud, X. Juárez, I.M. Rosas, A.M. Solano and A. Ramírez (2006) 'Examining the Role of Collective Action in an Informal Seed System: A Case Study from the Central Valleys of Oaxaca, Mexico', *Human Ecology* 34(2): 249–73.

Brockway, Lucile (1988) 'Plant Science and Colonial Expansion: The Botanical Chess Game', in J.R. Kloppenburg (ed.), *Seeds and Sovereignty: The Use and Control of Plant Genetic Resources*, Duke University Press, Durham, NC.

Byerlee, D., and H.J. Dubin (2010) 'Crop Improvement in the CGIAR as a Global Success Story of Open Access and International Collaboration', *International Journal of the Commons* 4(1): 1–19.

Commission on Genetic Resources for Food and Agriculture (CGRFA) (2011) *Thirteenth Regular Session of the Commission on Genetic Resources for Food and Agriculture*, Rome, Italy, 18–22 July 2011, Doc. CGRFA-13/11/Report, Food and Agriculture Organization, Rome.

Crosby, A.W. (1972) *The Columbian Exchange: Biological and Cultural Consequences of 1492*, Greenwood Press, Westport, CT.

Fowler, C., and T. Hodgkin (2004) 'Plant Genetic Resources for Food and Agriculture: Assessing Global Availability', *Annual Review of Environmental Resources* 29: 10.1–10.37.

Frankel, O.H. (1988) 'Genetic Resources: Evolutionary and Social Responsibilities', in J.R. Kloppenburg (ed.), *Seeds and Sovereignty: The Use and Control of Plant Genetic Resources*, Duke University Press, Durham, NC.

Garrison Wilkes, H. (1988) 'Plant Genetic Resources over Ten Thousand Years: From Handful of Seed to the Crop-Specific Mega-Gene Banks', in J.R. Kloppenburg (ed.), *Seeds and Sovereignty: The Use and Control of Plant Genetic Resources*, Duke University Press, Durham, NC.

Halewood, M. (2008) 'Use It or Lose It: Agricultural Biological Diversity', *Public Service Review: Science and Technology* 1: 109–11.

——(2010) 'Governing the Management and Use of Pooled Microbial Genetic Resources: Lessons from the Global Crop Commons', *International Journal of the Commons* 4(1): 404–36.

——(2012) 'Securing the global crop commons in support of agricultural innovation', in Mario Cimoli et al. (eds), *Intellectual Property Rights: Legal and Economic Challenges for Development*, Oxford University Press, Oxford.

Halewood, M., and K. Nnadozie (2008) 'Giving Priority to the Commons: The International Treaty on Plant Genetic Resources for Food and Agriculture', in G. Tansey and T. Rojotte (eds), *The Future Control of Food: A Guide to International Negotiations and Rules on Intellectual Property, Biodiversity and Food Security*, Earthscan, London.

Hammond, E. (2011) *How US Sorghum Seed Distributions Undermine the FAO Plant Treaty's Multilateral System: Overlap and Use of the CGIAR and US Sorghum Genebank Collections African Center for Biosafety Berne Declaration*, Development Fund, Melville, Zurich, Oslo, online: www.itpgrfa.net/International/sites/default/files/US_sorghum.pdf (last accessed 5 June 2011).

Harlan, J. (1992) *Crops and Man*, 2nd edition, American Society of Agronomy, Madison, WI.

Hess, C., and E. Ostrom (eds) (2007) *Understanding Knowledge as a Commons: From Theory to Practice*, MIT Press, Cambridge, MA.

Hodgkin, T., R. Rana, J. Tuxill, D. Balma, A. Subedi, I. Mar and I. Karamura (2007) 'Seed Systems and Crop Genetic Diversity in Agroecosystems', in D.I. Jarvis, C. Padoch and H.D. Cooper (eds), *Managing Biodiversity in Agricultural Ecosystems*, Columbia University Press, New York.

International Treaty on Plant Genetic Resources for Food and Agriculture (ITPGRFA) (2009) *Report of the Third Session of the Governing Body of the Treaty on Plant Genetic Resources for Food and Agriculture*, Tunis, 1–5 June 2009, Doc. IT/GB-3/09/Report, Food and Agricultural Organization, Rome, Italy, online: ftp://ftp.fao.org/ag/agp/planttreaty/gb3/gb3repe.pdf (last accessed 5 June 2011).

——(2008) *Norway Announces Annual Contribution to the Benefit-Sharing Fund of the International Treaty*, online: ftp://ftp.fao.org/ag/agp/planttreaty/news/noti005_en.pdf (last accessed 5 June 2011).

Jinna, S., and S. Jungcourt (2009) 'Could Access Requirements Stifle Your Research?' *Science* 323: 464.

Kanbur, R. (2002) *International Financial Institutions and International Public Goods: Operational Implications for the World Bank*, G-24 Discussion Paper Series, United Nations Conference on Trade and Development, New York and Geneva.

Kaul, I. (2010) *Global Public Goods and Responsible Sovereignty: The Broker*, online: www.stwr.org/the-un-people-politics/collective-self-interest-global-public-goods-and-responsible-sovereignty.html (last accessed 5 June 2011).

Kloppenburg, J. (1988a) *First the Seed: The Political Economy of Plant Biotechnology*, Cambridge University Press, Cambridge.

——(1988b) *Seeds and Sovereignty: The Use and Control of Plant Genetic Resources*, Duke University Press, Durham, NC.

Kloppenburg, J., and D.L. Kleinman (1988) 'Preface: Plant Genetic Resources: The Common Bowl', in J.R. Kloppenburg (ed.), *Seeds and Sovereignty: The Use and Control of Plant Genetic Resources*, Duke University Press, Durham, NC.

Linnemann, A.R., and J.S. Siemonsma (1987) 'Variety Choice and Seed Supply by Smallholders', *ILEIA Newsletter* 5(4): 22–23.

McGuire, S.J. (2007) 'Securing Access to Seed: Social Relations and Sorghum Seed Exchange in Eastern Ethiopia', *Human Ecology* 36(2): 217–29.

Mooney, P. (1983) 'The Law of the Seed: Another Development and Plant Genetic Resources', *Development Dialogue* 1–2: 1–172.

Morrissey, O., D. te Velde and A. Hewitt (2002) 'Defining International Public Goods: Conceptual Issues', in M. Ferroni and A. Mody (eds), *International Public Goods: Incentives, Measurements and Financing*, Kluwer Academic Publishers, Dordrecht, and World Bank, Washington, DC.

Nijar, G.S., Gan Pei Fern, Lee Yin Harn and Chan Hui Yun (2009) *Framework Study for Food Security and Access and Benefit Sharing for Genetic Resources for Food and Agriculture*, Background Study no. 42, United Nations Food and Agriculture Organization, Rome.

Olson, M. (1971) *The Logic of Collective Action*, Harvard University Press, Cambridge, MA.

Ostrom, E. (1990) *Governing the Commons: The Evolution of Institutions for Collective Action*, Cambridge University Press, Cambridge.

Pistorius, R. (1997) *Scientists, Plants and Politics: A History of Plant Genetic Movement*, International Plant Genetic Research Institute, Rome.

Ruiz, M., and R. Vernooy (eds) (2011) *The Custodians of Biodiversity: Sharing Access to and Benefits of Genetic Resources*, Taylor and Francis and International Development Research Centre, London.

Safrin, S. (2004) 'Hyperownership in a Time of Biotechnological Promise: The International Conflict to Control the Building Blocks of Life', *American Journal of International Law* 98: 641.

Schloen, M., S. Louafi and T. Dedeurwaerdere (2011) 'Access and Benefit-Sharing for Genetic Resources for Food and Agriculture: Current Use and Exchange Practices, Commonalities, Differences and User Community Needs, Report from a multi-stakeholder expert dialogue', Background Study Paper no. 59, Food and Agriculture Organization, Rome.

Shrestha, P., B.R. Sthapit, A. Subedi, D. Poudel, P. Shrestha, M.P. Upadhyay and B.K. Joshi (2008) 'Community Seed Bank: Good Practice for On-Farm Conservation of Agricultural Biodiversity', in B.R. Sthapit, D. Gauchan, A. Subedi and D.I. Jarvis (eds) *On-Farm Management of Agricultural Biodiversity in Nepal: Lessons Learned*, Proceedings of the National Symposium, 18–19 July 2006, Nepal Agricultural Research Council/Local Initiatives for Bioversity, Research and Development/International Development Research Centre/Sustainable Development Council, Kathmandu, Nepal.

Subedi, Anil, P. Chaudhary, B. Baniya, Ram Rana and R.K. Tiwari (2003) 'Who Maintains Genetic Diversity and How? Policy Implications for Agrobiodiversity Management', in D.I.D. Gauchan, B.R. Sthapit and A. Jarvis (eds), *Agrobiodiversity Conservation on-Farm: Nepal's Contribution to a Scientific Basis for National Policy Recommendations*, International Plant Genetic Research Institute, Rome.

System-Wide Genetic Resources Programme (SGRP) (2010) *Guidelines and Statements on Genetic Resources, Biotechnology and Intellectual Property Rights*, Booklet of CGIAR Centre Policy Instruments, Version 3, Bioversity International, online: www.sgrp.cgiar.org/sites/default/files/Policy_Booklet_Version3.pdf (last accessed 5 June 2011).

——(2011) *CGIAR Centres' Experience with the Implementation of Their Agreements with the Treaty's Governing Body, with Particular Reference to the Use of the SMTA for Annex 1 and Non-Annex 1 Materials*, online: www.itpgrfa.net/International/sites/default/files/gb4i05e.pdf (last accessed 5 June 2011).

Tapia, M.E., and A. Rosa (1993) 'Seed Fairs in the Andes: A Strategy for Local Conservation of Plant Genetic Resources', in W. de Boef, K. Amanor, K. Wellard and A. Bebbington (eds), *Cultivating Knowledge: Genetic Diversity, Farmer Experimentation and Crop Research*, ITDG Publishing, London.

United Nations Environment Programme (UNEP) (2006) *Outcomes and Recommendations of the Meeting of 'Biodiversity: The Megascience in Focus*, Doc. UNEP/CBD/COP/8/INF/46, online: www.cbd.int/doc/meetings/cop/cop-08/information/cop-08-inf-46-en.pdf (last accessed 5 June 2011).

van Overwalle, G. (ed.) (2008) *Gene Patents and Collaborative Licensing Models: Patent Pools, Clearinghouses, Open Source Models and Liability Regimes*, Cambridge University Press, Cambridge.

Vavilov, N.I. (2007) *Five Continents*, International Plant Genetic Resources Institute, Rome.

Weltzien, E., and K. Vom Brocke (2001) 'Seed Systems and Their Potential for Innovation: Conceptual Framework for Analysis', in *Targeted Seed Aid and Seed Systems Interventions: Strengthening Small-Farmer Seed Systems in East and Central Africa. Proceedings of a workshop held in Kampala, Uganda, 21–24 June 2000*, International Center for Tropical Agriculture, Cali, Colombia.

Whyte, R.O., and G. Julèn (eds) (1963) 'Proceedings of a Technical Meeting on Plant Exploration and Introduction', *Genetica Agraria* 17: 1–573.

PART I
Setting the scene

Countries' interdependence on plant genetic resources for food and agriculture and the imperative of international cooperation

2

DEMONSTRATING INTERDEPENDENCE ON PLANT GENETIC RESOURCES FOR FOOD AND AGRICULTURE

Marleni Ramirez, Rodomiro Ortiz, Suketoshi Taba, Leocadio Sebastián, Eduardo Peralta, David E. Williams, Andreas W. Ebert and Anne Vézina

Introduction

Plant genetic resources for food and agriculture (PGRFA) are strategic goods for crop breeding through farmer selection, conventional plant breeding and modern biotechnological techniques. Crop improvement enables their adaptation to biotic and environmental changes as well as the development of new foods and new uses. Currently, all countries depend in great measure on the PGRFA resources of plants domesticated, and subsequently developed, in other countries or regions for their food and sustainable agricultural development (Kloppenburg and Kleinman, 1987; Gepts, 2004). Even the world centres of crop diversity (Vavilov, 1926; Zeven and de Wet, 1982), which coincide with the centres of domestication, mostly rely on non-indigenous crop genetic resources to meet their food needs (Gepts, 2004). In her study of the regions, Ximena Flores Palacios (1998) showed that the southwest Asian centre of agricultural origin uses non-indigenous crops for 30 per cent of its production, while the Latin American and Chinese centres employ 60 per cent of their non-indigenous crops for their production. Furthermore, the new conditions that will be imposed by a changing climate will require that plant breeders and farmers have access to an even greater genetic diversity to attempt to adapt to the novel situations (Fujisaka et al., 2009).

The interdependence on plant genetic resources has been documented by looking at the pedigrees of crop varieties of worldwide importance for food security and by studying the flows of plant genetic resources. For an annotated bibliography on the subject of international pedigrees and flows of PGRFA, see the text by Christine Frison and Michael Halewood (2006). Crop varieties that are often cited to illustrate pedigree complexity include Sonalika, the most widely cultivated bread wheat variety in the world. Released in India in 1966, it has 17 generations in its pedigree, 420 parental combinations and 39 landraces, and breeders in 14 countries have contributed lines to its pedigree (Smale et al., 1996). The Veery cultivar, which was released in Mexico in 1977, includes an even larger pedigree with 23 generations, 3,169 parental combinations and 49 different landraces.

Between 1976 and 2000, the International Center for Tropical Agriculture (CIAT) bred and distributed, upon request, 205 bean varieties to 18 countries in Latin America. In most countries, local materials contributed to these released varieties with less than 50 per cent of the ancestors. In only two countries did more than half of the ancestors come from the same countries, while in five countries more than 90 per cent of the genetic material included in the released varieties was contributed from other countries (Johnson et al., 2003).

More than 60,000 samples of germplasm were reported to be distributed annually by the gene banks of the Consultative Group on International Agricultural Research (CGIAR) (S. Gaiji, International Plant Genetic Research Institute, unpublished information cited by Fowler and Hodgkin, 2004). Such data indicate that almost all countries are net recipients of such transfers (Fowler et al., 2001). More recent data for 2007–8 indicate that most (81 per cent) of the recipients of germplasm distributed by the International Agricultural Research Centres (IARCs) are developing countries, and the rest (19 per cent) are developed countries. Within these countries, 40 per cent of the germplasm goes to the national researchers of developing countries, 13 per cent to universities, 4 per cent to private breeders and 24 per cent to other individuals and organizations (SGRP, 2009). It is notable that in the first 19 months of operating under the International Treaty on Plant Genetic Resources for Food and Agriculture's (ITPGRFA) framework, the centres distributed approximately 550,000 samples of PGRFA all over the world using the Standard Material Transfer Agreement (SGRP, 2010).[1] A quarter of the materials were PGRFA as they were originally acquired, and the other three-quarters were materials that the centres had been involved in improving.

The flow of genetic resources from the IARCs to the National Agricultural Research Systems (NARS) breeding programmes are analysed in detail in the volume edited by Robert Eugene Evenson and Douglas Gollin (2003a). The authors note that in essentially all crops and throughout all regions the continuing diffusion of modern varieties to developing countries has depended to a large degree on germplasm coming from international research centres. Crop varieties being grown by farmers as well as those used in national breeding programmes appear to be based – in part or in whole – on genetic material supplied by these international centres (Evenson and Gollin, 2003b). The flows from the IARCs, which by design rely on access to an international genetic pool for the most important crops, therefore attest to the public benefits of having access to a wide array of genetic diversity.

In addition, the advantages of cooperation for germplasm improvement and exchange were recently demonstrated in response to the spread of wheat stem rust. The race Ug99 (*Puccinia graminis* Pers. f. sp.*tritici*), which is gradually moving out of Africa towards Yemen and West Asia, is highly virulent against the stem rust resistance genes that are commonly present in wheat germplasm and that have been effective in the control of the disease for over 30 years (Njau et al., 2010). Ug99 thus posed a serious threat to wheat production worldwide. In an effort to identify the sources of resistance, about 2,000 advanced and elite bread wheat lines from the International Centre for Maize and Wheat Improvement (CIMMYT) were evaluated in the field in Kenya from 2005 to 2007, while the seedlings were tested in St. Paul, Minnesota. This effort resulted in the identification of two race-specific resistance genes and three possibly new sources of resistance from synthetic, Chinese and other unknown origins, which represent very valuable genetic resources for breeding Ug99 resistance.

In the following pages, we present six cases that document broad interdependence on PGRFA between countries to supply their food needs and secure their access to materials for breeding and for other research. Rodomiro Ortiz and Suketoshi Taba outline maize's journey from its domestication in Meso-America to its current position as one of the primary foods of the world and how access to genetic diversity from multiple sources, including wild relatives, has allowed the crop to respond to the needs for increased grain yield, resistance to diseases and pests, and improved nutritional content. Leocadio Sebastián reminds us that rice would not have become the most important food crop in the world if it were not for germplasm exchange. Eduardo Peralta narrates the ascendance of chocho, an Andean legume, in the poorest provinces of Ecuador thanks to the availability of early materials from other countries, resulting in improved diets and sources of income for populations living in marginal conditions. David E. Williams underscores the high degree of countries' interdependence on peanut genetic resources, while Andreas W. Ebert demonstrates how access to genetic diversity is critical for developing resistance to diseases that have caused enormous losses for small-scale cacao farmers. Anne Vézina recounts the international collaboration on banana breeding that served the needs of small farmers and commercial producers and makes a call to take advantage of the multilateral system of access and benefit sharing (multilateral system) under the ITPGRFA for the benefit of banana growers, breeders and consumers.

Case 1: International dependence on maize germplasm to enhance global diets

In terms of total production, maize is the most important grain crop worldwide, at more than 800 million metric tonnes (FAO, 2010). Cultivated maize (*Zea mays* L. subsp. *mays* Iltis) was domesticated by early farmers in a region that is now occupied by southern Mexico from its wild relative teosinte (*Z. mays* ssp. *parviglumis* Iltis and Doebley) (Matsuoka et al., 2002). The direct maize ancestor was initially domesticated not for its grain but, rather, for its sugary pith (Smalley and Blake, 2003). Maize was subsequently selected for large cobs with many rows of kernels, thereby becoming an important food staple across Meso-America.

Maize was introduced to Europe with the shipments from the second voyage of Columbus and spread rapidly worldwide. Maize cultivation started in southern Spain at the end of the fifteenth century and spread to Africa in order to feed the growing numbers of slaves awaiting their transport to southwest Asia and the Americas. The multiple diversity flows from the American continent to the rest of the world served to widen the maize genetic base, which allowed additional selections to spread to all

Science, Policy & Law?

of the other continents (Ortiz et al., 2007). Today, maize is the most widely grown crop in the world based on number of countries and agro-economic zones, and it demonstrates a great culinary adaptation to diverse cultures worldwide.

In the early decades of the twentieth century, hybrid maize was developed using inbred lines from open-pollinated cultivars of North American origin, thus initiating one of the world's most profound changes in modern agriculture (Taba et al., 2005). Likewise, many tropical maize-breeding populations, largely based on dents from Mexico and flints from the Caribbean, further led to national cultivar releases – for example, the cultivar 'Eto' in Colombia (Chavarriaga, 1966), which, in turn, was used to breed exotic germplasm to broaden the genetic base of the US corn belt (Hallauer and Smith, 1979).

In the developing world, 82–92 per cent of crop yield is lost primarily due to abiotic and biotic stresses, with 17 per cent of loss due to drought, 20 per cent to infertile soils, 5 per cent each to leaf diseases and ear rots, 10 per cent or more to stem-boring insects, 10–20 per cent to insects that attack stored grain, and 15 per cent to the parasitic weed *Striga* (CIMMYT, 2004). The next four examples illustrate where it has been absolutely essential to get access to, and to use, maize genetic resources from a variety of countries and continents to address threats to maize crops and promote food security.

Tuxpeño maize: a truly global public good

The Tuxpeño landrace, a subset of the 250 maize landraces found in the New World, combines desirable traits such as resistance to important diseases and the ability to respond to high fertility conditions with some undesirable traits such as excessive plant height and a low harvest index. In the 1970s, CIMMYT researchers reduced the height of some of the Mexican Tuxpeño landraces with the aim of achieving greater per-plant grain yield (Taba, 1997). Bred populations were shared with partners across 43 countries in Asia, Africa and Latin America (CIMMYT, 1986). Furthermore, in the early 1980s, the CIMMYT and the International Institute of Tropical Agriculture (IITA) bred maize populations derived from La Posta (a Tuxpeño landrace), which showed host plant resistance to the maize streak virus (MSV), a significant disease in sub-Saharan Africa. With a reliable screening method and appropriate sources of resistance, IB32 – a streak-resistant line that was developed from this maize-breeding population – and La Revolution, a line from Reunion Island, gave rise to more than 100 cultivars and hybrids to encompass all of the relevant farming systems and agro-ecologies in sub-Saharan Africa.

Likewise, leading up to the 1990s, cooperating national programmes released 147 cultivars and hybrids that were bred from Tuxpeño populations, which went on to be grown in approximately 3.8 million hectares worldwide. Furthermore, Tuxpeño accessions are still in the pedigrees of many CIMMYT gene pools, populations and elite lines – for example, Thai Composite 1 (later called Suwan 1), which is a downy mildew resistant cultivar bred by Thai researchers and subsequently used in global maize breeding (Sriwatanapongse et al., 1993). The US-bred germplasm BS28 and BS29 were also derived from Tuxpeño Composite and Suwan 1, respectively (Hallauer, 1994).

Use of Teosinte for host plant resistance to *Striga hermonthica* in sub-Saharan Africa

Striga hermonthica infests millions of hectares of arable land in Africa and limits maize, pearl millet and sorghum production throughout the continent. IITA researchers bred maize lines with host plant resistance to this parasitic weed using genetically broad-based populations (including BC4, which is derived from *Zea diploperennis* Iltis, Doebley and Guzman, from Mexico) as well as synthetics that also possess resistance to MSV (Menkir et al., 2006). The BC4 population supported little or no *Striga hermonthica* emergence as a donor parent (Kling et al., 2000). The wild accession was crossed to an adapted maize germplasm, and the resulting F1 was backcrossed four times to four adapted maize genotypes (including Suwan 1-SR, which is resistant to downy mildew and bred by Thai researchers) under artificial infestation with *Striga hermonthica* in the screen house at Ibadan, Nigeria, to form a *Zea diploperennis* BC4 population. These *Striga*-resistant maize inbred lines have been useful sources of genes for developing germplasm with host plant resistance to *Striga hermonthica* and have been adapted to the lowland and mid-altitude areas of sub-Saharan Africa, where this weed is endemic.

Insect resistance from Central America to Eastern Africa

The larger grain borer, a native insect pest of Central America, was first observed in Tanzania in the late 1970s and early 1980s. CIMMYT researchers found host plant resistance to the borer in Caribbean accessions held in trust at the centre's gene bank. Conventional plant breeding techniques such as crossing those accessions with maize adapted to Eastern Africa led to combining the resistance of the Caribbean maize with key traits valued by Kenyan maize farmers. This newly bred maize was tested for resistance at the national programme research station in Kiboko, Kenya. New maize cultivars resulting from this research will clearly benefit farmers throughout Kenya (CIMMYT, 2007).

Opaque-2 and quality protein maize: the long journey of a Peruvian gene

A naturally occurring recessive mutant gene known as *Opaque-2*, which was first observed in a Peruvian maize landrace, gives a chalky appearance to the kernels and improves the protein quality of this crop by increasing levels of lysine and tryptophan (Crow and Kermicle, 2002). Since its discovery in 1963 (Mertz et al., 1964), the trait has been transferred through careful selection and meticulous breeding to quality protein maize breeding lines and cultivars (Vasal, 2000). Today, these lines grow from Latin America eastwards and westwards, reaching farmers' fields and users throughout sub-Saharan Africa (Krivanek et al., 2007) and Asia (Gupta et al., 2009), respectively. This quality protein maize improves the diets of people who depend on maize as a staple and also shows promise in animal feed.

Case 2: Interdependence on germplasm exchange for rice improvement

Rice is the most important staple food crop in the world. More than half of the world's population eats rice. Rice is also the one crop whose rapid development has been very dependent on the exchange of germplasm. The Green Revolution in rice-growing countries began with the introduction of IR8, a cultivar derived from a cross between the semi-dwarf variety Degeowoogen from Taiwan with the tall variety Peta. Peta was derived from a cross between Cina from China and Latisail from Pakistan. Subsequently, many other landraces have been used by the International Rice Research Institute (IRRI), the CIAT and national breeding programmes to increase yield, improve pest and disease resistance, confer tolerance to abiotic stresses and improve grain quality and other characteristics. The wide use of landraces from different countries as a source of desired traits has contributed to the increase in rice production in most rice-growing countries.

The availability of rice germplasm from the more than 125,000 accessions held at the IRRI has enabled a great deal of screening to uncover rice lines that carry the desired traits. It is important to note that the selected landraces were only discovered after screening thousands of accessions for different traits. Such tremendous effort is exemplified by the case of rice tungro virus resistance, which was identified in only a few accessions after more than 20 per cent of the IRRI rice germplasm collection was evaluated (Jackson, 1997; Angeles et al., 2008).

The pedigrees of lines developed by the IRRI and the national programmes evidence the use of several landraces coming from within the same country as well as from other countries. Major rice growing countries such as Bangladesh, Indonesia, India, Vietnam, Philippines and Thailand are using landraces that have been developed in their own breeding programmes as well as those obtained from other breeding programmes such as the IRRI. The use of IRRI breeding lines as the sources of desired traits has served to enrich local varieties with traits that help them cope with such diseases as bacterial leaf blight and grassy stunt virus as well as pests such as brown plant hopper and green leaf hopper (Hossain et al., 2003).

In the case of 78 Philippine cultivars released from 1968 (IR8) to 1995 by three rice breeding programmes, the cultivars can be traced to a total of 57 landraces coming from the Philippines and 15 other countries: Brazil, Burma, China, India, Japan, Korea, Malaysia, Nigeria, Pakistan, Senegal, Sri Lanka, Taiwan, Thailand, United States and Vietnam (Hossain et al., 2003). The combined traits from these landraces conferred the necessary characteristics that allowed the different cultivars to cope with changing pest and disease pressures, various soil and nutrient conditions and particular regional climatic conditions (Sebastian et al., 1998). None of the landraces, however, can be considered to have had a predominating influence in the genotype of the modern Philippine cultivars. Furthermore, molecular studies of these landraces reveal that they have come from very diverse genetic backgrounds.

Many of the improved lines that have been developed using these landraces have also spread back to many countries and have benefited thousands of farmers, including numerous rice lines developed by the IRRI that have been released as cultivars in several countries. The most widely planted rice variety in the world is IR64, which was first released in the Philippines but was subsequently released in 11 other countries. At its peak, it was estimated to have been planted over approximately 13 million hectares of land in Asia. It was most popular in Indonesia and the Philippines and was well liked by farmers for its high yield, by traders for its high milling recovery, and by consumers for its good quality as food. The pedigree of this variety comprises 20 landraces originating from nine countries: China, India, Indonesia, Japan, Korea, Philippines, Thailand, Vietnam and the United States. Another very widely planted IRRI variety is IR36, which is known as the first IR variety to be bred with multiple resistances to pests and diseases. Its pedigree includes 15 landraces coming from six countries and was released as a variety in 13 countries (Khush and Virk, 2005).

Most of the exchanged materials under the International Genetic Exchange in Rice (INGER) programme were in the form of advanced lines carrying desired traits derived from different countries. More than 48,000 nursery entries have been shipped to rice breeding programmes in 80 countries for adoptability tests or for use as possible parental material. The materials distributed include those from the IRRI breeding programmes as well as national breeding programmes nominated for inclusion in the breeding nurseries distributed by INGER. The exchange programme also facilitated the exchange of data and information on the performance of genetic materials, which has further enhanced the national and international breeding programsme.

This exchange of advanced lines has resulted in the release of more than 328 IRRI lines as varieties in 75 countries as well as many others that were used as parental lines in national breeding programmes (Khush and Virk, 2005). Many national programme entries in INGER have also been released as varieties in other countries or used as parental materials. Similar exchange and use of genetic materials has also been observed in Latin America. About 40 per cent of the approximately 300 varieties released in Latin America came from the CIAT breeding programme, and another 36 per cent came from international networks – mostly from INGER (Hossain et al., 2003). Many lines from other countries and sources (for example, materials from the Centre for Agricultural Research for Development (CIRAD)) were also released as varieties or used in national rice breeding programmes in the region (Chatel and Guimaraes, 2002).

The case of rice illustrates the great benefits of making available a wide range of genetic diversity for the benefit of humankind. The accessibility and availability of diverse sets of germplasm to rice breeders has enabled them to develop rice lines that can cope with various challenges in rice production. The reciprocal accessibility and availability of thousands of improved breeding lines has also allowed national breeding programmes to test for possible local releases or to use them as parental materials in developing locally adapted cultivars. Such activities have enabled a more rapid release of varieties as well as the development of locally adapted and improved cultivars.

Although it is difficult to attribute the reasons for the sharp increase in production, the world's production of rice has increased from about 215 million metric tonnes in 1961 to about 634 million metric tonnes in 2006 (a 296 per cent increase), while the area of land that is being harvested has increased from 115 million hectares to 154 million hectares (a 134 per cent increase) during the same period. Estimates indicate that the adoption of modern rice varieties in Bangladesh, West Bengal in India, Java in Indonesia, and Phillipines and Vietnam has resulted in an average net gain of 943 kilograms of rice per hectare or about $150 per hectare over what would have been harvested without these varieties (Hossain et al., 2003).

Future challenges will no longer be focused on meeting the increasing demand for rice but, rather, on developing ecologically sound, sustainable production systems as well as improved value chains under conditions involving less land, less water, fewer resources and the increasing effects of climate change. The need to develop varieties that can meet these challenges will require harnessing the available rice genetic resources. New traits will have to be discovered from cultivated and wild relatives, or existing traits will need to be combined in new ways to not only meet the need for increased production but also to cope with biotic (new pests and diseases or complex diseases) and abiotic (drought, flood and temperature changes) stresses, to use resources more efficiently (water and nutrients) and to meet the demand for high value rice (based on quality and speciality preferences). The new varieties will have to fit into the farming systems of the future.

Case 3: Harnessing introduced germplasm of chocho or lupin (*Lupinus mutabilis* Sweet) for Ecuador's benefit

In 1983, the first gene banks of Andean crops were established in Ecuador under the leadership of the Santa Catalina Experimental Station in Quito (INIAP, 2009). As a result of this initiative, a high percentage of the genetic variability of grains, tubers, roots and some native Andean fruits was collected. Among the first accessions to be collected were 257 accessions of Andean grain, chocho (*Lupinus mutabilis* Sweet).

Chocho or lupin is a legume that is native to the Andes of Bolivia, Ecuador and Peru. Its high protein content, which is even higher than soybeans, makes it a crop of great interest for human and animal nutrition. The National Department of Plant Genetic Resources and Biotechnology subjected the chocho collection of the Instituto Nacional Autónomo de Investigaciones Agropecuarias (INIAP) to an exhaustive morphological characterization and agronomic assessment. Many interesting traits were reported from this collection, but no early materials have been identified.

At the time, the research institutions in the Andean countries freely exchanged promising germplasm, and joint collecting trips were common practice. In 1992, as part of this cooperative framework and with the support of projects and regional and international institutions, a few sets of chocho populations from Peru and Bolivia were introduced to the INIAP and were evaluated at production locations in the Ecuadorian highlands. The best lines from this collection were stored at the INIAP gene bank. Subsequently, the gene bank material was assessed according to its adaptability and stability in different environments, and, based on this information, the accession ECU 2659, which is native to Peru, stood out as the most promising line for its adaptability, stability, high yield, grain quality and, especially, its earliness. Whereas local genotypes are usually harvested from 11 to 14 months after sowing, these varieties are harvested six to eight months after sowing. The lengthy growing period of the local genotypes results in many more risks, including frost, hailstorms, excess rain or drought. Obviously, by reducing this period of growth, farmers could substantially reduce their risks. In 1997, the responsibility for continuing this research was transferred from the Andean Crop Programme to the National Programme for Andean Legumes and Grains, which later distributed this material of chocho in Ecuador under the name INIAP-450 Andino (Caicedo and Peralta, 1999).

By growing the INIAP-450 Andino variety, farmers were able to improve their yield substantially – from 400 kilograms per hectare for the native local varieties to 1,350 kilograms per hectare with Andino. The earliness and pleasing appearance of the new variety were also beneficial and, combined with favourable agronomic characteristics and more efficient processing, encouraged a revival of chocho cultivation and consumption, which prior to 1997 had been very marginal in the rural areas and negatively affected by social prejudice in the cities (Peralta et al., 2009a).

Thanks to various efforts in public awareness, including the involvement of local agricultural research committees and individual training, the cultivation of the variety has spread successfully throughout the highlands. At the same time, its consumption has been promoted across all economic sectors by emphasizing its nutritional values, while the development of artisanal processing at the grassroots level has been encouraged and financially supported (Villacrez and Peralta, 2006). In recent years, small family enterprises have begun to offer high-quality chocho, packaged under proper sanitary conditions. The crop has gone from a product that was marginally consumed to a quality product that has a potential market abroad.

In poor communities throughout the state of Cotopaxi, which is located in the central high-Andean zone of the country, it is estimated that over 70 per cent of farmers are growing this variety, and, as a result, the amount of cultivated land has increased significantly each year (Peralta et al., 2009b). In 1999, in the communities of Ninin Cachipata and Chaluapamba, two families began to grow INIAP-450, and, today, this number has grown to 96, out of a total of 120 families. In the provinces of Chimborazo and Bolivar, the cultivation of this variety was re-started, and in three southern states of the country its consumption and cultivation was also introduced. In addition, four agro-industrial enterprises are producing high-quality chocho both in the form of processed branded products as well as the raw material that is of a very uniform quality (one of them uses only INIAP technology). Such production capabilities have enabled them to offer the product in different presentations and in places ranging from small bars to large food chains.

The chocho INIAP-450 Andino variety took Ecuador by storm. This case highlights the importance of having access to pooled materials from several countries to help in the search for varieties that may respond to local needs. Without access to precious chocho materials from neighbouring countries, this project would not have been possible. Today, INIAP-450 Andino and chocho cultivation and consumption have become emblematic of what is possible in the agricultural sector in the poorest and most marginal provinces of Ecuador.

Case 4: Worldwide interdependence on peanut genetic resources (*Arachis* spp.)

The peanut or groundnut (*Arachis hypogaea* L.) is the most widely cultivated grain legume in the world, with current annual production in excess of 35 million metric tonnes (FAO, 2010). The peanut's high protein (approximately 25 per cent) and oil (approximately 40 per cent) content make it a highly nutritious, easily processed and widely accepted food for millions of small-hold and subsistence farmers throughout the tropics.

It is postulated that the cultivated peanut originated about 7–8,000 years ago in the semi-arid region that is now southeastern Bolivia and northwestern Argentina. The cultigen is probably the result of a fortuitous naturally occurring hybridization event between two wild species of *Arachis*, giving rise to a larger, more vigorous species (*A. hypogaea* L.) that was subsequently propagated and ultimately domesticated by incipient Amerindian agriculturalists (Smartt, 1985). Over the next several millennia, an impressive amount of cultivated peanut diversity was generated as the crop was widely disseminated throughout the neotropics where it adapted to the different geographic, climatic and agricultural contexts in which it was grown by countless different ethnic groups.

The breadth of diversity within the cultivated peanut species is reflected in its infra-specific taxonomic classification (Krapovickas and Gregory, 2007). *Arachis hypogaea* is divided into two clearly distinguishable subspecies. The first (ssp. *hypogaea*) has a somewhat prostrate growth habit, with its branches often trailing along the ground, a long growth cycle and is often higher yielding; while the second (ssp. *fastigiata*) has a distinctly upright growth habit, a shorter growth cycle and tends to be more drought tolerant. The first subspecies is made up of two distinct botanical varieties, and the second subspecies is comprised of four botanical varieties. Furthermore, within each of these six botanical varieties, there are dozens of different local varieties, or 'landraces', each with unique traits and qualities. The result when viewed together is a colourful riot of different peanut shapes, sizes, seed colours, plant types and agronomic characteristics that few people other than peanut researchers have ever seen or imagined.

The great Columbian exchange that began in 1492 initiated the peanut's rapid diffusion to and widespread adoption in Africa, Asia and the Pacific and eventually to North America (Williams, 2004). Today, peanut production on other continents has far surpassed that of its South American home, and it now plays a more important role in the traditional diets of Asia, Africa and North America than it ever did in its region of origin. China, India and the United States are currently the three largest peanut-producing countries in the world. During the last three decades, improved peanut varieties developed in the United States and India have become widely adopted in Asia, Africa and Latin America, underscoring the pronounced global interdependence on *Arachis* genetic resources.

Nevertheless, the greatest concentration of peanut genetic diversity – including its related wild species – continues to survive in the fields of small-scale farmers and in unprotected natural areas in South America. The genetic diversity present within the cultivated peanut represents a wealth of options to better respond to the changing needs and preferences of peanut farmers, breeders and consumers. The important diversity of cultivated peanuts is complemented and greatly enriched by the even broader range of genetic diversity contained in the crop's wild relatives. Experts estimate that the genus *Arachis* comprises around 100 wild species, 80 of which are described, several more are in the process of being described and another dozen or so are believed to exist but remain undiscovered (Krapovickas and Gregory, 1994, 2007; Valls and Simpson, 2005). The wild species of *Arachis* are endemic to the grasslands and open-forested areas of lowland Bolivia, northwestern Argentina, Paraguay, Brazil and Uruguay. Of these wild species, around 30 are considered to be closely related to the cultivated peanut and therefore can serve as important sources of resistance to drought, pests and diseases for breeders, using conventional cross-breeding techniques without having to resort to genetic transformation.

Modern peanut-breeding efforts rely heavily on the large international peanut germplasm collections maintained by the International Crop Research Institute for the Semi-Arid Tropics (ICRISAT) and the National Plant Germplasm System in India, the US Department of Agriculture in the United States, the Agricultural Research Corporation in Brazil, the Instituto Nacional de Tecnologia Agropecuaria in Argentina, and the Chinese Academy of Agricultural Sciences. These important collections contain thousands of accessions of landrace materials, high-yielding improved varieties, experimental breeding lines and wild relatives, and they each have some unique accessions of peanuts collected from a different region of the world that are not duplicated anywhere else. However, these great collections do not hold all of the answers to our breeding needs because, even when considered together, the current gene bank collections do not yet have a representative sampling of the full scope of existing peanut diversity. Significant gaps remain to be filled by collecting missions before the full spectrum of peanut diversity can be known, studied and protected and before peanut breeders can have access to a safe and comprehensive source of the genetic resources they need to develop improved peanut varieties for the future.

Peanut breeders have already made use of peanut diversity obtained from many countries to produce agronomically and commercially successful improved varieties (Isleib and Wynne, 1992). Inter-subspecific hybrids are routinely employed to take advantage of the earliness of one subspecies combined with the higher-yielding qualities of the other. Through a complicated and painstaking process involving thousands of hand pollinations over a period of many years, conventional breeding techniques have been successfully employed to introduce nematode resistance from a wild *Arachis* species from Bolivia into a commercially acceptable peanut variety known as COAN, which has effectively solved a serious pest problem that had been plaguing peanut growers across a large area of southwestern United States (Simpson, 2001). Subsequently, other improved varieties have been released, building upon the nematode-resistance of

COAN, which are suited to the particular growing conditions of southeastern United States and elsewhere (Holbrook et al., 2008). Work at ICRISAT, where the world peanut collection of over 12,000 accessions is held, includes a peanut breeding programme that actively seeks out traits from South American landraces and wild species for breeding improved varieties that are better suited to the needs of peanut farmers in Africa and Asia. As part of a collaborative research programme between Bolivia and the United States, Bolivian peanut landraces of interest were made available to US researchers, while advanced breeding lines of improved US varieties with traits suitable for the Bolivian market and growing conditions are being evaluated in Bolivia for their potential for large-scale production (Peanut Collaborative Research Support Program, 2007).

Another important use of wild *Arachis* species is as forage, pasture, soil conservation and ornamental groundcover crops. Two species, *A. pintoi* and *A. glabrata*, which are both native to Brazil, have already been improved through selection, and commercial varieties of these species are now cultivated extensively in Australia, Central America, the United States and Africa. Sometimes referred to as 'tropical alfalfa', these 'forage peanuts' are valued for their vigorous growth, palatability for livestock, high protein content and tolerance to high temperatures – all traits that suggest their enormous potential for improving pastures and increasing livestock productivity in the tropics.

Drought tolerance is a general characteristic of the peanut, and valuable sources of resistance can be found in unimproved peanut landraces and closely related wild species, many of which have evolved in areas that are extremely arid during much of the year. This particular trait will be especially important as different crops and new varieties are sought that will provide viable alternatives to farmers and farming nations worldwide as they are faced with production constraints associated with climate change.

The global interdependence on peanut genetic resources is abundantly evident, perhaps even more than in most other crops. There are, however, two main obstacles that hinder the present and future exchange and use of peanut genetic resources. The first is that the peanut is not currently on the Annex 1 list of crops covered by the multilateral system of the ITPGRFA. This omission needs to be corrected as soon as possible to ensure that these globally important genetic resources can benefit from the same legal protection and facilitated access afforded by the multilateral system to the world's most important crops. Although the *Arachis* collection held by ICRISAT is subject to the terms and conditions of the multilateral system, thanks to the agreements between ICRISAT and the Governing Body of the ITPGRFA under Article 15 of the Treaty, this collection still lacks good representative coverage in some regions of known peanut diversity, particularly in regard to landrace materials and wild peanut species from South America. This situation is due in part to strict quarantine regulations in India (where ICRISAT's *Arachis* collection is located) that make it difficult to import germplasm from South America. Consequently, many unique accessions of both wild and cultivated *Arachis* are conserved *ex situ* only by national programmes in the Americas. The other main obstacle involves the extremely precarious state of conservation of wild *Arachis* species in their natural habitat where they are seriously threatened with extinction due to habitat destruction caused by the massive expansion of soybean cultivation, cattle ranching, urbanization and climate change (Jarvis et al., 2003). Many wild *Arachis* species are notoriously difficult to maintain in *ex situ* gene banks, and, as mentioned earlier, some wild species have yet to be discovered while many others have been scarcely collected.

Case 5: Searching for disease resistance in cacao (*Theobroma cacao* L.) in the gene bank of the Tropical Agricultural Research and Higher Education Centre (CATIE)

Latin America is the centre of origin and domestication of cacao (*Theobroma cacao* L.), which was introduced to Meso-America from South America in prehistoric times (Motomayor et al., 2002) and has been traded as far as North America since pre-Columbian times (Crown and Hurst, 2009). Cacao cultivation started in the lowlands of Mexico 2,000 years ago (Cope, 1976). It is a perennial crop cultivated in agroforestry systems under the shade of timber or fruit trees and, hence, is beneficial for the protection of tropical environments. Moreover, cacao has enormous socio-economic value as it is primarily cultivated by small-hold farmers worldwide and more than 20 million people depend directly on it for their livelihood.

Unfortunately, cacao is not included in the list of Annex I crops covered by the multilateral system of the ITPGRFA. By signing an agreement with the Governing Body of the Treaty, however, CATIE and the Cocoa Research Unit (hosted by the University of West Indies in Trinidad) have placed the international collections of cocoa germplasm that they maintain under the auspices of the ITPGRFA under the mandate of the multilateral system and have committed to make them available according to the Treaty. However, these two international gene banks conserve only a portion of the existing diversity of cacao, much of which is in national gene banks and in farmers' fields. To facilitate scientific cooperation and knowledge sharing aimed to optimize the conservation and use of cacao genetic resources, a global network entitled CacaoNet was established in 2006 (Engels, 2006).

About 30 per cent (810,000 metric tonnes) of the world's cacao production is lost every year due to three major diseases: (1) 'witches broom' caused by *Moniliophthora* (= *Crinipellis*) *perniciosa;* (2) frosty pod rot caused by *Moniliophthora roreri*; and (3) black pod disease caused by *Phytophthora palmivora* and/or *P. capsici* (Guiltinan, 2007). Among these three pathogens, frosty pod rot is the most devastating since it can lead to the abandonment of cacao cultivation. Frosty pod, which is believed to have originated in northeastern Colombia, has spread to ten countries over a period of 200 years. Starting in Colombia in 1817, damage was reported in Ecuador in 1917 and in Venezuela in 1941 (Phillips-Mora, 2003). In 1956, frosty pod rot was reported in Panama and subsequently spread to Costa Rica in 1978, Nicaragua in 1979, Honduras in 1997 and Guatemala in 2002. In 2004, the disease was detected in Belize and in 2005 it was detected in Mexico (Phillips-Mora et al., 2006a) where it destroyed up to 80 per cent of cacao production in infected areas.

Cultural practices and plant protection measures to combat the pathogen are inefficient and costly, and, hence, small-scale farmers generally abandon their cacao fields once the disease has spread, leading to huge harvest losses. The most economic and long-lasting solutions are resistant cacao cultivars that reduce the need for chemical treatments, thus rendering cacao production once again attractive for farmers and beneficial for the environment.

The cacao breeding programme undertaken by CATIE extensively evaluated the international cacao collection comprising in 2006 of a total of 942 accessions of different origins (Phillips-Mora et al., 2006b). This programme aimed at identifying genotypes with high productivity, resistance to frosty pod rot and black pod disease and outstanding industrial quality. Tolerance to frosty pod rot is a rare trait, as only five genotypes out of over 600 accessions evaluated showed resistance. These five genotypes were acquired from diverse sources: UF-273 and UF-712 from Costa Rica in 1960; EET-75 from Ecuador between 1965 and 1966; ICS-95 from Trinidad in 1959; and PA-169 from Peru in 1961.

Recent breeding results are very promising as numerous superior genotypes have been developed, whose industrial quality is being evaluated in collaboration with international chocolate manufacturers. The best genotypes of the CATIE breeding programme were established in clonal gardens for multiplication and subsequent validation in multilocational trials under a wide range of agro-ecological conditions in different regions of Latin America.

Testing started in Costa Rica in 2005 with the transfer of elite material to cacao farmers in different strategic locations. With the introduction of a regional cacao project for Central America in 2007, a massive multiplication and dissemination process began, which was aimed at assisting at least 2,500 indigenous families and small-scale farmers in six countries (Panama, Costa Rica, Nicaragua, Honduras, Belize and Guatemala). Five clonal gardens of one hectare each were established in these countries for local multiplication of the planting material, thus avoiding the cost of long distance transportation of the grafted plants. Five superior frosty pod rot resistant breeding materials were selected for multiplication, together with three outstanding local and/or international cacao lines and four lines that will serve as rootstock. In addition, 22 superior lines, the majority of which are frosty pod rot resistant, will be validated alongside ten locally selected and/or international varieties in multilocational trials in these countries.

The CATIE breeding programme is also sending frosty pod rot resistant lines to national breeding programmes in Nicaragua, Honduras, Mexico, Trinidad, Ecuador, Peru and Brazil as well as to the Intermediate Cocoa Quarantine Facility at the University of Reading in the United Kingdom for disease testing and subsequent worldwide distribution if found safe.

The accelerated dissemination of frosty pod rot in Central America in recent years suggests active human participation. If the disease was to spread to Africa and Asia, which are the main production areas of cacao, the livelihood of tens of thousands of poor farmers would be at risk and the entire chocolate industry worldwide would be threatened. The deployment of frosty pod resistant genotypes to Africa and Asia, through the Intermediate Cocoa Quarantine Facility at the University of Reading, could ensure that these countries could cope with the possible arrival of the devastating disease, without suffering the enormous production losses experienced in Central America.

The successful completion of the breeding programme at CATIE would not have been possible without easy access to its international germplasm collection, composed of numerous valuable genotypes from different Latin American countries. The dissemination of superior genotypes developed by this and other breeding programmes has enabled cacao producers worldwide to cope with major disease threats, providing substantial benefits to small-hold farmers, the chocolate industry, consumers and the environment.

Case 6: The importance of accessing *Musa* genetic resources for Latin America and the Caribbean

Bananas originate from the tropical and sub-tropical humid forests that extend from India to Papua New Guinea. Domestication of its seedy fruits is believed to have started more than 7,000 years ago with the selection by farmers of fruits with more pulp than seeds (Denham et al., 2003). Dispersal from the centre of origin resulted in the development of additional groups of bananas arriving in Polynesia between 3,500 and 4,500 years ago (De Langhe and De Maret, 1999) and to East Africa 4,500 years ago where they diversified into East African highland bananas and plantains in west and central Africa (Lejju et al., 2006). Introduction to the Americas happened much later and has not resulted in the evolution of significant diversity since most domesticated varieties are, for all practical purposes, sterile, and genetic diversity is generated by occasional, naturally occurring mutations. The Portuguese brought the banana from West Africa to the Canary Islands some time after 1402, and they were later transported on the ships sailing to the New World during the Spanish conquest (Simmonds, 1966). Plantains were introduced to the Americas via the Caribbean in the seventeenth century and are now widely grown for local consumption.

Bananas are a staple in many developing countries.[2] The major banana-growing regions of the world – Latin America and the Caribbean, Asia-Pacific and Africa – each produce about one-third of the annual production, which was estimated at 95 million metric tonnes in 2009 (FAO, 2010). The majority of producers are small-hold farmers who grow the crop for either home consumption or for local markets. Only about 15 per cent of the global production is exported. Even in Latin America and the Caribbean, which produce most of the bananas for the export market, only 30 per cent are exported.

Bananas are unusual among major crops in that most of the types grown, either for export or local consumption, are farmer-selected varieties rather than improved hybrids produced by breeding programmes. Current banana-breeding efforts are inadequate to address the scale of the problems faced by small-hold producers. Meanwhile, the genetic base on which solutions to these problems depend – either through genetic improvement or a better use of diversity in production systems – is being eroded (Karamura and Mgenzi, 2004; Sharrock and Frison, 2004). Market forces, in particular, have encouraged small-hold farmers to focus on the commercial varieties, leading to the loss of traditional cultivars and, thus, making the crop even more vulnerable to pests and diseases.

Although a significant share of the crop's genepool is conserved in gene banks, the coming into force in 1993 of the Convention on Biological Diversity and the concurrent uncertainty and politization of issues surrounding genetic resources has severely limited the acquisition of new materials, especially of wild *Musa*.[3] The situation is especially challenging for Latin American and Caribbean countries, which are dependent on banana genetic resources from beyond their borders.

The commercial potential of bananas was first exploited in Central America and the Caribbean. By the 1900s, the market was increasingly reliant on the dessert variety Gros Michel, and this variety was planted over a widespread area, thereby creating a favourable environment for the spread of diseases. Indeed, a disease caused by a fungus thought to have originated in Southeast Asia was reported in Panama in 1890 (Simmonds, 1966). By 1930, Fusarium wilt, or Panama disease, had spread to Costa Rica, Surinam, Honduras, Guatemala and Jamaica. Since the soil-dwelling fungus cannot be controlled with chemical pesticides, the only way to grow susceptible cultivars such as Gros Michel on a large scale was to clear forests to set up new plantations and move on as soon as the disease arrived.

Panama disease triggered the first breeding efforts to produce a disease-resistant Gros Michel-like export banana. In the 1920s, breeding programmes were started in Trinidad and Jamaica by the British government and in Panama by the United Fruit Company, which was later moved to Honduras (Rowe and Richardson, 1975). Progress to produce a disease-resistant Gros Michel variety using pollen from wild bananas resulted in inferior progenies. Breeders then decided to concentrate on breeding 'improved' male parents that possessed both disease resistance and good agronomic characteristics (Rowe and Richardson, 1975). Meanwhile, the banana companies had solved their Panama disease problem by switching to Cavendish varieties that were resistant to the disease, although they are susceptible to other pests and diseases that could be controlled using chemical pesticides. By the beginning of the 1980s, with no new commercial cultivars in sight, the main breeding programmes were facing closure. At the same time, donors were considering creating an independent networking organization to encourage cooperation at the international level regarding, among other things, the safe exchange of *Musa* germplasm for the development of disease-resistant varieties (Nestel, 1984).

Progress towards international cooperation was made when the International Network for the Improvement of Bananas and Plantains (INIBAP) was created in 1985. One of its first actions was to establish a collection of *Musa* germplasm at the International Transit Centre (ITC) in Leuven, Belgium. It addressed quarantine issues by setting up virus indexing centres to screen germplasm and started channelling financial support to the Fundación Hondureña de Investigación Agrícola (FHIA). The FHIA soon delivered disease-resistant hybrids, many of which, along with improved male parents, are available for international distribution through the ITC. In 1994, the International *Musa* Germplasm Collection at the ITC became part of the international network of *ex situ* gene banks following the signature of an agreement between Bioversity International, which INIBAP had joined, and the Food and Agriculture Organization (FAO).

However, the legal and policy uncertainties surrounding the access to genetic resources has led several countries to withhold the export of PGRFA, inaugurating a period of drastically reduced access to *new* germplasm for subsequent global distribution, even after the adoption of the ITPGRFA in 2001. The ITC was not exempted. The number of accessions acquired in the last few years has been reduced to nearly zero except for African cultivars (Vézina, 2008). Meanwhile, over the same period of time, the number of samples from the ITC's existing collection that have been distributed around the world for research, field evaluation and breeding purposes has risen dramatically.

The genetic pedigree of the FHIA hybrids (the most important improved varieties currently being distributed internationally) demonstrates the importance of being able to draw upon a wide range of genetic resources and the value of sharing resources across regions. For example, Jonathan Robinson (2000) traced the genetic resource history that led to the making of FHIA-03, a cooking banana bred by Phillip Rowe and Franklin Rosales, as part of an exercise to examine the impact of genetic resources. In total, some 14 crosses involving 11 wild types and two triploid landraces took place to bring about FHIA-03. Crosses between four wild types from Papua New Guinea, Java, Malaysia and the Philippines gave birth to a vigorous diploid (SH-2095), which produces large bunches weighing up to 30 kilograms. SH-2095 is a parent of many of the FHIA varieties.

Although the FHIA hybrids are highly productive and disease-resistant, and tens of thousands of plantlets of these varieties have been distributed to farmers through various projects in Latin America, Africa and Asia, the factors that favour adoption are still not fully understood. In Tanzania, growing FHIA banana hybrids substantially and significantly reduced expected yield losses from pests and diseases (Nkuba et al., 2006). Furthermore, since hybrid cultivars yield larger bunches, fewer plants are required to meet the consumption needs of the household, leaving more land available for the production of other crops or pasture. In Uganda, however, adoption rates have been lower, perhaps because of the differences in the severity of pest and disease pressures, the cooking qualities of the hybrids and the dissemination strategies that are being used (Nkuba et al., 2006).

The most enthusiastic adoption of the FHIA's hybrids has been in Cuba, with more than 14,000 hectares planted since 1991. This enthusiastic adoption came about at a time when the cost of controlling black leaf streak disease on the Cavendish varieties had quadrupled, shortly after the arrival of the disease on the island (Perez Vicente et al., 2003).

In the meantime, public sector support for the FHIA ended in 2004, and the banana breeding programme is back to being funded by commercial interests, with its objectives more aligned with the needs of the export industry and its products no longer readily available for distribution to small-hold farmers and public sector research programmes. Other programmes that have been established by other organizations, such as the Agricultural Research Corporation in Brazil, are continuing to produce hybrids for small-hold farmers, but despite the fact that these efforts are being made by the public sector their efforts are essentially targeted towards serving national producers. Still, the main factor limiting the anticipated benefits of future breeding efforts is access to PGRFA. For example, the demand for varieties that are tolerant to cold, excessive moisture or drought in order to help banana growers cope with climate change or increase the amount of arable land is unlikely to be met unless more wild species are collected and conserved in gene banks. One of the priorities of the ITC is to boost its holding of wild species so that the benefits that could be derived from them are spread more widely.

Concluding remarks

The cases reviewed in this chapter amply demonstrate the importance of access to as broad a range of diversity of PGRFA as possible to respond to the needs of farmers. They constitute a compelling argument in favour of a system that guarantees access to a wide range of germplasm to overcome the food, nutrition and livelihood constraints faced by tens of thousands of small-scale farmers and their families. It is expected that the increased use of molecular tools will lead researchers to make even greater use of the diversity available in wild and cultivated crops in improving crop varieties, thus strengthening the case for the need to improve the availability and accessibility of genetic resources (Hajjar and Hodgkin, 2007).

The multilateral system of the ITPGRFA provides the legal framework within which genetic resources of the most relevant crops and forages (many of which are already included in vast *ex situ* collections) can be virtually 'pooled' and accessed and used for the purposes of training, research and breeding for food and agriculture. The multilateral system also includes mechanisms for sharing benefits associated with the use of genetic resources accessed from the system. The criteria for crops or forages to be included in the Treaty's multilateral system include their importance to 'food security and interdependence' (Article 11.1).

All six of the crops examined in this chapter are important for food security, and each of them is greatly depended upon by their countries. (The number of countries interdependent on lupin are clearly fewer, and all are located within the same region.) Yet three of these crops – peanuts, lupin and cacao – are not included in the ITPGRFA's multilateral system. The case of lupin (chocho), which is a locally important crop in the Andean subregion, shows how a national programme can make use of materials obtained through collaborative collecting beyond its frontiers in order to undertake important research on its own. The authors of the sections on cacao and peanuts, which are both crops of worldwide importance, warn us that much-needed research on these crops can be hindered by the lack of facilitated access to their genetic diversity. Subsequent chapters in this book illustrate how it came to be that some of these important crops were not included in the Treaty's multilateral system. Politics, rather than the significance of these crops for food security and interdependence, has led to this outcome. There are clearly scientifically sound arguments to make – regardless of whether or not there is political will to make them – that these crops, and others like them, should be included in the multilateral system.

The ITPGRFA and its multilateral system are components of the global system of conservation and use of PGRFA as envisaged by the FAO. The multilateral system is under implementation, and although there is still quite a way to go it remains the best hope to deal with the challenges of increasing productivity in a sustainable manner by expanding the use of agricultural biodiversity. At its heart lies the principle of access to the broadest range of crop diversity.

Notes

1 International Treaty on Plant Genetic Resources for Food and Agriculture, 29 June 2004, www.planttreaty.org/texts_en.htm (last accessed 30 March 2011). Standard Material Transfer Agreement, 16 June 2006, ftp://ftp.fao.org/ag/agp/planttreaty/agreements/smta/SMTAe.pdf (last accessed 30 March 2011).
2 Here we are using the term banana broadly to encompass both dessert and cooking varieties that belong to the genus *Musa*.
3 Convention on Biological Diversity, 31 I.L.M. 818 (1992).

References

Angeles, E.R., R.C. Cabunagan, R.E. Tabien and G.S. Khush (2008) 'Resistance to Tungro Vectors and Viruses', in E.R. Tiongco, E.R. Angeles and L.S. Sebastian (eds), *The Rice Tungro Virus Disease: A Paradigm in Disease Management*, Philippines Rice Research Institute, Science City of Munoz, Nueva Ecija, Philippines, and Honda Research Institute, Japan.
Caicedo, C., and E. Peralta (1999) 'INIAP-450 Andino: Variedad de Chocho (*Lupinus mutabilis* Sweet)', Programa Nacional de Leguminosas. Estación Experimental Santa Catalina, Instituto Nacional Autónomo de Investigaciones Agropecuarias, Quito. Ecuador.
Centro Internacional de Mejoramiento de Maiz y Trigo (CIMMYT) (1986) *Improving on Excellence: Achievements of Breeding with the Maize Race Tuxpeño*, CIMMYT, Mexico City, Mexico.
——(2004) *Adding Value for Development: CIMMYT Annual Report 2003–2004*, CIMMYT, Mexico City, Mexico.
——(2007) 'Body Blow to Grain Borer', *CIMMYT E-News*, vol. 4, no. 9, online: www.cimmyt.org/english/wps/news/2007/sep/borers.htm (last accessed 23 March 2011).
Chatel, M., and E.P. Guimaraes (2002) 'International Partnership for Rice Improvement in Latin America: CIRAD, a Case Study', Brazilian Society of Plant Breeding, *Crop Breeding and Applied Bulletin* 165(2): 639–48.
Chavarriaga, E. (1966) 'Corn ETO, una variedad producida en Colombia', *Revista Instituto Colombiano Agropecuario* 1: 5–30.
Cope, F.W. (1976) 'Cacao', in N.W. Simmonds (ed.), *Evolution of Crop Plants*, Longman, London.
Crow, J.F., and J. Kermicle (2002) 'Oliver Nelson and Quality Protein Maize', *Genetics* 160: 819–21.
Crown, P., and J. Hurst (2009) 'Evidence of Cacao Use in the Prehispanic American Southwest', *Proceedings of the National Academy of Sciences* 106(7): 2110–13.
De Langhe, E., and P. De Maret (1999) 'Tracking the Banana: Its Significance in Early Agriculture', in C. Gosden and J. Hather (eds), *The Prehistory of Food: Appetites for Change*, Routledge, London.
Denham, T.P., S.G. Haberle, C. Lentfer, R. Fullagar, J. Field, M. Therin, N. Porch and B. Winsborough (2003) 'Origins of Agriculture at Kuk Swamp in the Highlands of New Guinea', *Science* 301(5643): 189–93.
Engels, J. (2006) 'Towards the Establishment of a Cacao Genetic Resources Network – CacaoNet', *Consultative Board on the World Cocoa Economy* Modica, Italy (6 March 2006).

Evenson, R.E., and D. Gollin (eds) (2003a) *Crop Variety Improvement and Its Effect on Productivity: The Impact of International Agricultural Research*, CABI Publishing, Wallingford, UK.

——(2003b) 'Crop Genetic Improvement in Developing Countries', in R.E. Evenson and D. Gollin (eds), *Crop Variety Improvement and Its Effect on Productivity: The Impact of International Agricultural Research*, CABI Publishing, Wallingford, UK.

Flores Palacios, X. (1998) 'Contribution to the Estimation of Countries' Interdependence in the Area of Plant Genetic Resources', Report no. 7, Rev. 1, Committee on Genetic Resources for Food and Agriculture (CGRFA), FAO, Rome, Italy.

Food and Agriculture Organization (FAO) (2010) *Production Indices* (September 2010), online: http://faostat.fao.org/site/567/default.aspx#ancor (last accessed 23 March 2011).

Fowler, C. M. and T. Hodgkin (2004) 'Plant Genetic Resources for Food and Agriculture: Assessing Global Availability', *Annual Review of Environment and Resources* 29: 143–79.

Fowler, C., M. Smale and S. Gaiji (2001) 'Unequal Exchange? Recent Transfers of Agricultural Resources and Their Implications for Developing Countries', *Development Policy Review* 19: 181–204.

Frison, C., and M. Halewood (2006) *Annotated Bibliography Addressing the International Pedigrees and Flows of Plant Genetic Resources for Food and Agriculture*, System-wide Genetic Resources Programme, online: www.sgrp.cgiar.org/?q=node/486 (last accessed 23 March 2011).

Fujisaka, S., D. Williams and M. Halewood (eds) (2009) 'The Impact of Climate Change on Countries' Interdependence on Genetic Resources for Food and Agriculture', Background Study paper no. 48, CGRFA, FAO, Rome, Italy.

Gepts, P. (2004) 'Who Owns Biodiversity and How Should the Owners Be Compensated?' *Plant Physiology* 134: 1295–1307.

Guiltinan, M.J. (2007) 'Recent Advances and Future Directions in the Applications of Biotechnology to the Improvement of *Theobroma cacao* L.: The Chocolate Tree', in E. Pua and M. Davey (eds), *Biotechnology in Agriculture and Forestry: Transgenic Crops*, V. Springer Verlag, Berlin.

Gupta, H.S., P.K. Agrawal, V. Mahajan, G.S. Bisht, A. Kumar, P. Verma, A. Srivastava, S. Saha, R. Babu, M.C. Pant and V.P. Mani (2009) 'Quality Protein Maize for Nutritional Security: Rapid Development of Short Duration Hybrids through Molecular Marker Assisted Breeding', *Current Science* 96: 230–37.

Hajjar, R., and T. Hodgkin (2007) 'The Use of Wild Relatives in Crop Improvement: A Survey of Developments over the Last Twenty Years', *Euphytica* 156: 1–13.

Hallauer, A.R. (1994) 'Registration of BS28 and BS29 Corn Germplasm', *Crop Science* 34: 544–45.

Hallauer, A.R., and O.S. Smith (1979) 'Registration of BS13(S2)C1 and BS16 Corn Germplasm', *Crop Science* 19: 755.

Holbrook, C.C., P. Timper, A.K. Culbreath and C.K. Kvien (2008) 'Registration of "Tifguard" Peanut', *Journal of Plant Registrations* 2: 92–94.

Hossain, M., D. Gollin, V. Cabanilla, E. Cabrerra, N. Johnson, G.S. Khush and G. McLaren (2003) 'International Research and Genetic Improvement in Rice: Evidence from Asia and Latin America', in R.E. Evenson and D. Gollin (eds), *Crop Variety Improvement and Its Effect on Productivity: The Impact of International Agricultural Research*, CABI Publishing, Wallingford, UK.

Instituto Nacional Autónomo de Investigaciones Agropecuarias (INIAP) (2009) *INIAP Cincuenta años aportando al país*, INIAP, Quito, Ecuador.

Isleib, T.G., and J.C. Wynne (1992) 'Use of Plant Introductions in Peanut Improvement', in H.L. Shands and L.E. Weisner (eds), *Use of Plant Introductions in Cultivar Development – Part 2*. Crop Science Society of America (CSSA) Special Publication no. 20, CSSA, Madison, WI.

Jackson, M.T. (1997) 'Conservation of Rice Genetic Resources: The Role of the International Rice Genebank at IRRI', *Plant Molecular Biology* 35: 61–67.

Jarvis, A., M.E. Ferguson, D.E. Williams, L. Guarino, P.G. Jones, H.T. Stalker, J.F.M. Valls, R.N. Pittman, C.E. Simpson and P. Bramel (2003) 'Biogeography of Wild Arachis: Assessing Conservation Status and Setting Future Priorities', *Crop Science* 43(3): 1100–108.

Johnson, N.L., D. Pachico and O. Voysest (2003) 'The Distribution of Benefits from Public International Germplasm Banks: The Case of Beans in Latin America', *Agricultural Economics* 29: 277–86.
Karamura, D., and B. Mgenzi (2004) 'On Farm Conservation of Musa Diversity in the Great Lakes Region of East Africa', *African Crop Science Journal* 12(1): 75–83.
Khush, G.S., and P.S. Virk (2005) *IR Varieties and Their Impact*, International Rice Research Institute (IRRI), Los Baños, Philippines.
Kling, J.G., J.M. Fajemisin, B. Badu-Apraku, A. Diallo, A. Menkir and Melake-A. Berhan (2000) '*Striga* Resistance Breeding in Maize', in B.I.G. Haussmann, D.E. Hess, M.L. Koyama, L. Grivet, H.F.W. Rattunde and H.H. Geiger (eds), *Breeding for Striga Resistance in Cereals*, Margraf Verlag, Weikersheim, Germany.
Kloppenburg, J., and D.L. Kleinman (1987) 'Analyzing Empirically the Distribution of the World's Plant Genetic Resources: The Plant Germplasm Controversy', *Bioscience* 37(3): 190–98.
Krapovickas, A., and W.C. Gregory (1994) 'Taxonomía del género *Arachis* (Leguminosae)', *Bonplandia* 8: 1–186.
——(2007) 'Taxonomy of the Genus *Arachis* (Leguminosae)', *Bonplandia* 16(Sup. L): 1–205 [English translation of this Spanish-language monograph is published in D.E. Williams and C.E. Simpson (1994), translators, *Bonplandia*, Instituto de Botánica del Nordeste, Corrientes, Argentina].
Krivanek, A., F.H. De Groote, N.S. Gunaratna, A.O. Diallo and D. Friesen (2007) 'Breeding and Disseminating Quality Protein Maize for Africa', *African Journal of Biotechnology* 6: 312–24.
Lejju, B.J., P. Robertshaw and D. Taylor (2006) 'Africa's Earliest Bananas?' *Journal of Archaeological Science* 33: 102–13.
Matsuoka, Y., Y. Vigouroux, M.M. Goodman, J. Sanchez Garcia, E. Buckler and J. Doebley (2002) 'A Single Domestication for Maize Shown by Multilocus Microsatellite Genotyping', *Proceedings of the National Academy of Sciences* 99: 6080–84.
Menkir, A., J.G. Kling, B. Badu-Apraku and O. Ibikunle (2006) 'Registration of 26 Tropical Maize Germplasm Lines with Resistance to *Striga hermonthica*', *Crop Science* 46: 1007–9.
Mertz, E.T., L.S. Bates and O.E. Nelson (1964) 'Mutant Gene That Changes Protein Composition and Increases Lysine Content of Maize Endosperm', *Science* 145: 279–80.
Motomayor, J.C., A.M. Risterucci, P.A. Lopez, C.F. Ortiz, A. Moreno and C. Lanaud (2002) 'Cacao Domestication 1: The Origin of the Cacao Cultivated by the Mayas', *Heredity* 89: 380–86.
Nestel, B. (1984) *An International Network for the Improvement of Bananas and Plantains*, Discussion paper presented by International Development Research Council to a donor group meeting in Rome, 22 May 1984, online: http://idl-bnc.ca/dsapace/bitstream/10625/7516/1/64968.pdf (last accessed 30 March 2011).
Njau, P.N., Y. Jin, J. Huerta-Espino, B. Keller and R.P. Singh (2010) 'Identification and Evaluation of Sources of Resistance to Stem Rust Race Ug99 in Wheat', *Plant Disease* 94: 413–19.
Nkuba, J., S. Edmeades and M. Smale (2006) 'Gauging Potential Based on Current Adoption of Banana Hybrids in Tanzania', International Food Policy Research Institute Policy (IFPRI) Brief no. 21, IFPRI, Washington, DC.
Ortiz, R., M. Perez Fernandez, J. Dixon, J. Hellin and M. Iwanaga (2007) 'Specialty Maize: Global Horticultural Crop', *Chronica Horticulturae* 47(4): 20–25.
Peanut Collaborative Research Support Program (2007) *Development and Use of Multiple-Pest Resistance to Improve Production Efficiency of Peanut*, Project no. UFL16P, Peanut Collaborative Research Suort Program, online: http://peanutcrsp.org/home.cfm (last accessed 23 March 2011).
Peralta, E., N. Mazon, A. Murillo, M. Rivera and C. Monar (2009a) 'Manual de granos andinos. CHOCHO, QUINUA, AMARANTO Y ATACO. Cultivos, variedades y costos de producción', Manual no. 69 (segunda edición), Programa Nacional de

Leguminosas y Granos Andinos, Estación Experimental Santa Catalina, INIAP, Quito, Ecuador.
Peralta, E., N. Mazon, A. Murillo, E. Villacres, M. Rivera and C. Subia (2009b) 'Catálogo de variedades mejoradas de granos andinos: chocho, quinua y amaranto, para la sierra del Ecuador', Publicación miscellanea no. 151, November 2009, Programa Nacional de Leguminosas y Granos Andinos. Estación Experimental Santa Catalina, INIAP, Quito, Ecuador.
Perez Vicente, L., J.M. Alvarez, and M. Perez (2003) 'Economic Impact and Management of Black Leaf Streak Disease in Cuba', in *Mycosphaerella Leaf Spot Disease of Bananas: Present Status Outlook* (Proceedings of the Second International Workshop on Mycosphaerella Leaf Spot Diseases, San Jose, Costa Rica, 20–23 May 2002), International Network for the Improvement of Banana and Plantain, Montpellier, France.
Phillips-Mora, W. (2003) *Origin, Biogeography, Genetic Diversity and Taxonomic Affinities of the Cacao Fungus* Moniliophthtera Roreri *as Determined Using Molecular, Phytopathological and Morpho-Physiological Evidence*, PhD dissertation, University of Reading, Reading.
Phillips-Mora, W., C.F. Ortiz and M.C. Aime (2006a) 'Fifty Years of Frosty Pod Rot in Central America: Chronology of Its Spread and Impact from Panamá to Mexico', in *Proceedings of the Fifteenth International Cocoa Research Conference* (Cocoa Producers' Alliance in collaboration with CATIE, 9–14 October 2006, San José, Costa Rica).
Phillips-Mora, W., A. Mora, E. Johnson and C. Astorga (2006b) 'Recent Efforts to Improve the Genetic and Physical Conditions of the International Cacao Collection at CATIE', in *Proceedings of the Fifteenth International Cocoa Research Conference* (Cocoa Producers' Alliance in collaboration with CATIE, 9–14 October 2006, San José, Costa Rica).
Robinson, J. (2000) *Genetic Resources Impact Tracing Study: A Report for the System-Wide Genetic Resources Programme*, Bioversity International, Rome.
Rowe, P., and D.L. Richardson (1975) 'Breeding Bananas for Disease Resistance, Fruit Quality and Yield', *Tropical Agriculture Research Services*, Bulletin no. 2, La Lima, Honduras.
Sebastian, L.S., L.R. Hipolito and J. Garcia (1998) 'Molecular Contribution of Landraces in the Development of Philippine-bred Modern Rice Varieties', *SABRAO Journal of Plant Breeding and Genetics* 30(2): 73–82.
Sharrock, S.L., and E.A. Frison (2004) 'Prospects and Challenges of Biodiversity in Small-Holder Systems', *African Crop Science Journal* 12(1): 51–57.
Simmonds, N.W. (1966) *Bananas*, 2nd edition, Longman, London.
Simpson, C.E. (2001) 'Use of Wild *Arachis* Species/Introgression of Genes into *A. hypogaea* L.', *Peanut Science* 28: 114–16.
Smale, M., with contributions from P. Aquino, J. Crossa, E. del Toro, J. Dubin, T. Fisher, P. Fox, M. Khairallah, A. Mujeeb-Kazi, K.J. Nightingale, I. Ortiz-Monasterio, S. Rajaram, R. Singh, B. Skovmand, M. van Ginkel, G. Varughese and R. Ward (1996) *Understanding Global Trends in the Use of Wheat Diversity and International Flows of Wheat Genetic Resources*, Economics Working Paper no. 96–02, CIMMYT, Mexico City, Mexico.
Smalley, J., and M. Blake (2003) 'Sweet Beginnings: Stalk Sugar and the Domestication of Maize', *Current Anthropology* 44: 675–703.
Smartt, J. (1985) 'Evolution of Grain Legumes: The Oilseeds', *Experimental Agriculture* 21: 305–20.
Sriwatanapongse, S., S. Jinahyon and S.K. Vasal (1993) *Suwan-1: Maize from Thailand to the World*, CIMMYT, Mexico City, Mexico.
System-wide Genetic Resources Programme (SGRP) (2009) 'Reaping the Benefits of Multilateral Exchange: The CGIAR Centres and the Implementation of Article 15: Some Case Studies', Third Session of the Governing Body, Tunis, 1–5 June 2009, International Treaty on Plant Genetic Resources, SGRP, Rome.
—— (2010) 'The Importance of Recognizing the International Treaty in the CBD's Protocol on Access and Benefit-Sharing', Policy Brief, July 2010, SGRP, Rome.
Taba, S. (1997) 'A. Maize', in D. Fuccilo, L. Sears and P. Stapleton (eds), *Biodiversity in Trust*, Cambridge University Press, Cambridge.

Taba, S., H.L. Shands and S.A. Eberhart (2005) 'The Growth of CIMMYT's Maize Collection with the Introduction of Latin American Maize Landrace Accessions through the Cooperative Regeneration Project', in S. Taba (ed.), *Latin American Maize Germplasm Conservation: Regeneration, in Situ Conservation, Core Subsets and Prebreeding*, Proceedings of a Workshop held at the CIMMYT, 7–10 April 2003, CIMMYT, Mexico City, Mexico.

Valls, J.F.M., and C.E. Simpson (2005) 'New Species of *Arachis* L. (Leguminosae) from Brazil, Paraguay and Bolivia', *Bonplandia* 14: 35–64.

Vasal, S.K. (2000) 'The Quality Protein Maize Story', *Food Nutrition Bulletin* 21: 445–50.

Vavilov, N.I. (1926) 'Studies on the Origin of Cultivated Plants', *Bulletin of Applied Botany* 16(2): 1–248.

Vézina, A. (2008) 'Importancia del acceso a recursos genéticos de Musa en América Latina y el Caribe', *Recursos Naturales y Ambiente* 53: 72–80.

Villacrez, E., and E. Peralta (2006) 'El aporte de la investigación a la agroindustria y transformación del chocho (*Lupinus mutabilis* sweet)', in *Memoria XII Congreso Internacional de Cultivos Andinos*. INIAP and Pontificia Universidad Católica del Ecuador, Quito, Ecuador.

Williams, D.E. (2004) 'Columbian Exchange: The Role of Analogue Crops in the Adoption and Dissemination of Exotic Cultigens', in *Encyclopedia of Plant and Crop Science*, Marcel Dekker, New York.

Zeven, A.C., and J.M.J. de Wet (1982) *Dictionary of Cultivated Plants and Their Regions of Diversity*, Centre for Agricultural Publishing and Documentation, Wageningen, Netherlands.

3

FLOWS OF CROP GERMPLASM RESOURCES INTO AND OUT OF CHINA

Fuyou Wang

Introduction

China is not only a major grain producer but also a major food consumer. What is more, food production is related to national economic and social stability. Therefore, food production is considered to be highly important in China and ranks among the highest priorities of the government at all levels. In March 2010, Premier Jiabao Wen (2010), in the *Report on the Work of the Government*, proposed that:

[w]e will comprehensively implement the plan to increase grain production by 50 million tons nationwide, with the emphasis on major grain-producing regions. With the focus on breeding improved crop varieties, we will accelerate innovation in agricultural technology, widespread adoption of scientific and technological advances, and carry out a major science and technology project to create new crop varieties using gene transfer technology.

According to expert analysis, the average yields of rice, wheat and maize obtained in China during 1995–97 were significantly lower than the highest experimental yields of these grains (Lin, 1998). Ten years later in 2005–7, they were still low, even assuming that the maximum yield of experimental crops had not increased during that time (Table 3.1).

In recent years, the investment of central and local governments in agricultural research has increased rapidly, and substantial breakthroughs in agricultural technology have been made: the highest experimental yields have certainly increased. What this tells us is that there is great potential for improving the yields of these crops in China. The question is, what needs to be done to accomplish this? According to a survey on crop yields, about 50 per cent of the constraints can be attributed to crop characteristics (Lin, 1998). Therefore, in order to improve crop yields in China, it is essential to obtain germplasm with the characteristics of high yields, the ability to adapt to a wide variety of conditions and resistance to pests and diseases.

TABLE 3.1 Average yields of rice, wheat and maize in China compared to yields of experimental crops, 1995–97 and 2005–7

Crop	1995–97		2005–7	
	Yield (tonnes per hectare)	Percentage of experimental crop yields	Yield (tonnes per hectare)	Percentage of experimental crop yields
Rice	6.19	35	6.32	35.8
Wheat	3.79	28	4.49	33.2
Maize	4.84	25.7	5.26	27.9

Source: Lin (1998); National Bureau of Statistics of China (2008).

Although Vavilov identified China as one of the eight centres of geographical diversity in the world (Moore and Tymowski, 2005, 23), China is still highly dependent on exotic crop genetic resources: 46–55 per cent of China's food production is based on foreign germplasm (Palacios, 1998, 14).

Overview of the introduction of crop genetic resources into China

A large number of crop genetic resources have been introduced into China from around the world. This has benefited China's agricultural industry, resulting in food self-sufficiency being largely achieved in the country, which has, in turn, contributed to food security in the world.

Before the establishment of the People's Republic of China in 1949, Chinese scientists brought in many elite crop genetic resources. The famous wheat variety Nanda 2419 (Mentana) was introduced from Italy in 1942 and was grown around the Yangtze River and in southwest and southern China on a large scale. Since then, it has been extended farther south and north in wheat-growing areas, and with more than 4.7 million hectares a year devoted to it, it is one of the most widely grown wheat varieties in China. In 1930, the high-yielding maize variety Golden Queen, which originated in the Midwest of the United States, was introduced and extended rapidly in Shanxi Province. The success of Golden Queen promoted the development of China's maize breeding, and after half a century this variety is still a main contributor to hybrid materials (Shen, 2004, 27). The sweet potato variety Nancy Hall was introduced from the United States in 1938, and Okinawa 100 from Japan in 1941. These two varieties have made a significant impact on China's sweet potato breeding programmes for over half a century (Ma et al., 1998, 1).

After the establishment of the People's Republic of China, much more effort went into the introduction of crop genetic resources. From 1949 to 1995, more than 80 introduced wheat varieties were being grown without further improvement or adaptation (Ministry of Agriculture of China, 1996, 36). Of these wheat varieties, 30 covered more than 33,000 hectares, 19 were grown on more than 66,700 hectares, and six (which included the Italian wheat varieties Mentana, Funo, Abbondanza, St1472/506, the United States variety CI12203 and the Australian variety Quality) were grown on more than 666,700 hectares (Tong and Zhu, 2001, 48). These introduced varieties had a major impact on wheat breeding and production in China (He et al., 2001, 12). Six potato varieties (Mira, Berlihingen and Aquila from Germany, Epoka from Polland, Aquahbanbma and K-495 from Hungary) were introduced during this period and, since the 1950s, have been extended without improvement after characterization, selection and trial. Mira is still the main variety planted in the mountain areas of southwest China (Tian et al., 2001, 249).

In 1984, China formally became a member country of the Consultative Group on International Agricultural Research (CGIAR). China had begun to introduce crop genetic resources – such as the high-yield rice variety IR-8 from the International Rice Research Institute (IRRI) and Siete Cenos from the International Maize and Wheat Improvement Center (CIMMYT) – from the International Agricultural Research Centres (IARCs) under the CGIAR much earlier, but since joining the CGIAR, China had acquired a large number of crop genetic resources, such as rice, wheat, maize, sweet potato, potato and pigeon pea, from the IARCs (CGIAR, 2005a). Between 1973 and 1983, nearly 300 accessions were introduced from the IARCs every year. However, this number rose to 406 accessions in 1984 and has been more than 500 since 1985, according to the CGIAR's System-Wide Information Network for Genetic Resources (SINGER), which is a relational database that links the databases of the 11 IARCs having gene banks supported by the CGIAR. After 2005, the IRRI alone was providing more than 2,000 samples to China every year, according to the IRRI's records.

After the Convention on Biological Diversity (CBD) came into force in 1993, many countries set up laws and regulations to protect their plant genetic resources, restricting the export and exchange of such resources.[1] This has resulted in increasing difficulties in introducing foreign plant genetic resources (Ministry of Agriculture of China, 2008, 123). Crop genetic resources have mainly been exchanged through bilateral or cooperative agreements between China and other countries on agricultural science and technology. For the period 1996 to 2007, the China Ministry of Science and Technology and the United States Department of Agriculture (USDA) signed a Protocol on Cooperation in Agricultural Science and Technology, promoting the exchange and sharing of plant genetic resources. In addition, the Chinese Academy of Agricultural Sciences (CAAS) has signed a series of agreements with its counterparts in other countries, including Russia, Brazil, Argentina, Australia, France, Uruguay and the Association of Southeast Asian Nations (ASEAN), in which the exchange of plant genetic resources is one of the core areas of cooperation (Ministry of Agriculture of China, 2008, 123).

Overall status of China's introduction and conservation of exotic germplasm

At present, more than 661 species (excluding forest species) are cultivated in China, including 35 grain crops, 74 cash crops and 163 vegetable crops. Of these, only about 300 originated in China (Ministry of Agriculture of China, 2008, 60) – rice originated in China; wheat originated from the Fertile Crescent of the Near East; maize, potato and sweet potato originated in the Americas. Cotton is one of the most important cash crops in China, and its species of upland cotton (which has the highest yield, best quality and is grown in the widest area at present) was introduced from the United States. Sugarcane, which is the most important source of sugar in China, is indigenous to tropical South Asia and Southeast Asia. It was probably introduced into China around 110 BC from India (Sharpe, 1998). Among the legumes, soybean and adzuki bean originated in China, but rice bean, kidney bean, common bean, lima bean, vegetable pea, chickpea, pigeon pea and winged bean all originated outside China.

There are 163 cultivated vegetable species now in China, of which only 41 originated in China (including secondary origin) – 85 per cent were introduced. The majority of fruit varieties cultivated in China at present, such as apple, grape, orange, papaya, mango, pomegranate, walnut, pineapple and strawberry, were introduced. Sweet potato, potato, broad bean, cucumber, tomato, sunflower and tobacco, which are crucial to Chinese daily life, are all introduced species as well (Tong and Zhu, 2001, 49).

Great progress has been made in the collection and introduction of crop genetic resources in China, but there is still a big gap, compared to developed countries such as the United States. In 2000, 370,000 accessions were collected and conserved in China's National Long-term Genebank of Crop Germplasm (NLGCG) and a variety of nurseries (Tong and Zhu, 2001, 48); In December 2007, the total collections of plant genetic resources numbered 391,919 accessions, including 351,332 accessions of seeds and 40,587 accessions of plants and *in vitro* seedlings (Ministry of Agriculture of China, 2008, 81–82). By the end of 2009, there were only 400,000 accessions (CAAS-Bioversity International, 2009). In contrast, in the year 2000, there were 550,000 accessions in the Germplasm Resources Information Network (GRIN) of the USDA's National Plant Germplasm System (NPGS), with 70 per cent of these collected from other countries worldwide. In April 2010, the number of plant introductions (as determined by the plant introduction (PI) number) had reached 658,000 in the NPGS (USDA, 2010). In ten years, USDA accessions have increased by 19.6 per cent, with an average of 10,000 accessions added annually, while the accessions of the NLGCG increased by only 8.11 per cent, with an average of 3,000 accessions added annually.

The rate of growth in China's collection of plant genetic resources has continued to lag far behind that of the United States. In addition, only 18 per cent of the accessions in China's national conservation system have been collected from abroad (Ministry of Agriculture of China, 2008, 82), the result of which is not only an obvious lack of geographical distance but also of genetic variation. So in order to enrich the genetic diversity of its crop genetic resources, China *must* promote the introduction of germplasm.

In the Database on Germplasm Introduced from Abroad, which is managed by the CAAS, there are only 31,576 accessions, among which 23.8 per cent (or 7,054 accessions) are from the IARCs. The IRRI has contributed the largest share of rice germplasm (70.9 per cent). CIMMYT and the International Institute of Tropical Agriculture (IITA) have the second largest share of germplasm: sorghum, which accounts for

58.4 per cent. The International Crops Research Institute for the Semi-Arid Tropics (ICRISAT) has contributed 56.9 per cent of groundnut germplasm; CIMMYT, 29.1 per cent of wheat germplasm; and CIMMYT and IITA, 16.0 per cent of maize (Agridata, 2010). In a word, it is very clear from the data that most of the germplasm introduced from abroad has come through the IARCs, and it is not surprising that it particularly involves the crops that are the focus of the CGIAR, such as rice, wheat, maize, sorghum, food legumes, groundnut and so on.

Germplasm contributed by the United States accounts for 31.2 per cent of the total accessions, which is higher than the proportion obtained from the IARCs. However, the germplasm introduced into China from the United States consists mainly of coarse cereals, tobacco, fibre crops, flowers, soybean, cotton, forages and green manure, which are not covered by Annex I of the International Treaty on Plant Genetic Resources for Food and Agriculture (ITPGRFA), meaning that the genetic resources of the IARCs play a more important role for China's food security (Agridata, 2010).[2]

Sweet potato germplasm is conserved in the Xuzhou Sweet Potato *in Vitro* Seedling Bank in Jiangsu Province, and potato germplasm is collected in the Keshan Potato *in Vitro* Seedling Bank in Heilongjiang Province. According to the data provided by the Keshan Potato *in Vitro* Seedling Bank on 24 May 2010, 62.6 per cent of its potato genetic resources have been introduced from abroad, some from the International Potato Center (CIP). However, there are no sweet potato accessions from CIP in the Xuzhou *in Vitro* Seedling Bank (CAAS, 2010a). Moreover, there are 144 accessions introduced in the National Sweet Potato Nursery in Guangzhou – 13.07 per cent of the total amount conserved – which include genetic resources from the CIP.

China's introduction of genetic resources of staple crops

Rice, wheat, maize, soybean, sweet potato and potato are the main staple crops in China. The IARCs that conserve the germplasm of these crops (the IRRI, CIMMYT, the CIP and the World Vegetable Centre (AVRDC), where soybean is mainly conserved) were contacted for this study, and they provided data on the flow of the corresponding crop genetic resources to and from China before 2007. The Standard Material Transfer Agreement (SMTA) used in conjunction with the ITPGRFA has been adopted and has been used by the IARCs to distribute genetic resources since 2007.[3] Data from the SMTAs from 2007 to 2009 were used in this study and are accessible on the IARCs' databases.

Rice

Rice originated in China some 7,000 years ago (IPGRI-CAAS, 2010a). It is the biggest and most important food crop in China, averaging 29 million hectares annually and producing more than 180 million tons of grain in recent years (National Bureau of Statistics of China, 2008). Along with other institutions, the Chinese National Rice Research Institute participates in the International Network for

Genetic Evaluation of Rice (INGER), which is a global model for the exchange, evaluation, release and use of genetic resources through SMTAs under the ITPGRFA. From 1996 to 2007, China obtained more than 6,000 accessions of improved rice germplasm from other countries through INGER. Through the use of these rice genetic resources, research institutes in China have developed a number of rice varieties, which are now grown on more than 15 million hectares and have increased yields by 5.5 million tons (Ministry of Agriculture of China, 2008, 112).

High-yielding rice varieties (including IR-8) developed by the IRRI were used by Chinese researchers well before the formal relationship was established with the IRRI in 1982. Since the opening of an IRRI liaison office in Beijing in 1997, the cooperation between IRRI and Chinese scientists has been strengthened and there has been collaboration on many projects. Overall, the IRRI's impact on Chinese agriculture has been extensive: about 90 per cent of Chinese hybrid rice varieties, which account for about half of China's rice production, have IRRI parentage. Thirty-seven modern varieties that have been shared via the IRRI's breeding network have been released in China (CGIAR, 2005b).

Samples of the IRRI's rice germplasm have been introduced into China relatively consistently, at a rate of 2,430 annually during the period from 1985 to 2002. In 2003, there was a sharp rise in acquisitions when three rice experts from the IRRI were hired by the CAAS's Institute of Crop Sciences. They brought 13,036 samples with them in order to continue their research. In 2004, the number of introduced samples dropped back down to 3,630. In 2006–9, there was a gradual increase in annual samples introduced.

The number of samples of rice genetic resources used by China in breeding programmes has been rising year by year (13.0 per cent in 2007, 13.6 per cent in 2008 and 16.6 per cent in 2009), putting China second for the number of rice samples shipped by the IRRI during 2007–9, and first in the world for the total number of samples shipped, followed by India, the United States and the Philippines. In addition, as of 10 September 2010, 10,460 accessions of rice were introduced into China from the United States, according to the shipping records of the USDA-GRIN. In total, 108,272 samples of rice are housed in the IRRI's Genetic Resources Center (SINGER, 2010). While 106,486 samples were shipped into China from the IRRI during 1984–2009, only 4,609 samples have been entered into the database on germplasm introduced from abroad. This implies that many samples have been introduced repeatedly and also that the rice germplasm introduced into China is not readily shared.

Wheat

Wheat originated in the Fertile Crescent region of the Near East 10,500 years ago (Tanno and Willcox, 2006, 1886) and spread into China around 2500 BC (Wang and Zhao, 2010). After rice, wheat used to be the second largest food crop in China, but with the increase in maize acreage, it has been the third since 2002. Its average annual yield has been around 105 million tons in recent years, and acreage is about 23 million hectares (National Bureau of Statistics of China, 2008).

Wheat genetic resources introduced from abroad, such as Mentana, Funo, Abbondanza, St1472/506, CI12203 and Quality, have made a great contribution to enriching China's wheat genetic resources and promoting the breeding of new varieties. For example, more than 260 new varieties have been bred directly or indirectly from Orofen of Chile (Tong and Zhu, 2001, 49), and a series of rust-resistant varieties using Jubileina I, Jubileina II, Virgilio and Lovrin 10 has been cultivated and grown on a large scale in different wheat zones in China (Li et al., 2009, 778). More than 80 per cent of China's cultivated wheat varieties have parentage from introduced wheat germplasm. Most of the materials with resistance to diseases and pests, in particular, have been introduced from abroad (Tong and Zhu, 2001, 49). Of the more than 2,000 wheat varieties that were bred during 1949–2000, half originated from 16 cornerstone parents, five of which were introduced (Li et al., 2009, 779).

From 1989 to 2009, China introduced 52,750 samples of wheat from CIMMYT, most of which (85.29 per cent) was bread wheat because of a lack of bread wheat varieties. Good quality flour for bread still has to be imported. From 1982 to 2010, 265 CIMMYT wheat varieties were released in China; 65 of them have become leading varieties. CIMMYT's wheat varieties and derivatives are well suited to conditions in Yunnan, Xinjiang and Sichuan. CIMMYT varieties have played a leading role in wheat cultivation in Xinjiang and Yunnan since the 1980s and in Sichuan since the mid-1990s. According to the shipping records of the USDA-GRIN, 14,798 accessions of wheat have been introduced from the United States as of 10 September 2010.

Maize

Maize originated in Mexico and Central America and was introduced into China at the beginning of the sixteenth century (Tong, 2001, 231). From 1900 to 1936, maize was planted all over China and was among the top six crops (*ibid.*). Yields have been increasing since 1978, and maize has become the second biggest food crop in China since 2002, covering about 28.0 million hectares and yielding 150 million tons annually in recent years (National Bureau of Statistics of China, 2008).

Introduced varieties have made significant contributions to the maize industry in China. Contributions of germplasm from the United States have been on an upward trend, even surpassing the contribution of China's domestic germplasm by more than 50 per cent. Each 1 per cent contribution of United States germplasm has resulted in an increase of 0.01 tons per hectare in maize yields. As of 10 September 2010, China had introduced 1,173 accessions of maize from the United States, according to the USDA-GRIN shipping records.

Although the genetic contribution of germplasm from the IARCs (mainly CIMMYT) averages no more than 3 per cent annually, it has been increasing, rising especially fast since the 1990s. The yield increase that follows each 1 per cent increase in germplasm contributed by the IARCs is 0.025 tonnes per hectare, which is higher than that of the United States (Li et al., 2005).

During 1986–2009, China introduced 7,626 samples of maize genetic resources from CIMMYT, most of which (85.73 per cent) were lines (maize is developed through crosses, with lines being mainly the parents of crossed hybrids). Only 577 gene bank accessions were introduced. Before 2000, fewer than 100 samples were introduced every year; there were five (non-consecutive) years during this time when no samples were introduced. In 2002, 2007 and 2008, more than 1,000 samples were requested and shipped, but, after 2009, there was a sharp drop. According to Thomas Payne, head of the CIMMYT Germplasm Bank, the drop was mainly due to customs regulations, which differ by province in China and are often very strict. It was therefore very difficult for CIMMYT to send seed into China. In addition, the provincial departments of agriculture (or the respective provincial academies of agricultural science) do not always cooperate with the central Chinese Academy of Agricultural Sciences in Beijing. CIMMYT would send just one set of seed to Yunnan, for example, to have it regenerated there with subsequent redistribution through its Beijing liaison office.

Soybeans

Soybeans originated in central China, and their cultivation history can be traced back 5,000 years (Zhao and Gai, 2004, 954). According to the report of the Ministry of Agriculture on the *State of Plant Genetic Resources for Food and Agriculture in China (1996–2007)*, there are four species of soybeans, 24,931 accessions of cultivated soybean and 6,644 accessions of wild soybeans in the NLGCG (Ministry of Agriculture of China, 2008, 83). Calculated by acreage, soybeans are the fourth biggest food crop in China, grown on 9.3 million hectares in 2000 (although this decreased to 8.8 million hectares in 2009 because of competition from the United States and Brazil) (FAO, 2008).

Although the soybean originated in China, as of 10 September 2010, 1,861 soybean samples (1,495 accessions) had been introduced into China from the United States, according to the USDA-GRIN shipping records, and as of 11 November 2010 ten gene bank accessions had been acquired from the AVRDC (originating from Taiwan, the United States and Hungary), with 369 breeding lines resulting from the AVRDC's breeding programme.

Potatoes

Potatoes originated in the mountains of Peru and Bolivia and were domesticated by Native Americans more than 8,000 years ago. In 1650, they were introduced into China, where they have been grown ever since (IPGRI-CAAS, 2010c). It is the fifth biggest food crop in China, grown on 4.8 million hectares in 2009 (FAO, 2008).

From CIP, 837 samples of potato germplasms have been introduced, averaging 36 samples a year between 1984 and 2006. No accession was introduced in 2007, but 168 samples were introduced in 2008 and 303 in 2009. This increase is probably due to closer relations with the CIP and the joint founding of the CIP-China Center for Asia and the Pacific (CCCAP). Moreover, China has been focusing on developing the potato industry in recent years, and a further 186 potato accessions had been

TABLE 3.2 Ratio of International Potato Center parentage in released and adopted potato varieties, 1997 and 2007

Released	Total	CIP parents	Ratio (percentage)
1997	112	12	10.7
2007	257	37	14.4
Adopted			
1997	63	8	12.7
2007	162	21	13.0

Source: Thiele et al. (2008, 11 and 18).

introduced from the United States as of 10 September 2010, according to the USDA-GRIN shipping records. There have been several new potato varieties developed, including the following:

- CIP-24, which is resistant to disease and was bred through cooperation between China and CIP in 1978. It is grown on about 70,000 hectares, mainly in the dry areas of China's north provinces.
- Cooperation 88, which was also co-bred, is grown on more than 100,000 hectares in Yuan Province (CGIAR, 2005a).

The proportion of varieties with CIP parents has been growing faster among released varieties than among adopted varieties (Table 3.2). This shows that, while CIP germplasm has wider adaptability, it is not used well: the extension of varieties with CIP parentage should be given much more prominence.

Sweet potatoes

Sweet potatoes are native to the tropical areas of South and Central America and were domesticated there more than 5,000 years ago (CGIAR, 2005c). In the mid-1500s, they were introduced into China, which is now the largest producer of sweet potatoes in the world, accounting for 61 per cent of the world's production (IPGRI-CAAS, 2010b). It is the sixth biggest food crop in China, grown on 3.9 million hectares in 2009 (FAO, 2008).

While many new varieties have been developed, from Nancy Hall to Okinawa 100, for half a century there was little progress in introducing sweet potato germplasm, resulting in a high degree of inbreeding. In recent years, however, through international cooperation a good deal of elite sweet potato germplasm has been introduced, especially from the CIP, and this has broadened the breeding base. Furthermore, the CIP's breeding strategy has improved China's yields, which has made China's breeders pay much more attention to the introduction of genetic resources (Ma et al., 1998, 1 and 3). In total, 525 samples of sweet potato were introduced from the CIP from 1990 to 2009, with an average of 20 samples introduced annually before 2006. Sixty-five accessions of sweet potato had been introduced from the United States as of 10 September 2010, according to USDA-GRIN shipping records.

Outflow of plant genetic resources from China

The Chinese government provided access to plant genetic resources for other countries through regional and international cooperation. From 1996 to 2007, China provided over 40,000 samples of plant genetic resources to other countries, of which the CAAS provided 11,288 samples of 120 species to over 100 countries or international organizations, such as the United States and United Kingdom, the IRRI and CIMMYT (Ministry of Agriculture of China, 2008, 123 and 111). While this number appears small at first glance, it only includes those samples provided by the CAAS. In China, there are more than 30 agricultural academies at the provincial level, all of which also cooperate in the exchange of plant genetic resources with foreign institutes.

The exchange of genetic resources has played a major role in plant breeding and agricultural production all over the world. According to the GENESYS[4] Internet site: (http://www.genesys-pgr.org/), 29,490 accessions of plant genetic are held in genebanks located in CGIAR centres, European countries, and the USA (as part of the USDA collections) maintained in Europe and the USDA-GRIN collections. Rice germplasm collected from China accounts for 6.59 per cent of those rice collections (the United States ranks first, followed by India, the Lao People's Democratic Republic and the Philippines). Barley, wheat and sorghum follow rice in the proportion of accessions from China.

Soybeans were introduced into North America in 1765 by Samuel Bowen, a sailor who had visited China (Brachfeld and Choate, 2007, 275). In 1893 and 1906, the USDA sent researchers to China to look for and collect soybean samples, and they collected more than 60 varieties (Simons, 1987, 338–42). As of 16 May 2010, there are 6,359 accessions (two species) collected by USDA-GRIN. According to its database, the AVRDC has collected 721 accessions of soybean from China, and, according to SINGER, the International Livestock Research Institute (ILRI) had collected four accessions as of May 2009.

In a strategic move, Brazil started introducing the entire soybean germplasm collection of the United States in 2006 in order to broaden its soybean genetic base and ensure much-needed genetic variability (Mariante et al., 2009, 64). The Brazilian Enterprise for Agricultural Research (Embrapa) and the Agronomic Institute of Campinas (IAC) have collected and introduced 13,300 accessions of soybean (*ibid.*, 159). According to the National Report on the Status of Genetic Resources for Food and Agriculture in 2004, there are 680 accessions of soybean conserved in the Genebank of the National Institute of Agricultural Technology (INTA), Argentina (Argentina, 2008, 21). Soybeans have made a great contribution to world agriculture. The United States, Brazil and Argentina have been the top three producers and exporters since 1990, with China being the biggest soybean importer since 1999 (FAO, 2008).

Benefits to China from the flow of crop genetic resources

China has obtained 169,656 samples of germplasm from the IARCs in 16 crops covered by the GENESYS from 1984 to 2009, which is over 11 times the amount of germplasm it has provided to the IARCs and over 5 times the amount of germplasm that has been acquired from China by USDA-GRIN, EURISCO and the IARCs together (Table 3.3). In comparing the flow of genetic resources to and from the IARCs for individual crops, the number of introduced rice samples is approximately 12 times the number of samples provided to the IARCs, the number of wheat samples is over 60 times, maize is over 293 times and sweet potato is 75 times. Although China did not provide any samples of potato germplasm to IARCs, 837 samples of the CIP's potato genetic resources have been introduced (Table 3.3). However, it must be noted that these numbers are probably inflated due to the quantity of duplicate samples of one accession shipped into China. Based on the SINGER data, there are 293 duplicates in the top 1,000 samples of rice introduced to China, 162 duplicates among 365 potato samples and 217 duplicates among 419 sweet potato samples, meaning that the percentage of samples flowing into and out of China must be discounted by 30–50 per cent.

As of 16 May 2010, 70,000 samples of PGRFA had been introduced through USDA-GRIN into China, and 21,388 accessions of PGRFA from China had been collected and stored in USDA-GRIN, meaning that China received 3.27 times the number of plant genetic resources from the United States than it provided. In terms of single crops, the ratio of rice is 5.0 times more received than provided, wheat is 8.0, maize

TABLE 3.3 Flow of germplasm out of/into China

Crop	Export GENESYS	SINGER	Import	Ratio of (Import/GENESYS)	(Import/SINGER)
Rice	11195	8839	106486	9.51	12.05
Wheat	5715	757	47921	8.39	63.30
Barley	7045	3104	2490	0.35	0.80
Maize	841	26	7626	9.07	293.31
Sorghum	2090	666	922	0.44	1.38
Sweet potato	51	7	525	10.29	75.00
Cultivated potato	687	0	837	1.22	
Faba bean	675	571	156	0.23	0.27
Bean	550	190	998	1.81	5.25
Cow pea	523	149	713	1.36	4.79
Chick pea	70	62	161	2.30	2.60
Lentil	26	15	76	2.92	5.07
Pigeon pea	12	10	84	7.00	8.40
Cassava	2	2	230	115.00	115.00
Pearl millet	8	1		0.00	
Forage	–	278	431	–	1.55
Total	29490	14677	169656	5.75	11.56

Source: GENESYS (last accessed 17 Sep 2011); Data on Samples into China from the IARCs other than mentioned above are from the SINGER Database (last accessed 30 April 2010).
Note: Because maize is not included in the GENESYS, its data are adjusted according to the data provided by CIMMYT. The flows of crop genetic resources into/out of Hong Kong and Taiwan of the Peoples Republic of China are not included by this Table.

is 8.1 and sweet potatoes are 2.3 times more. Soybeans are only 0.29, which is less than the amount exported to the United States. Again, based on the PI identifier, there are 567 duplicate samples in the top 1,000 samples, so these ratios should be discounted accordingly. Despite the problems with duplicate accessions, the introduced germplasm of rice, wheat, maize, sweet potatoes and potatoes has played an important role in China's breeding programmes, which, in turn, promote the development of corresponding industries and make an invaluable contribution to China's food security. Therefore, China has benefited enormously from the total flow of genetic resources.

Suggestions for policies on the flow of plant genetic resources adopted in China

Since the 1980s, China's crop breeding has been either stagnating or developing at a very low speed, which has mainly been caused by a narrow genetic base according to expert analysis (Liu, 1999, 2). For instance, hybrid rice has been planted widely, but most of its sterile lines are derived from one common Chinese wild rice. The hybrid maize, which accounts for about 90 per cent of the maize area cultivated in the country, has parents of only six major inbred lines in its pedigrees. Over 50 per cent of the wheat varieties have Mentana, Funo, Abbondanza and Orofen in their pedigrees (Ministry of Agriculture of China, 2008, 72). There has also been serious inbreeding in the sweet potato. As a result, many of China's experts have called for further increasing the introduction of crop genetic resources from abroad. In addition, the management and introduction of crop genetic resources should be strengthened in order to reduce duplicate accessions as well as controlling the serious problem of illegal outflow.

Further cooperation with the IARCs

As shown earlier, the crop genetic resources from the IARCs make an extremely important contribution – and remain of great significance – to China's breeding programmes, especially in light of the fact that the IARCs hold and manage genetic resources for the benefit of the international community. Their holdings are public goods in nature and can be introduced much more easily than accessions from individual countries. In addition, the complete information and high phytosanitary level of the IARCs' germplasm guarantee China's biological safety. Therefore, China should strengthen its cooperation with the IARCs, facilitate access to PGRFA for the IARCs and even provide more funding to the IARCs if national fiscal resources permit.

Research institutions in China should look for opportunities to cooperate with the IARCs. They can participate in shuttle breeding and crop improvement programmes, implement research projects or establish joint laboratory or regional centres. Joint laboratories have already been set up between the CAAS and the IRRI, the ILRI, the International Center for Agricultural Research in the Dry Areas (ICARDA), ICRISAT and CIMMYT. In February 2010, the Chinese Ministry of Agriculture and the CIP signed an agreement to launch CCCAP (CIP, 2010). The development of tropical areas is a particular focus of the IARCs, so sub-centres of tropical agriculture could be jointly established in southern China with the International Center for Tropical Agriculture (CIAT), the IITA and Bioversity International.

Considering accession to the ITPGRFA

China is not yet a contracting party of the ITPGRFA, but in the report on the *State of PGRFA in China (1996–2007)*, it was clearly stated that the Chinese government had recognized the importance of the ITPGRFA and agreed with the targets it sets in regard to promoting the conservation and sustainable use of PGRFA, fairly and equitably sharing the benefits arising out of their use and, finally, realizing sustainable agriculture and food security. The Chinese government supports the ITPGRFA in establishing a multilateral system on access and benefit sharing (multilateral system) and in the use of SMTAs for accessing PGRFA and sharing the benefits. China is considering joining the ITPGRFA in order to have more opportunities to share resources (Ministry of Agriculture of China, 2008, 115–16 and 123).

China should organize experts on genetic resources, law and agricultural economics to analyse the ITPGRFA systematically, particularly what benefits and costs would be incurred under the ITPGRFA and how to amend national laws to be in harmony with the ITPGRFA. Under the provisions of the ITPGRFA, every contracting party should facilitate access to PGRFA under the multilateral system covered by Annex I, without the need for bilateral negotiations, which can be very difficult and costly. This would benefit China in accessing abundant crop genetic resources at an international level, which would help address the problem of the narrow genetic base in breeding. Furthermore, once it has acceded to the ITPGRFA, China can share technology about PGRFA with other contracting parties, and it can get technical and financial resources from international entities, such as the Governing Body of the ITPGRFA, the IARCs and the Global Crop Diversity Trust. This would benefit China in helping to protect its domestic PGRFA and to prevent the biopiracy of biotechnology entities in developed countries.

Certainly, China could also make a great contribution to the food security of the world. China owns abundant PGRFA. If China accedes to the ITPGRFA, the PGRFA covered by the multilateral system would be broadened. In recent years, China has been investing highly in agricultural sciences and technologies and has made great progress. Therefore, China could play an important role in international technology transfer.

Enhancing the integrated management of plant genetic resources and regulating their flow

In order to implement the CBD and exercise national sovereignty over genetic resources, China's State Council has declared that the Ministry of Environmental Protection is responsible for coordinating the protection and management of all national resources of biological species. As a result, the Inter-Ministerial Joint Meeting of Biological Species Resources, which is made up of 17 ministries, was founded in August 2003. The National Committee of Crop Genetic Resources, which will be responsible for coordinating the management of crop genetic resources and studying strategy and policy in relation to them, was to be founded under the Ministry of Agriculture, but it is still not yet established.

The outflow and introduction of crop genetic resources should be managed jointly. Inventorying the germplasm introduced and deciding on the provision of germplasm to entities outside the country should be managed in an integrated way. This could be done by authorizing the CAAS and the Chinese Academy of Tropical Agriculture Sciences (CATAS) to carry out this mandate and by building the capacity to provide crop genetic resources. In the meantime, the General Administration of Customs of the People's Republic of China could regulate and facilitate the import and export of PGRFA.

Developing a mechanism for sharing plant genetic resources

A meta-database of national crop genetic resources has been established in China and can be accessed through the Internet. However, there are still some limitations, such as registration. The information on crop genetic resources should be available to all users, which would benefit breeders and facilitate breeding programmes. In the *Regulation of Crop Germplasm Management*, access should be accorded expeditiously, free of charge. If a fee is charged, it should not exceed the minimal cost involved. In addition, controls such as intellectual property rights on new plant varieties should not be applied directly for the germplasm requested. Unfortunately, there are still problems surrounding the sharing of intellectual property rights and benefits from varieties developed from germplasm accessions: most mid-term gene banks and nurseries are reluctant to share their holdings. China needs to carry out a nationwide survey and develop detailed policies on using the holdings of national mid-term gene banks and nurseries.

Notes

Many scientists made contributions to this chapter. Michael Halewood, Head of the Policy and Support Unit of Bioversity International, discussed and guided the framework of this study and introduced me to the gene bank managers of the IARCs. He reviewed my draft chapter carefully and gave me many helpful comments. My thanks also go to Skofic A. Milko, manager of SINGER; Mohamed F. Nawar, database manager of GENESYS; Patria G. Gonzales, manager of the Seed Health Unit of the IRRI; Thomas Payne, head of the Germplasm Bank of CIMMYT; Zhonghu He, director of the CIMMYT Office in Beijing; Andreas W. Ebert, gene bank manager and head of Genetic Resources and the Seed Unit of AVRDC; Yu Li, maize expert of CAAS; Zhikang Li and Jianlong Xu, rice experts of CAAS; Xicai Liu, manager of the Keshan Potato *in Vitro* Seedling Bank of China; Mark A. Bohning, manager of the GRIN Database Management Unit, ARS of the USDA; and Francesca Giampieri, Librarian, Bioversity International.

1 Convention on Biological Diversity, 31 I.L.M. 818 (1992).
2 International Treaty on Plant Genetic Resources for Food and Agriculture, 29 June 2004, www.planttreaty.org/texts_en.htm (last accessed 30 March 2011).
3 Standard Material Transfer Agreement, 16 June 2006, ftp://ftp.fao.org/ag/agp/planttreaty/agreements/smta/SMTAe.pdf (last accessed 30 March 2011).
4 GENESYS (http://www.genesys-pgr.org/) is a global portal developed by Bioversity International (with support from the Global Crop Diversity Trust and the Secretariat of the ITPGRFA) that acts as a one-stop access point to the information about PGRFA accessions. GENESYS currently includes information provided from EURISCO about materials held in European genebanks, from the CGIAR System-wide Information

Network for Genetic Resources (SINGER) and the Genetic Resources Information Network of the USDA.

References

Agridata (2010) *Database on Germplasm Introduced from Abroad*, Sub-Centre of Crop Sciences Information of National Agricultural Data Sharing Centre, online: http://crop.agridata.cn/A010104.ASP (last accessed 20 April 2010) [in Chinese].

Argentina (2008) *Informe nacional sobre el estado de los recursos fitogenéticos para la agricultura y la alimentación: Argentina* [*National Report on the Status of Genetic Resources for Food and Agriculture: Argentina*], online: www.fao.org/docrep/013/i1500e/Argentina.pdf (last accessed 10 November 2010) [in Spanish].

Brachfeld, A., and M. Choate (2007) *Eat Your Food! Gastronomical Glory from Garden to Gut*, Coastalfields Press, Arvada, CO.

CAAS-Bioversity International (2009) *Minutes of CAAS-Bioversity High Level Meeting*, 9 September 2009, Beijing.

Chinese Academy of Agricultural Sciences (CAAS) (2010a) *Chinese Crop Germplasm Resources Information System*, online: http://icgr.caas.net.cn/32pu/xzgssgm.htm (last accessed 20 April 2010) [in Chinese].

——(2010b) *National Sweet Potato Nursery in Guangzhou*, Chinese Crop Germplasm Information Network, online: http://icgr.caas.net.cn/32pu/gzgs.htm (last accessed 20 April 2010) [in Chinese].

Consultative Group on International Agricultural Research (CGIAR) (2005a) *China and the CGIAR*, online: www.cgiar.org/china/index-CHINESE.html (last accessed 28 April 2010) [in Chinese].

——(2005b) *CGIAR in Action: China*, online: www.cgiar.org/inaction/china.html (last accessed 20 April 2010).

——(2005c) *Sweet Potato*, online: www.cgiar.org/impact/research/sweetpotato.html (last accessed 13 September 2010).

Food and Agriculture Organization (FAO) (2008) *FAOSTAT*, online: http://faostat.fao.org/site/342/default.aspx (last accessed 9 November 2010).

He, Z-H., S. Rajaram, Z-Y Xin and G-Z. Huang (eds) (2001) *A History of Wheat Breeding in China*. CIMMYT, Mexico, online: www.cimmyt.org/english/docs/book/historywbchina/histwbchina.html (last accessed 20 April 2010).

International Plant Genetic Research Institute (IPGRI)-CAAS (2010a) online: http://icgr.caas.net.cn/Ipgri/foodCrop/shuidao.HTM (last accessed 26 May 2010) [in Chinese].

International Potato Center (CIP) (2010) *China and the International Potato Center Launch New Center to Boost Potato and Sweet Potato Capacity across China, Asia, and the Pacific*, online: www.cipotato.org/pressroom/press_releases_detail.asp?cod=77 (last accessed 28 April 2010).

IPGRI-CAAS (2010a) http://icgr.caas.net.cn/Ipgri/foodCrop/shuidao.HTM (last accessed 26 May 2010) [in Chinese].

——(2010b) online: http://icgr.caas.net.cn/Ipgri/foodCrop/ganshu.HTM (last accessed 26 May 2010) [in Chinese].

——(2010c) online: http://icgr.caas.net.cn/Ipgri/foodCrop/malingshu.HTM (last accessed 26 May 2010) [in Chinese].

Li, H-M., R-F. Hu, S-H. Zhang and Y-M. Zhang (2005) 'The Impacts of US and CGIAR Germplasm on Maize Production in China', *Scientia Agricultura Sinica*, 38(11): 2189–97.

Li, X-J., X. Xu, W-H. Liu, X-Q. Li, X-M. Yang and L-H. Li (2009) 'Genetic Contribution of Introduced Varieties to Wheat Breeding in China Evaluated Using SSR Markers', *Acta Agronomica Sinica* 35(5): 778–85 [in Chinese].

Lin, J.Y. (1998) 'How Did China Feed Itself in the Past? How Will China Feed Itself in the Future?' Second Distinguished Economist Lecture, CIMMYT, Mexico, online: www.cimmyt.org/english/docs/special_publ/del/2del/2distecon_contents.htm (last accessed 20 April 2010).

Liu, X. (1999) 'Origin and Development of Germplasm Enhancement', *Crop Variety Resources* 2: 2–5 [in Chinese].

Ma, D-F., H-M. Li and I.G. Mok (1998) *Introduction and Use of Exotic Germplasm in the Chinese Sweetpotato Breeding Program*, CIP Program Report 1997–98, online: www.cipotato.org/market/PgmRprts/pr97–98/37china.pdf (last accessed 20 April 2010).

Mariante, A. da S., M.J.A. Sampaio and M.C.V. Inglis (2009) *State of the Brazil's Plant Genetic Resources: Second National Report (1996–2007)*, online: www.fao.org/docrep/013/i1500e/Brazil.pdf (last accessed 28 April 2010).

Ministry of Agriculture of China (1996) *Country Report to the FAO International Technical Conference on Plant Genetic Resources*, Leipzig, 1996, online: www.fao.org/ag/AGP/AGPS/pgrfa/pdf/china.pdf (last accessed 27 February 2010).

——(2008) *State of Plant Genetic Resources for Food and Agriculture in China (1996–2007)*, online: ftp://ftp.fao.org/ag/agp/countryreports/ChinaFINALCRAug2008.pdf (last accessed 28 April 2010).

Moore, G., and W. Tymowski (2005) *Explanatory Guide to the International Treaty on Plant Genetic Resources for Food and Agriculture*, International Union on the Conservation of Nature, Gland, Switzerland.

National Bureau of Statistics of China (2008) *China Statistical Yearbook (2008)*, online: www.stats.gov.cn/tjsj/ndsj/2008/indexch.htm (last accessed 20 April 2010) [in Chinese].

Palacios, X.F. (1998) *Contribution to the Estimation of Countries' Interdependence in the Area of Plant Genetic Resources*, Background Study Paper no.7, Rev. 1, FAO, Rome.

Sharpe, P. (1998) *Sugar Cane: Past and Present*, Southern Illinois University, Carbondale, IL, online: www.ethnoleaflets.com//leaflets/sugar.htm (last accessed 6 September 2010).

Shen, Z-Z. (2004) *A Study on the Sino-U.S. Exchange and Cooperation in Agricultural Science and Technology during the Modern Times*, PhD dissertation, Nanjing Agricultural University, China) [in Chinese; unpublished].

Simons, N. M. (ed.) (1987) *Crop Evolution*, translated by W. Zhao, China Agricultural Press, Beijing.

System-Wide Information Network for Genetic Resources (SINGER) (2010) online: http://singer.cgiar.org/index.jsp?page=showkeycount&search=cuc=cuchn (last accessed 18 February 2011).

Tanno, K., and G. Willcox (2006) 'How Fast Was Wild Wheat Domesticated?' *Science* 311 (5769): 1886.

Thiele, G., G. Hareau, V. Suarez, E. Chujoy, M. Bonierbale and L. Maldonado (2008) *Varietal Change in Potatoes in Developing Countries and the Contribution of the International Potato Centre: 1972–2007*, Working Paper no. 2008–6.46, CIP, Lima, Peru.

Tian, Z-M., Y-C. Zhao and Q. Cheng (2001) 'Introduction, Selection and Use of Exotic Potato Germplasm', *China Potato* 15(4) [in Chinese].

Tong, D-X., and Z-H. Zhu (2001) 'Crop Introduction for Agriculture Development in China,' *Review of China Agricultural Science and Technology* 13(3) [in Chinese].

Tong, P.-Y. (2001) 'A Brief Review of the Historical Development of Chinese Maize Production in 1900–1948', *Agricultural Archeology* 1 [in Chinese].

US Department of Agriculture (USDA) (2010) *NPGS of GRIN*, online: www.ars-grin.gov/npgs/acc/acc_queries.html (last accessed 12 April 2010).

Wang, W., and H. Zhao (2010) 'Main Results of the Project on Origin of Chinese Civilization', *Guangming Daily*, 23 February 2010, online: http://news.guoxue.com/article.php?articleid=24487 (last accessed 11 November 2010) [in Chinese].

Wen, J.-B. (2010) *Report on the Work of the Government*, paper delivered at the Third Session of the Eleventh National People's Congress on 5 March 2010, online: http://news.xinhuanet.com/politics/2010–03/15/content_13174348.htm (last accessed 18 October 2010) [in Chinese].

Zhao, T.-J., and J-Y. Gai (2004) 'The Origin and Evolution of Cultivated Soybean', *Scientia Agricultura Sinica* 37(7): 954–62 [in Chinese].

4

CROP AND FORAGE GENETIC RESOURCES

International interdependence in the face of climate change

Julian Ramirez-Villegas, Andy Jarvis, Sam Fujisaka, Jean Hanson and Christoph Leibing

Introduction

Efficient use of agricultural diversity and genetic resources of both crops and forages will be needed to maintain current levels of food production in the face of future challenges under future conditions. The world's population is expected to grow to 9.1 billion by 2050, with increasing consumption of dairy and meat products (United Nations, 2005). This increased demand means that production will need to be increased without the option of increasing the amount of arable land. The added expected impacts of climate change suggest that we face worsening multiple challenges and decreasing options to address these challenges. Plant genetic resources for food and agriculture (PGRFA) will play a crucial role in providing the genes to help confront these challenges.

Considerable interdependence exists among the different agro-ecosystems. Clear examples are livestock systems that depend on fodder species that originate elsewhere as well as sites where crops are grown often but which are not their centres of diversity. The conservation and flow of genetic diversity between and within countries is therefore critical for sustaining livelihoods, but it will also be necessary to intensify production. However, the intensification of cropping systems often leads to genetic erosion (Heal et al., 2004). Currently, only some 150 plant species are now being cultivated, and mankind primarily depends on no more than 12 of these (Esquinas-Alcazar, 2005). Despite this reality, there are calls for diversifying agricultural production to adapt to climate change, to enhance nutritional security, and to service an increasingly complex global market for agricultural goods (Cavatassi et al., 2006; Cleveland et al., 1994; Reidsma and Ewert, 2008). Trade-offs between intensification, conservation and diversification further complicate decision-making processes.

PGRFA have been widely exchanged over the past 10,000 years among farmers and, recently, by collection, *ex situ* conservation, and use by research organizations. The globalization of plant genetic resources is evident in the expansion of crops outside their centres of origin, with near global coverage of many crops whose origins were once geographically restricted (Vavilov, 1926). The global interdependence on PGRFA has given rise to policies to facilitate the access and exchange of plant genetic resources across the planet (Palacios, 1998). However, no comprehensive knowledge on the current and future interdependence on PGRFA exists in order to determine the degree to which policies need to be improved or maintained during the coming years.

Studies using a variety of approaches have tried to quantify the interdependence on plant genetic resources (see, for example, Palacios, 1998). It is important in the face of climate change to re-appraise interdependence and identify the changes in demand for PGRFA to help create policies to address future challenges. This chapter focuses on the effect that climate change will have on the international interdependence on PGRFA. Rather than quantify current demand, we look for evidence of how climate change might enhance, reduce or shift patterns of demand in the future. Analysis is based on literature review and numerical analysis of climate data. We structure the chapter in the following way: (1) review of current patterns of interdependence for PGRFA (to set the baseline); (2) review of the expected impacts of climate change on agriculture; (3) quantitative analysis of the changes in 'climatic interdependence' among countries for a number of crops; (4) discussion of likely changes in interdependence based on evidence from previous chapters; and (5) discussion of policy implications to address future interdependence patterns.

Current international interdependence on PGRFA: establishing the baseline

Virtually all countries depend on PGRFA that were originally found in other countries. Today's improved varieties have resulted from innumerable crosses between materials from different countries (Zeven and De Wet, 1982). The globally popular Veery wheat, for example, is the product of 3,170 crosses involving 51 parents from 26 countries. In addition, many countries rely on non-indigenous crops and thus need to import germplasm to grow them. Countries in southern and central Africa rely on crops that originated outside the region for 50–100 per cent of their food, with the majority of these countries exceeding 80 per cent dependence on 'foreign' germplasm (Palacios, 1998). Such dependence never falls below 80 per cent in the Andean countries. Crops such as cassava, maize, groundnut and beans originated in South America but have become extremely important staples in sub-Saharan Africa. Cassava is a major food source for 200 million Africans in 31 countries, with a farm-gate value of over US $7 billion (FAO, 1997). And Africa – with its indigenous millets and sorghums – makes a considerable contribution to other areas such as South Asia (13 per cent) and Latin America (8 per cent) (Kloppenburg and Kleinman, 1987). In many cases, problems with disease and/or pest susceptible varieties have only been resolved by breeding resistant varieties using genetic diversity from other countries (National Research Council, 1972).

The case of forages is similar. Over 90 per cent of the major cultivated forage grasses in the world are indigenous to sub-Saharan Africa (Boonman, 1993). These grasses have been used to improve extensive cattle pastures in Latin America. Several examples of the use of pastures in non-native environments exist, and the following are just a few.

- By 1996, over 40 million hectares were sown to *Brachiaria* in Brazil (Miles et al., 1996).
- Ruzi grass (*Brachiaria ruziensis*) was introduced into Thailand and is increasing in demand (Phaikaew et al., 1993).
- *Cenchrus ciliaris* is now grown in 31 countries (Cox et al., 1988), with four million hectares planted in the United States, over six million hectares in Mexico, and 7.5 million hectares in Australia (Humphreys, 1967).
- Other African grasses (for example, *Panicum maximum*, *Chloris gayana* and *Pennisetum purpureum*) are now also widely distributed throughout the tropics.
- Alfalfa (*Medicago sativa*) is extensively cultivated in the warm temperate, cool sub-tropical regions and tropical highlands and is native to southwest Asia. Today, it is the major forage legume crop, covering 79 million hectares worldwide, including 13 million hectares in the United States, where it is the third most important crop in value (Putnam et al., 2007).
- Vetch originated in the Near East and is now grown on close to one million hectares worldwide (FAOSTAT, 2009).
- Red and white clover (*Trifolium pratense* and *Trifolium repens* respectively) originated in Europe and North Africa and are now widely cultivated in sub-tropical areas in North America, southern South America, Australia and New Zealand.
- *Stylosanthes*, from Latin America, is now used in India (Ramesh et al., 2005), Thailand (Phaikaew and Hare, 2005) and Australia for improving pastures and producing leaf meal for monogastric feed.

Crop wild relatives provide researchers with genes that are useful for developing biotic and abiotic resistance (Maxted et al., 2008; Gurney et al., 2002; Lane and Jarvis, 2007). The use of crop wild relatives has increased dramatically over the past decade and will continue to increase thanks to biotechnology tools (Hajjar and Hodgkin, 2007). A number of crops such as sugar cane, tomatoes and tobacco could not be grown on a substantial commercial scale in countries that are for the most part far from the native habitats of their wild relatives were it not for the contribution made by these wild relatives to disease resistance (FAO, 1997). Current interdependence is therefore highly related to the extent to which PGRFA contributes to crop improvement for sustaining agriculture and food security, but it is also very dependent on environmental compatibility. The actual flow of PGRFA between countries and regions is difficult to monitor and so is the actual knowledge of their usefulness and importance. Hence, an actual quantification of the current interdependence needs to be done in order to improve policies and decision-making processes.

Climate change impacts on agriculture and PGRFA

Climate change will likely bring increases in temperature of between 2–6 degrees Celcius, both increases and decreases in precipitation and increased frequency of droughts and floods (IPCC, 2007). These changes will especially affect rain-fed agriculture, making adaptation necessary. Climate change will also impact agricultural biodiversity by increasing the genetic erosion of landraces (Mercer and Perales, 2010) and shifting wild species niches, including crop wild relatives (Jarvis et al., 2008). Severe pest outbreaks may increase with climate change, profoundly affecting agro-ecosystems and global food availability, although this also may vary geographically (World Bank, 2008).

Modelling indicates that yields from rain-fed agriculture in some regions of Africa could be reduced by up to 50 per cent by 2020 (Challinor et al., 2007; Fischer et al., 2001; IPCC, 2007). Food production and access in many African countries will likely be severely compromised, exacerbating food insecurity and malnutrition (IPCC, 2007). The most problematic areas are southern Africa, where land that is suitable for maize, the major staple, will likely disappear by 2050, and South Asia, where productivity of groundnut, millet, sorghum, wheat and rapeseed will be heavily reduced (Lobell et al., 2008).

Even under the most conservative of scenarios, climate change will cause shifts in suitable areas for cultivation of a wide range of crops. Changes will include a general trend of loss in suitable areas in sub-Saharan Africa, the Caribbean, India and northern Australia as well as gains in the northern United States, Canada and most of Europe for a number of staple crops (Fischer et al., 2002; Fischer et al., 2005; Lane and Jarvis 2007). Developed nations will see an expansion of suitable arable land to higher latitudes and a potential increase in production if those lands are brought under cultivation (Fisher et al., 2001). By the 2080s, rain-fed cereal production in the developing world could decrease by 3 per cent in negatively impacted countries and increase by 6 per cent in positively impacted countries. Production losses from climate change could worsen hunger in developing countries beyond the current level of one billion going hungry (FAO, 2008; Fischer et al., 2002).

Although farmers have always had to adjust and adapt their cropping systems to changing climatic and environmental conditions, the speed and complexity of current climate change poses problems on a much bigger scale (Adger et al., 2007). Rising temperatures and changing rainfall regimes will not decrease the global suitability for crops *per se*, but they will cause geographic shifts in suitable cropping areas. For certain areas, there is high likelihood that currently adapted crops will become mal-adapted (IPCC, 2007). The use of plant genetic resources is thus critical for the improvement of the resilience of crop varieties and the sustainability of the production of agricultural goods. New within-crop diversity and/or different and better-adapted crops will be needed to respond to future conditions. The negative impacts of climate change can be mitigated if farmers can adapt by changing to more suitable varieties and, if necessary, crops (Lane and Jarvis, 2007). Barriers to such adaptation, however, are discussed in the following sections of this chapter.

The areas that are currently most food-insecure will be most affected by climate change. These areas have the greatest need for new crops and varieties that are tolerant to extreme conditions, such as drought, heat, flooding, submergence and salinity. The need for adapted germplasm will require the characterization, evaluation and availability of germplasm. Some of the more important traits to be found in varieties and genotypes for responding to climate change include: drought tolerance, extreme events tolerance, resistance to very hot and humid conditions and pest and disease resistance. Most of these traits are present in traditional cultivars or wild species. Unfortunately, with modern crop intensification, many such traditional varieties have became under-utilized and, in certain instances, lost completely (Shewayrga et al., 2008; van Heerwaarden et al., 2009; Chaudhary et al., 2004; Hammer et al., 1996; Parzies et al., 2000). The US Department of Agriculture indicates that out of the 7,098 apple varieties that were in use between 1804 and 1904, 86 per cent have been lost. Similarly, 95 per cent of cabbage, 91 per cent of maize, 94 per cent of pea and 81 per cent of tomato varieties no longer exist (Fowler, 1994). Agricultural expansion also accounts for wild habitat loss and for the genetic erosion of many wild gene pools that could provide both biotic and abiotic resistances for traditional and modern crop varieties (FAO, 1998). Most of the raw materials for adaptation are likely to be found only in gene banks, and while some gene banks lack appropriate characterization and distribution mechanisms, most of the genetic resources that are useful for adaptation remain inaccessible to the world.

Genetic erosion is likely to increase due to changes in the climate system. Recent changes in the climate system have already led to shifts in species distributions (Chen et al., 2009; Parmesan and Yohe, 2003; Root et al., 2003). The magnitude of these changes includes potential extinctions, and it is likely to increase with the growing levels of climate change (Carnaval and Moritz, 2008; IPCC, 2007; Nogués-Bravo et al., 2008; Thomas et al., 2004). Many ecosystems are projected to become more depauperate, and many communities are projected to start unravelling as the ecology of individual species is affected by the changing climate at different rates (IPCC, 2007). Many of these ecosystems hold several wild species, which in some cases are not properly conserved either *in situ* (Bass et al., 2010; Kim and Byrne, 2006) or *ex situ* (Ramirez et al., 2010).

The Food and Agriculture Organization (FAO) estimates that a large proportion of the gene pool of the major crops has been sampled and placed in *ex situ* collections. Much of the diversity found in farmers' fields today, and a great amount of diversity that no longer exists on-farm, can now be accessed solely through gene banks (Fowler and Smale, 2000). However, improved mechanisms of *in situ* conservation are required in order to maintain evolutionary dynamics. *In situ* conservation has the advantage of maintaining interactions between species and is therefore also likely very important for maintaining evolutionary dynamics. However, its implementation typically involves traditional farming communities, indigenous territories and/or protected areas whose monitoring can be very difficult in some cases. One proposed approach is the preservation of a number of valuable crop species and varieties in selected areas of traditional agriculture (Brush, 2000). One example of this approach is the Globally Important Agricultural Heritage System Initiative (CGRFA, 2000). Conservation of forests and pristine sites valued for their wildlife or ecological value is a means to achieve the *in situ* conservation of wild species and wild crop relatives. However, to date, no inventories exist on how many crop wild relatives might be preserved *in situ*, turning the issue of the *in situ* conservation of crop wild relatives into a more complicated issue.

Forages, however, are a different case, and they are more vulnerable to land use change than other crops. Forages in small-hold systems in the tropics are often inter-cropped and planted around fences or used in crop fallow periods or for vegetation on degraded lands, where competition for crop land, water and fertile soil are high and resources are limited. Climate change will require new management options and the use of alternative forage species or the increased use of existing forages in many areas. Use of specific forage species in small-hold livestock systems is limited by the length of the growing season and temperature (Cox et al., 1988; Thornton et al., 2006b). The predicted temperature increases of 2–6 degrees Celcius by the end of the century in Africa could cause the spread of tropical species from countries in other agro-ecological regions into new environments where water is not the limiting factor (Collier et al., 2008; Hoffman and Vogel, 2008; IPCC, 2007). Many of the important forage grasses, including Napier, Rhodes and *Brachiaria*, are not frost tolerant. As temperatures increase and the likelihood of frost decreases, these tropical species could be introduced into sub-tropical regions, including the tropical highlands.

Changes in climate in Africa are predicted to be more severe than in other regions (Collier et al., 2008; IPCC, 2007). Impacts will be substantial on the use of cultivated forages and on indigenous forage diversity in grasslands and natural pastures. Some areas of East Africa are predicted to receive 10–20 per cent more rainfall (Collier et al., 2008). Current grazing or marginal lands may be converted to crops, leading to the loss of forage diversity and to increasing degradation. Climate change could also allow (or suggest) the introduction of species from Latin America such as *Centrosema* and *Stylosanthes guianensis*, which are adapted to sub-humid areas. In areas that are predicted to get drier, continued cropping can result in more rapid degradation through the loss of land cover. The use of drought tolerant forages in these areas, such as buffel grass or *Stylosanthes scabra*, could, on the other hand, increase soil cover and reduce soil degradation (Batjes and Sombroek, 1997). Appropriate adaptation strategies need to be developed on a site-specific basis and transferred appropriately to analogous areas.

The importance of grasslands increases under the context of climate change, particularly for mitigation. Natural grasslands act as an important carbon sink (Morgan, 2005). However, increasing levels of carbon dioxide will increase overall biomass production and may result in reduced forage quality and digestibility due to lignification (Thornton et al., 2006a). Changes in the level of carbon dioxide could result in changes in species richness in natural pastures because some legumes are more responsive to increased levels of carbon dioxide as a result of biological nitrogen fixation (Aguiar, 2005). As a result, shifts in particular areas due to climate change will lead to changes in both the distribution of wild species and grasslands, producing a complex mixture of responses that could lead (in some cases, at best) to a substantial degradation of genetic resources.

Numerical analysis of the changes in climatic similarity between countries for selected crops (methods and results)

We performed a numerical analysis of the effect of climate change predictions on crop distributions to identify the extent to which climate change will impact climatic compatibility of genetic resources among countries. We analysed the climatic similarity between countries under current and future conditions and interpreted it as a proxy for genetic resource interdependence. We assumed that both crop distributional ranges and climatic compatibilities at the same time could provide an indication of whether similar PGRFA can be used in two sites, regardless of how much actual exchange has occurred between these two sites in the recent past. Mining current and future climates and detecting interdependencies between current and future climates, we identified the potentially useful links between countries today and in the future and determined whether interdependencies play (or could play) an important role in both the conservation and utilization of plant genetic resources now and in the future.

We selected 17 staple food and cash crops that account for 75 per cent of the global harvested area (FAOSTAT, 2009). There were a number of different climatic responses within these crops because they cover a range of different ecologies. Crops included cereals (that is, wheat, maize, millet, rice, sorghum and barley), legumes (that is, beans, groundnuts and soybeans), roots and tubers (such as potatoes, sugar beets, sweet potatoes and cassava), fruit (bananas and plantains), fibres (cotton), and high value industrial and cash crops (such as sugarcane and coffee). 'Climatic similarity' was described in terms of the following four dimensions:

1. current climatic similarities among countries (the baseline);
2. climatic similarity between countries in the future, which shows the likely changes in interdependence with respect to the baseline;
3. climatic similarity between each country's current conditions and future conditions, which provides the extent to which the current climate upon which agriculture is built and for which genetic resources are adapted could be important for the future of other countries; and
4. climatic similarity between each country's future conditions and current conditions, which provides the extent to which a country in the future could depend on other countries for genetic resources conserved presently.

A set of metrics (described later on) are used to assess each of these dimensions. Both dimensions (3) and (4) include the concept of self-dependence, which is useful to investigate the likelihood of a country being able to hold the necessary genetic resources for its own adaptation. The analysis looks at the similarity between countries for these four situations with respect to each of the 17 crops under study. Current harvested areas of each crop were used to define the production environments for each crop within each country (FAOSTAT, 2009). The extent of each crop within each country of the world was determined using a spatial allocation model (You and Wood, 2006; You et al., 2006). For each country and for cropped lands (per crop) within each country, the current distribution of climates was derived from WorldClim (Hijmans et al., 2005). We downloaded monthly minimum, maximum and mean temperatures as well as monthly total rainfall gridded datasets at 30 arcs per second (approximately 1 kilometre in the Equator) and used them to derive 19 bioclimatic indices (Busby, 1991). The average of all cells within a country was taken as the country's current climate.

Future climates were derived from the results of 18 global circulation models (GCMs) from the third and fourth IPCC assessment reports (IPCC, 2001, 2007) for the decade beginning 2050 and for the Special Report on Emission Scenarios (SRES) A2 emission scenario (business as usual). GCM minimum, maximum and mean monthly temperatures and total monthly precipitation outputs were downscaled using the delta method and were used to derive the same 19 bioclimatic indices as for current climate (Ramirez and Jarvis, 2008, 2010). The average of all cells within each country was again taken as an indicator of the country's future climate.

For each crop's distributional range, the 19 bioclimatic indices were standardized to have a mean of 0 and a standard deviation of 1. Three 19-dimensional Euclidean distance (ED) matrices of *n-by-n* (*n* varying on a crop basis) countries were computed:

1. ED between current climates to measure the current climatic compatibility between countries;
2. ED between future climates to measure the likely change in climatic similarity between countries and determine whether current compatibilities are to be preserved or lost in the future; and
3. ED between current and future (and *vice versa*) climates to determine opportunities for sharing genetic resources from one period to the other and from one country to the other.

Both matrices (1) and (2) provide a geographical perspective of the potential links between countries, and (3) provides both a geographical and temporal perspective of the potential links that can be established between countries. Since all of the variables were standardized, the ED is a non-dimensional and scale-independent measure. Therefore, direct comparisons between countries, crops and periods can be done with no bias.

With the aim of (1) investigating the likely changes in self-dependence; (2) comparing interdependencies of one country with the others; (3) investigating the relevance of current and future potential links; and (4) detecting key potential links in the future that are not suspected currently, we calculated a range of metrics aiming to provide a numerical basis for interpreting the results:

- average ED under current conditions (AEDC) – the ED assessed using current conditions for each pair of countries averaged on a country basis;
- average ED under future conditions (AEDF) – the ED assessed using future conditions for each pair of countries averaged on a country basis;
- average change in country self-dependence (ACCSD) – the average changes in ED for each country and its future;

- average change in country inter-dependence (ACCID) – the average changes in ED between each country and all of the others;
- average current-future (ACFI) and future-current (AFCI) interdependence – the average ED from one country's current climates to future climates of all of the others and *vice versa*;
- number of gained (NGR) and lost (NLR) relationships – country rates of gained and lost climatic similarities; and
- new key climatic similarities (NNKS) – number of key climatic similarities that may arrive within the context of each crop when climatic conditions change.

Despite the above specifics, metrics are not reported given the number of countries and crops analysed here. Instead, we provide a summary of our findings. Our analysis shows that the current climatic similarity among countries for all crops was high. Regions were not only capable of providing genetic resources to other regions and importing/exporting food crops to supply the basic needs of their populations, but they also showed high climatic similarity. Indeed, they were probably able to provide genetic resources because of their climatic similarity. On average, a country's croplands was significantly similar to more than 80 per cent of other countries' croplands for all crops. Globally, 116 out of the 188 countries (62 per cent) analysed will likely decrease their ACCID, indicating an increased similarity among the global areas of these countries (see Table 4.1 and Figure 4.1). Of the remaining 38 per cent of countries, 94 per cent will likely increase their ACCID by only less than 5 per cent. For all crops, the similarities are likely to strengthen for more than 40 per cent of countries (with sugar beets showing the most similarity at 91 per cent and cassava showing the least at 41 per cent). Generally, moist environments are more likely to increase in climatic similarity, although no significant latitudinal trends were found as there were no separations among individual countries of different geographic areas (that is, Europe, sub-Saharan Africa, Latin America and Australia) (see Figure 4.1a).

Considering the entire area of the country and pooling results by region, climatic similarity is likely to increase with changing climates. On average, all regions showed reduced Euclidean distances in 2050s, with North America presenting the greatest decrease and Latin America presenting the least decrease. These values, however, may disguise a lot of variation within the country. Country self-distances (distance from a country's current position to its future position) are the greatest for North Africa and North America, indicating a lower level of climatic similarity than in regions such as the Caribbean, where countries are relatively climatically similar (see Figure 4.1).

Although the ACCID ranged from between minus 24 per cent (Faroe Islands) and 54 per cent (Grenada), 97 per cent of the countries presented ACCID between 10 and minus 10 per cent. For 11 per cent of the countries (that is, Serbia and Montenegro, United Arab Emirates, Sweden, Romania, Macedonia, Lithuania, Iran, Iceland, Grenada, Georgia, France, Faroe Islands, Egypt, Canada, Bulgaria, Bosnia, Belgium and Azerbaijan), changing climate conditions were found to add at least one key link that was even closer than they would be themselves in the future.

TABLE 4.1 Likely changes in interdependence for 17 crops globally for 2050

Crop	A1	Decreases in ED (%)	Change in ED (%)	Maximum loss of interdependence	Maximum gain of interdependence	% Countries with key current actors to take PGR for future	% Analogues that may be more important	% Analogues that may appear
Bananas*	Y	52.4	−0.12	Solomon Islands	Pakistan	2.9	53.3	1.0
Barley	Y	70.4	−0.65	Finland	Yemen	16.8	69.6	5.6
Beans	Y	59.7	−0.31	Colombia	Saudi Arabia	14.2	67.2	3.0
Cassava	Y	40.6	0.33	Solomon Islands	Guyana	0.0	40.6	0.0
Coffee	N	55.3	−0.21	Kazakhstan	Yemen	1.2	49.4	0.0
Cotton	N	45.4	0.08	Uganda	Saudi Arabia	8.3	65.7	2.8
Groundnut	N	90.2	−2.73	Iraq	Yemen	30.1	74.0	15.4
Maize	Y	57.4	−0.23	Solomon Islands	Yemen	9.7	60.6	1.3
Millet	Y**	86.4	−2.99	Germany	Albania	42.1	73.9	22.7
Potatoes	Y	68.8	−0.65	Finland	Oman	13.6	66.2	2.6
Rice	Y	52.2	−0.16	Iraq	Somalia	8.8	59.6	0.0
Sorghum	Y	63.3	−0.34	Peru	Oman	10.2	67.2	2.3
Soybeans	N	46.8	0.07	Colombia	Turkmenistan	12.1	66.9	2.4
Sugar beets	Y	61.1	0.27	Finland	Uzbekistan	31.5	90.7	27.8
Sugar cane	N	43.6	0.04	Colombia	Turkmenistan	6.4	52.7	4.5
Sweet potatoes	Y	41.9	0.12	Bhutan	Saudi Arabia	1.7	51.3	0.0
Wheat	Y	74.0	−1.15	Finland	Yemen	42.0	83.2	22.9
Total	–	61.7	−0.25	Faroe Islands	Grenada	11.7	69.1	3.7

Notes: * bananas and plantain
** Pearl millet: Y, finger millet: Y, any other millet: N

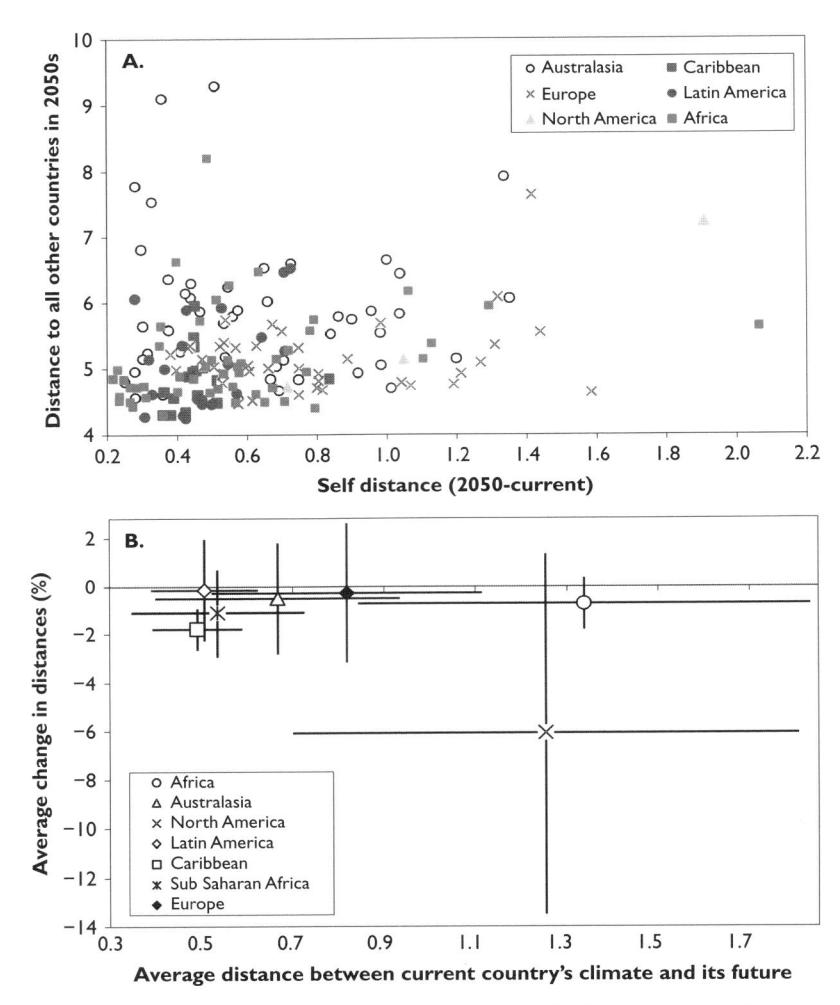

FIGURE 4.1 Overview of changes in interdependence: (a) Changes in interdependence within country land areas. Self-distance (X axis) is the ED between each country's current and its future conditions and average distance (Y axis) is the average ED between the country's current climates and all of the others' future climates. (b) Average changes in ED pooled by continental zone (Y axis) confidence interval versus the average distance between current and future climates of each country plus a 95 percent confidence interval, pooled by continental zone (X axis)

On the other hand, 98 per cent of the countries have ACFI that is lower than their AEDC, meaning that they are closer to most other countries' future conditions than they are to these countries' current conditions. While 76 per cent of the climatically 'far' countries (that is, those presenting very high AEDC) will decrease their ED with some countries (thus, will become more similar to at least one country), 52 per cent presented the opposite pattern. However, significant within-country variability is still

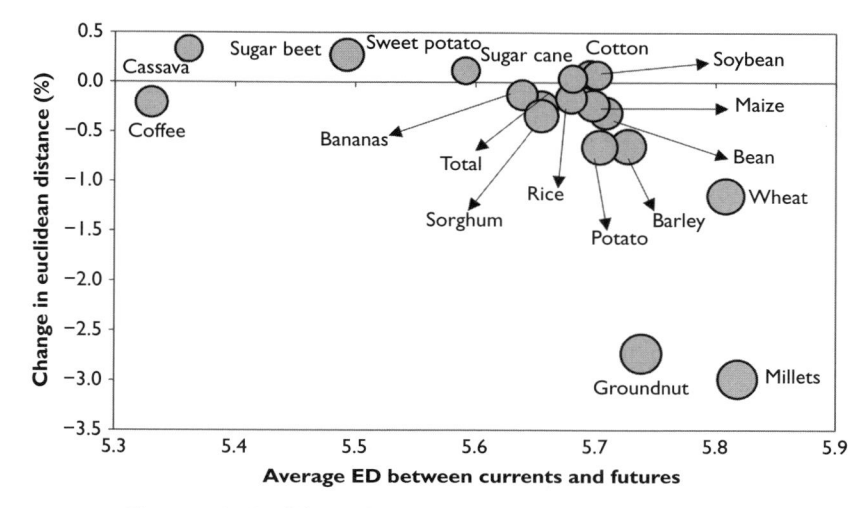

FIGURE 4.2 Changes in Euclidean distances pooled by crop for all countries having one or more hectares on each crop versus the average Euclidean distance between current climates and future climates of each country pooled by crop. Bubble size is the percentage of countries having one or more hectares on each crop and decreasing climatic distances (that is, proxy of a higher interdependency)

present within these figures due to the different orographic and landscape features of the various countries.

There are different observed trends in crop-based climatic interdependence. The proportion of countries increasing their 'closeness' (that is, ACCID < 0) is above 50 per cent for 12 of the 17 crops considered (see Figure 4.2), and the average ED decreases (that is, AEDF < AEDC) for 11 out of the 17 crops (such as bananas, barley, beans, coffee, groundnuts, maize, millets, potatoes, rice, sorghum and wheat). Cassava was the only crop for which no country showed a closer analogy than itself. Wheat, maize and millet showed 42, 30 and 42 per cent of their current country-level cropping environments under analysis being nearer to at least one other future cropping environment respectively. This outcome may indicate that environments included in these percentages may have at least one key country with which they should consider reviewing existing and past flows of plant genetic resources and that strengthening their current treaties with such countries could improve their current levels of PGRFA conservation and utilization.

Coffee had the least distance from current to future environments, probably due to strong climatic specialization. ACFI values for all crops, however, occurred within quite a small range (from 5.3 to 5.9). Most of the changes in the different environments of crops are near to the total area, suggesting that all of these crops hold relatively similar patterns of interdependencies and that their environments may be relatively similar now and in the future. Cassava, coffee, sugar beets and sweet potatoes from one side, and wheat, millet and groundnuts from the other are the only crops that showed significant differences in behaviour in comparison with the other crops.

Loss of similarity occurs when a cropping environment moves further away from all of the climates of other countries (that is, AEDF > AEDC). The environments of grain legumes (that is, beans and soybeans) are predicted to reduce their overall average climatic similarity (that is, ACCID > 0) in Latin American countries with croplands in the Andes, signifying that these countries tend to become more 'climatically' isolated and very likely to require very specific germplasm for adaptation to future conditions. Cold environments such as those in Finland are also likely to lose climatic similarity in their croplands. Finland is actually the most frequent 'loser' (of climatic similarity) in the whole world for the selected set of crops (that is, the country presenting an ACCID > 0 most frequently). On the other hand, Asian countries tend to increase their climatic similarity (that is, ACCID < 0) at greater rates than other countries (Table 4.1, Figure 4.1a), which may be due to their sub-tropical conditions and different within-region climate change patterns. The same thing happens in some sub-Saharan African countries such as Somalia and others nearby, especially in rice cropping. All of these results indicate that Asia and certain countries in sub-Saharan Africa could take more advantage of strengthened and globalized treaties for sharing PGRFA.

Novel climate conditions (that is, conditions not observed in any country under current conditions) appear in all of the current croplands except for cassava, coffee, rice and sweet potatoes. This outcome may be due either to a very high similarity in current conditions (that is, specialization of cropping environments) or because they are cropped in countries that become 'climatically isolated', or both. All cropping environments of all of the crops analysed show a significant number of areas (that is, > 30 per cent), with any other current-future and/or future-future ED being less than their self-future ED – meaning there is another place that is more compatible for them than their own place. The fact that the rates of lost interdependencies are far less significant than the rates of gained or strengthened dependencies should be also taken into account when analysing the patterns of exchange of future potential PGRFA. For all crops, at least 30 per cent of the cropping environments have lost climatic similarity with at least one other country. Due to the significant climatic and orographic variability within some countries, specific analyses within different cropping environments are required to further develop a collaborative and dynamic network that will allow fluxes of relevant genetic materials from one country to another.

Expected changes in interdependence as a result of climate change

Two clear issues may lead to greater interdependence as a result of climate change:

1. Novel climatic conditions for countries will mean that currently adapted landraces and varieties may become maladapted, requiring the import of new materials with novel and more appropriate traits. The identification of environments from which these materials can be imported is therefore critical for adaptation.
2. Climate change will bring about new types of and increased demand for PGRFA globally, requiring greater volume and variety of genetic materials. Identifying the

appropriate channels through which these resources might flow from one country to the other will facilitate the utilization and conservation of PGRFA and will thus aid adaptation.

Almost 35 per cent of the global land area may experience 'novel climates' – essentially climatic conditions that are currently not experienced anywhere – and the geography of the world's climate may shift significantly (Williams et al., 2007). In regard to agriculture, temperatures in the growing season in 2100 in the tropics and sub-tropics are likely to be hotter than experienced over the past century, particularly in the highlands, where cropping areas are more reduced and are subjected to other constraints. Current local planting materials are unlikely to withstand such conditions (Battisti and Naylor, 2009). In summary, our numerical analysis shows that, globally:

- regions are able to provide genetic resources because of their climatic similarity and/or compatibility with other regions;
- similarity among global areas will increase;
- some 30 per cent or more of the countries will benefit more from other countries' PGRFA than from their own PGRFA;
- 98 per cent of the countries will increase their average climatic compatibility;
- 76 per cent of the currently climatically 'isolated' countries will become more similar to at least one other country;
- more than 50 per cent of the countries for 70 per cent of the crops analysed will become more interdependent; and
- at least 30 per cent of the cropping environments will lose at least one potential key link.

Between regions and countries:

- increased compatibility by 2050s was observed, with North America being the most benefited and Latin America the least benefited;
- large regions (e.g., North Africa and North America) tend to have lower climatic similarity compared to relatively small areas (e.g., the Caribbean);
- Serbia and Montenegro, United Arab Emirates, Sweden, Romania, Macedonia, Lithuania, Iran, Iceland, Grenada, Georgia, France, Faroe Islands, Egypt, Canada, Bulgaria, Bosnia, Belgium and Azerbaijan will have a key link with another country that is even closer than themselves in the future;
- Andean countries and high-latitude countries (that is, Finland) will become more 'climatically isolated'; and
- climate and climate change geographical variability is high among regions and within them, and this factor will lead to the need for more site-specific analyses.

Crop level:

- similar ecologies often mean similar patterns of changes in interdependencies;
- adaption to climate change and flows of PGRFA between countries and regions might depend on both shifts of species distributions, land use changes and current and future conservation actions;
- sugar beet farmers would benefit the most from improved future PGRFA exchange mechanisms and cassava farmers the least;
- all areas cropped under cassava, coffee, rice and sweet potato will not present any 'novel climate' (a climate not existing currently). In fact, cassava was the only crop for which no country showed a closer analogy than the current;
- bananas, barley, beans, coffee, groundnuts, maize, millet, potatoes, rice, sorghum and wheat environments are more compatible in the future than they are currently;
- coffee was the most climatically specialized (and isolated) crop; and
- 42, 30 and 42 per cent of the environments where wheat, maize and millet are cropped (respectively) may need to review existing and past flows of plant genetic resources and strengthen their current treaties with at least one other country.

Developing countries have provided the biological basis for agriculture both in developed countries and for each other (Fowler and Smale, 2000). The dimension and direction of flow of PGRFA (for example, south to south and north to south) is notoriously difficult to track, monitor and quantify. The limited information available on flows seems to indicate that we are in a period of reverse flow in which material is no longer exported from its centres of origin. Rather, the opposite is occurring: farmers and research institutes in the developing world have become net recipients of both local varieties and of improved materials (Fowler and Smale, 2000; Visser et al., 2003). The quantification of PGRFA flows and utilization is needed in order to revise current projects and develop future actions.

M.B. Burke, D.B. Lobell and L. Guarino (2009) found that shifts in crop climates (that is, of maize, millet and sorghum) leading up to 2050 indicate that many countries will experience novel climates not currently found within their borders in 2050 and that 75 per cent of these will have analogues in at least five other countries. Our figures confirm that these crops are likely to present novel climates. The international movement of germplasm is essential to enable adaptation (*ibid.*). Overall, crops will become maladapted in the face of climate change (Lobell et al., 2008). Given the speed of climate change, farmers will most likely be unable to adapt rapidly enough through traditional selection practices (Burke et al., 2009). What will be needed is the facilitated exchange of exotic varieties and landraces from analogous sites elsewhere.

Thus, climate change will bring new and enhanced demand for genetic resources in order for farmers to adapt to the new climatic situation. National and international breeding programmes for a number of crops are already targeting new varieties with adaptations for future climatic stresses (see, for example, Ortiz et al., 2008), including heat, drought and waterlogging tolerance. In addition, the effort to breed for traits valued both today and for the future is likely to increase the general demand for PGRFA.

Demand will also likely increase for the genetic resources of crop wild relatives to address biotic and abiotic constraints, many of which are being exacerbated by climate change (Lane and Jarvis, 2007). While demand for such genetic resources is global, their natural distribution is generally restricted. For example, no wild relatives of the cultivated peanut exist outside Brazil, Paraguay, Bolivia, Uruguay or Argentina (Ferguson et al., 2005; Jarvis et al., 2003). The increased demand for these resources implies increased interdependence among countries, possibly encouraging a greater flow of germplasm from south to north since many centres of origin occur in the south (Fowler and Smale, 2000). The opposite can occur for beans or maize, for which the centres of diversity are in the northern hemisphere (that is, in Mexico and the United States). Since crop wild relatives are poorly conserved in *ex situ* collections (Ramirez et al., 2010), policies must facilitate the access to wild gene pools through targeted collection to fill specific gaps (Maxted et al., 2008; Maxted and Kell, 2009; Ramirez et al., 2010).

Biotechnology will also affect the demand for PGRFA. On the one hand, new tools and methods mean that more accessions can be screened and potentially used. On the other hand, transgenics, marker-assisted selection and other biotechnology tools may reduce the amount of diversity required in breeding programmes, as individual genes, rather than collections of traits, become the target. Although increased demand for new varieties to confront climate change will rely heavily on biotechnology, it is difficult to foresee the outcome of these developments on interdependence (Hajjar and Hodgkin, 2007; Ortiz et al., 2008).

Climate change may also increase the importance of otherwise minor or under-utilized crops and plant species. These include species suitable for biofuel production (biodiesel, ethanol and second and third generation biofuel technologies) as well as hardy crops and species that until now have had only local or regional significance but which may in the future provide valuable alternatives to confront climate change, especially in marginal environments.

Conclusions

This chapter has reviewed the interdependence of PGRFA and discussed how climate change might change interdependence patterns and levels. The numerical analysis of climatic similarity is presented as a proxy for potential genetic resource interdependence. The analysis shows that similarity between countries will be greater for 75 per cent of the global croplands. Regardless of the level of ecological specialization of a crop, there will be a significant shift in the climates of these crops, which may bring novel climates that are often nearer to the current climates of other countries (that is, 98 per cent of countries increase their climatic compatibility, and 76 per cent of the crops present novel climates). The exchange of PGRFA among countries will thus continue as a key issue in the face of climate change, although some changes in the mechanism may be required (that is, a clear definition of the key providers and key receivers). Although some special cropping environments will decrease their climatic similarity, the gaining and strengthening rates will markedly overcome climatic interdependence loss rates in all regions and crops. There will, therefore, be significant opportunities for setting collaborative networks to conserve, characterize, improve and share PGRFA. There is high likelihood of a change in the types of demand for PGRFA in the future and a likely increase in overall demand.

Changes in the types of demand are expected in the following ways:

- an increased demand for PGRFA with characteristics that will help adapt agriculture to future climates (heat, drought and waterlogging-tolerant materials, among others);
- an increasing demand for crop wild relatives to address biotic and abiotic constraints; and
- an increased demand for 'minor' crops (including neglected and under-utilized crops) that might help communities adapt to climate change in marginal environments and/or contribute to climate change mitigation through biofuel or a similar combustible.

Increases in demand are expected for the following reasons:

- shifting geography of climate, leading to shifts in crop distribution among and within countries;
- the appearance of globally and regionally novel climates making locally adapted genetic resources less suitable or no longer suitable beyond the next 20–30 years; and
- increasing global population and expected negative impacts of climate change on agriculture, leading to more and more need for new seed technologies to produce more food in less area with greater water productivity.

Facilitated access to PGRFA will be needed. While facilitated access and benefit sharing exists for Annex 1 crops from the International Treaty for Plant Genetic Resources for Food and Agriculture, there are few options for benefit sharing of non-Annex 1 crops.[1] Twelve of the crops we analysed are in Annex 1 (that is, bananas, barley, beans, cassava, maize, millet, potatoes, rice, sorghum, sugar beets, sweet potatoes and wheat). Exclusions include major staples (including some of those studied in this chapter's climatic similarity analyses such as groundnut, soybeans and coffee), numerous forage species, crop wild relatives, and many minor and/or neglected and under-utilized crops. Thousands of farmers depend on these non-Annex 1 crops, and adaptation to climate change without a facilitated exchange of genetic resources will be difficult. A more facilitated germplasm exchange for these crops would enable *ex situ* conservation through new collecting (conservation status is incomplete for many of these crops and species) and would make the genetic resources themselves available for countries to adapt to the future challenges of climate change.

Note

1 International Treaty on Plant Genetic Resources for Food and Agriculture, 29 June 2004, www.planttreaty.org/texts_en.htm (last accessed 11 June 2011).

References

Adger, W., S. Agrawala, M.M.Q. Mirza, C. Conde, K.L. O'Brien, J. Pulhin, R. Pulwarty, B. Smit and K. Takahashi (2007) *Assessment of Adaptation Practices, Options, Constraints, and Capacity. Climate Change: Impacts, Adaptation and Vulnerability*, Contribution of Working Group II to the Fourth Assessment Report of the Intergovernmental Panel on Climate Change, Cambridge, UK, and New York.

Aguiar, M.R. (2005) 'Biodiversity in Grasslands: Current Changes and Future Scenarios', in S.G. Reynolds and J. Frame (eds), *Grasslands: Developments Opportunities Perspectives*, FAO, Rome.

Bass, M.S., M. Finer, C.N. Jenkins, H. Kreft and D.F. Cisneros-Heredia (2010) *Global Conservation Significance of Ecuador's Yasuní National Park*. PLoS One 5(1): e8767, doi:10.1371/journal.pone.0008767.

Batjes, N.H., and W.G. Sombroek (1997) 'Possibilities for Carbon Sequestration in Tropical and Subtropical Soils', *Global Change Biology* 3: 161–73.

Battisti, D., and R. Naylor (2009) 'Historical Warnings of Food Insecurity with Unprecedented Seasonal Heat', *Science* 323(5911): 240–44.

Boonman, J.G. (1993) *East Africa's Grasses and Fodders: Their Ecology and Husbandry*, Kluwer Academic Publishers, Dordrecht, The Netherlands.

Brush, S. (ed.) (2000) *Genes in the Field: On-farm Conservation of Crop Diversity*, International Development Resources Centre International Plant Genetic Resources Institute, Boca Raton, FL.

Burke, M.B., D.B. Lobell and L. Guarino (2009) 'Shifts in African Crop Climates by 2050, and the Implications for Crop Improvement and Genetic Resources Conservation', *Global Environmental Change* 19: 316–25.

Busby, J.R. (1991) 'BIOCLIM: A Bioclimatic Analysis and Prediction System', in C.R. Margules and M.P. Austin (eds), *Nature Conservation: Cost Effective Biological Surveys and Data Analysis*, Commonwealth Scientific and Industrial Research Organization, Canberra, Australia.

Carnaval, A.C., and C. Moritz (2008) 'Historical Climate Change Predicts Current Biodiversity Patterns in the Brazilian Atlantic Rainforest', *Journal of Biogeography* 35: 1187–1201.

Cavatassi, R., L. Lipper and J. Hopkins (2006) *The Role of Crop Genetic Diversity in Coping with Agricultural Production Shocks: Insights from Eastern Ethiopia*, Working Papers no. 06–17, Agricultural and Development Economics Division, FAO, Rome, online: http://ideas.repec.org/p/fao/wpaper/0617.html (last accessed 15 May 2011).

Challinor, A.J., T.R. Wheeler, P.Q. Craufurd, C.A.T. Ferro and D.B. Stephenson (2007) 'Adaptation of Crops to Climate Change through Genotypic Responses to Mean and Extreme Temperatures', *Agriculture, Ecosystems and Environment* 119: 190–204.

Chaudhary, P., D. Gauchan, R.B. Rana, B.R. Sthapit and D.I. Jarvis (2004) *Potential Loss of Rice Landraces from a Terai Community in Nepal: A Case Study from Kachorwa, Bara.* PGR Newsletter, volume 137, International Plant Genetic Research Institute, Rome.

Chen, I.C., H.J. Shiu, S. Benedick, J. Holloway, V.K. Chey, H.S. Barlow, J.K. Hill, C.D. Thomas (2009) 'Elevation Increases in Moth Assemblages over 42 Years on a Tropical Mountain', *Proceedings of the National Academy of Sciences* 106(5): 1479–83.

Cleveland, D.A., D. Soleri and S.E. Smith (1994) 'Do Folk Crop Varieties Have a Role in Sustainable Agriculture?' *Bioscience* 44(11): 740–51.

Commission on Genetic Resources for Food and Agriculture (CGRFA) (2000) *Progress Report on the Development of a Network of in situ Conservation Areas*, Globally Important Agriculture Heritage Systems, online: ftp://ext-ftp.fao.org/ag/cgrfa/cgrfa9/r9w13e.pdf (last accessed 8 July 2011).

Collier, P., G. Conway and T. Venables (2008) 'Climate Change and Africa', *Oxford Review of Economic Policy* 24: 337–53.

Cox, J.R., M.H. Martin, F.A. Ibarra, J.H. Fourie, N.F.G. Rethman and D.G. Wilco (1988) 'The Influence of Climate and Soils on the Distribution of Four African Grasses', *Journal of Range Management* 41: 127–39.

Esquinas-Alcazar, J. (2005) 'Protecting Crop Genetic Diversity for Food Security: Political, Ethical and Technical Challenges', *Nature Reviews Genetics* 6: 946–53.
Food and Agriculture Organization (FAO) (1997) *The State of the World's Plant Genetic Resources for Food and Agriculture*, FAO, Rome, online: www.fao.org/ag/AGP/AGPS/Pgrfa/pdf/swrfull.pdf (last accessed 2 June 2006).
——(1998) *The Conservation and Sustainable Utilization of Mahogany Genetic Resources*, Forest Genetic Resources no.26, FAO, Rome.
——(2008) *The State of Food Insecurity in the World*, FAO, Rome, Italy.
FAOSTAT (2009) United Nations Food and Agriculture Organization, Statistical Database, online: http://faostat.fao.org/site/567/default.aspx#ancor (last accessed 11 June 2011).
Ferguson, M.E., A. Jarvis, A. H.T. Stalker, D. Williams, L. Guarino, J.F.M Valls, R.N. Pittman, C.E. Simpson, and P. Bramel (2005) 'Biogeography of Wild Arachis (Leguminosae): Distribution and Environmental Characterisation,' *Biodiversity and Conservation* 14(7): 1777–98.
Fischer, G., M. Shah, F.N. Tubiello and H. van Velhuizen (2005) 'Socio-Economic and Climate Change Impacts on Agriculture: An Integrated Assessment, 1990–2080', *Philosophical Transactions of the Royal Society* 1463: 2067–83.
Fischer, G., M. Shah and H. van Velthuizen (2002) 'Climate Change and Agricultural Vulnerability', Special Report as a contribution to the World Summit on Sustainable Development, International Institute for Applied Systems Analysis, Laxenburg, Austria.
Fischer, G., H. van Velthuizen, M. Shah, and F. Nachtergaele (2001) *Global Agro-ecological Assessment for Agriculture in the Twenty-First Century*, International Institute for Applied Systems, FAO, Austria.
Fowler, C. (1994) *Unnatural Selection: Technology, Politics and Plant Evolution*, International Studies in Global Change no. 6, Gordon and Breach Science Publishers, Langhorne, PA.
Fowler, C. and M. Smale (2000) *Germplasm Flows between Developing Countries and the CGIAR: An Initial Assessment*, Economics Working Paper, Gordon and Breach Science Publishers, Yverdon, Switzerland.
Gurney, A.L., M.C. Press and J.D. Scholes (2002) 'Can Wild Relatives of Sorghum Provide New Sources of Resistance or Tolerance against Striga Species?' *Weed Resources* 42: 317–24.
Hajjar, R., and Hodgkin, T. (2007) 'The Use of Wild Relatives in Crop Improvement: A Survey of Developments over the Last Twenty Years', *Euphytica* 156: 1–13.
Hammer, K., H. Knüpffer, L. Xhuveli and P. Perrino (1996) 'Estimating Genetic Erosion in Landraces: Two Case Studies', *Genetic Resources and Crop Evolution* 43: 329–36.
Heal, G., B. Walker, S. Levin, K. Arrow, P. Dasgupta and G. Daily (2004) 'Genetic Diversity and Interdependent Crop Choices in Agriculture', *Resource and Energy Economics* 26: 175–84.
Hijmans, R.J., S.E. Cameron, J.L. Parra, P.G. Jones and A. Jarvis (2005) 'Very High Resolution Interpolated Climate Surfaces for Global Land Areas', *International Journal of Climatology* 25: 1965–78, online: www.worldclim.org (last accessed 15 May 2011).
Hoffman, M.T., and C. Vogel (2008) 'Climate Change Impacts on African Rangelands', *Rangelands* 30: 12–17.
Humphreys, L.R. (1967) 'Buffelgrass (*Cenchrus ciliaris*) in Australia', *Tropical Grasslands* 1: 123–34.
Intergovernmental Panel on Climate Change (IPCC) (2001) *Third Assessment Report: Climate Change 2001*, IPCC, Geneva.
——(2007) *Fourth Assessment Report: Climate Change 2007*, IPCC, Geneva.
Jarvis, A., A. Lane and R.J. Hijmans (2008) 'The Effect of Climate Change on Crop Wild Relatives', *Agriculture, Ecosystems and Environment* 126: 13–23.
Jarvis, A., M. Ferguson, D. Williams, L. Guarino, P. Jones, H. Stalker, J. Valls, R. Pittman, C. Simpson and P. Bramel (2003) 'Biogeography of Wild Arachis: Assessing Conservation Status and Setting Future Priorities', *Crop Science* 43: 1100–108.
Kim, K.C., and L.B. Byrne (2006) 'Biodiversity Loss and the Taxonomic Bottleneck: Emerging Biodiversity Science', *Ecological Resources* 21: 794–810.

Kloppenburg, J.R., and D.R. Kleinman (1987) 'Plant Germplasm Controversy Analyzing Empirically the Distribution of the World's Plant Genetic Resources', *Bioscience* 37: 3–190.

Lane, A., and A. Jarvis (2007) 'Changes in Climate Will Modify the Geography of Crop Suitability: Agricultural Biodiversity Can Help with Adaptation', *Journal of Semi-arid Tropics* 4(1), online: www.icrisat.org/Journal/specialproject.htm (last accessed 15 May 2011).

Lobell, D., M.B. Burke, C. Tebaldi, M.D. Mastrandea, W.P. Falcon and R.L. Naylor (2008) 'Prioritizing Climate Change Adaptation Needs for Food Security in 2030', *Science* 319: 607–10.

Maxted, N., E. Dulloo, B.V. Ford-Lloyd, J.M. Iriondo and A. Jarvis (2008) 'Gap Analysis: A Tool for Complementary Genetic Conservation Assessment', *Diversity and Distributions* 14(6): 1018–30.

Maxted, N., B. Ford-Loyd, J. Irinodo, S.P. Kell and E. Dulloo (2008) *Crop Wild Relative Conservation and Use*, CABI Publishing, Wallingford, United Kingdom.

Maxted, N., and S. Kell (2009) *Establishment of a Global Network for the In Situ Conservation of Crop Wild Relatives: Status and Needs*, FAO Commission on Genetic Resources for Food and Agriculture, Rome.

Mercer, K.L., and H.R. Perales (2010) 'Evolutionary Response of Landraces to Climate Change in Centers of Crop Diversity', *Evolutionary Applications* 3(5–6): 480–93.

Miles, J.W., B.L. Maass and C.B. do Valle (eds) (1996) *Brachiaria: Biology, Agronomy and Improvement*, International Center for Tropical Agriculture, Cali, Colombia.

Morgan, J.A. (2005) 'Rising Atmospheric CO2 and Global Climate Change: Responses and Management Implications for Grazing Lands', in S.G. Reynolds and J. Frame (eds), *Grasslands: Developments, Opportunities and Perspectives*, FAO, Rome.

National Research Council (1972) *Genetic Vulnerability of Major Crops*, National Academy of Sciences, Washington, DC.

Nogués-Bravo, D., M.B. Araújo, T. Romdal and C. Rahbek (2008) 'Scale Effects and Human Impact on the Elevational Species Richness Gradients', *Nature* 453: 216–20.

Ortiz, R., D. Kenneth, B. Sayre, R. Govaerts Gupta, G.V. Subbarao, B. Tomohiro, D. Hodson, J.M. Dixon, J.I. Ortiz-Monasterio and M. Reynolds (2008) 'Climate Change: Can Wheat Beat the Heat?' *Agriculture, Ecosystems and Environment* 126: 46–58.

Palacios, X.F. (1998) *Contribution to the Estimation of Countries' Interdependence in the Area of Plant Genetic Resources*, Background Study Paper no. 7, Rev. 1, Commission on Genetic Resources for Food and Agriculture, Rome.

Parmesan, C., and G. Yohe (2003) 'A Globally Coherent Fingerprint of Climate Change Impacts across Natural Systems', *Nature* 421: 37–42.

Parzies, H.K., W. Spoor and R.A. Ennos (2000) 'Genetic Diversity of Barley Landrace Accessions (Horderum vulgare ssp. vulgare) Conserved for Different Lengths of Time in Ex Situ Gene Banks', *Heredity* 84: 476–86.

Phaikaew, C., and M.D. Hare (2005) 'Stylo Adoption in Thailand: Three Decades of Progress', in F.P. O'Mara, R.J. Wilkins, L. Mannetje, D.K. Lovett, P.A.M. Rogers and T.M. Boland (eds), *Twentieth International Grassland Congress*, Wageningen Academic Publishers, Wageningen, The Netherlands.

Phaikaew, C., C. Manidool and P. Devahuti (1993) *Ruzi Grass (Brachiaria ruziziensis) Seed Production in North-east Thailand*, Proceedings of the Twenty-Seventh International Grassland Congress, Palmerston North and Rockhampton.

Putnam, D.H., C.G. Summers and S.B. Orloff (2007) 'Alfalfa Production Systems in California', in C.G. Summers and D.H. Putman (eds), *Irrigated Alfalfa Management for Mediterranean and Desert Zones*, University of California Agriculture and Natural Resources Publication, Berkeley, CA.

Ramesh, C.R., S. Chakraborty, P.S. Pathak, N. Biradar and P. Bhat (2005) 'Stylo in India – Much More Than a Plant for the Revegetation of Wasteland', in F.P. O'Mara, R.J. Wilkins, L. t'Mannetje, D.K. Lovett, P.A.M. Rogers and T.M. Boland (eds), *Twentieth International Grassland Congress*, Wageningen Academic Publishers, Wageningen, The Netherlands.

Ramirez, J., and A. Jarvis (2008) *High Resolution Statistically Downscaled Future Climate Surfaces*, International Centre for Tropical Agriculture, online: http://gisweb.ciat.cgiar.org/ GCMPage (last accessed 11 June 2011).
—— (2010) *Downscaling of Global Circulation Model Outputs: The Delta Method*, Decision and Policy Analysis Working Paper no. 1, International Center for Tropical Agriculture, Colombia.
Ramirez, C., Khoury, A. Jarvis, D.G. Debouck and L. Guarino (2010) *A Gap Analysis Methodology for Collecting Crop Gene Pools: A Case Study with Phaseolus Beans*. PLoS One, 5(10): e13497, doi: 10.1371/journal.pone.0013497.
Reidsma, P., and F. Ewert (2008) 'Regional Farm Diversity Can Reduce Vulnerability of Food Production to Climate Change', *Ecology and Society* 13(1): 38.
Root, T.L., J.T. Price, K.R. Hall, S.H. Schneider, C. Rosenzweig and J.A. Pounds (2003) 'Fingerprints of Global Warming on Wild Animals and Plants', *Nature* 421: 57–60.
Shewayrga, H., D.R. Jordan and I.D. Godwin (2008) 'Genetic Erosion and Changes in Distribution of Sorghum (Sorghum bicolor L. (Moench)) Landraces in North-Eastern Ethiopia', *Plant Genetic Resources* 6: 1–10.
Thomas, C.D., A. Cameron, R.E. Green, M. Bakkenes, M. Beaumont and Y.C. Collingham (2004) 'Extinction Risk from Climate Change', *Nature* 427: 145–48.
Thornton, P., M. Herrero, A. Freeman, O. Mwai, E. Rege, P. Jones and J. McDermott (2006a) *Vulnerability, Climate Change and Livestock: Research Opportunities and Challenges for Poverty Alleviation*, International Livestock Research Institute, Nairobi, Kenya.
Thornton, P., P.G. Jones, T. Owiyo, R.L. Kruska, M. Herrero, P. Kristjanson, A. Notenbaert, N. Bekele and A. Omolo (2006b) *Mapping Climate Vulnerability and Poverty in Africa*, International Livestock Research Institute, Nairobi, Kenya.
United Nations (2005) *World Population Prospects: The 2004 Revision*, United Nations, Rome.
Van Heerwaarden, J., J. Hellin, R.F. Visser and F.A. van Eeuwijk (2009) 'Estimating Maize Genetic Erosion in Modernized Smallholder Agriculture', *Theoretical Applied Genetics* 119: 875–88.
Vavilov, N.I. (1926) 'Centres of Origin of Cultivated Plants', *Bulletin of Applied Botany and Plant Breeding* (Leningrad) 16: 139–248.
Visser, B., D. Eaton, N. Louwaars and J. Engels (2003) 'Transaction Costs of Germplasm Exchange under Bilateral Agreements', in Global Forum on Agricultural Research (GFAR) / International Plant Genetic Research Institute (eds), *Strengthening Partnerships in Agricultural Research for Development in the Context of Globalisation* (Proceedings of the GFAR 2000 Conference, 21–23 May 2000, Dresden, Germany), GFAR / International Plant Genetic Research Institute, Rome.
Williams, J.W., S.T. Jackson and J.E. Kutzbach (2007) 'Projected Distributions of Novel and Disappearing Climates by 2100 AD', *Proceedings of the National Academy of Sciences* 104(14): 5738–42.
World Bank (2008) *World Development Report 2008: Agriculture for Development*, World Bank, Washington, DC.
You, L., and S. Wood (2006) 'An Entropy Approach to Spatial Disaggregation of Agricultural Production', *Agricultural Systems* 90(1–3): 329–47.
You, L., S. Wood and U. Wood-Sichra (2006) 'Generating Global Crop Distribution Maps: From Census to Grid', Selected paper at International Atomic Energy Agency Conference, Brisbane, Australia.
Zeven, A.C., and J.M.J. De Wet (1982) *Dictionary of Cultivated Plants and Their Regions of Diversity Excluding Most Ornamentals, Forest Trees and Lower Plants*, 2nd edition, Pudoc, Wageningen, The Netherlands.

5

CHANGING RATES OF ACQUISITION OF PLANT GENETIC RESOURCES BY INTERNATIONAL GENE BANKS

Setting the scene to monitor an impact of the International Treaty

M. Halewood, R. Sood, R. Sackville Hamilton, A. Amri, I. Van den Houwe, N. Roux, D. Dumet, J. Hanson, H.D. Upadhyaya, A. Jorge and D. Tay

Introduction

The international crop and forage collections conserved and hosted by the International Agricultural Research Centres (IARCs), which are supported by the Consultative Group on International Agricultural Research (CGIAR), have long played a key supportive role in national and international public agricultural research and plant breeding. The genetic resources in these collections are an important source of diversity for plant breeders and farmers to develop crops and forages with the ability to resist pests and diseases, withstand climactic stresses and grow in degraded soils. It is widely predicted that access to such diversity will be increasingly important as countries struggle to meet the challenges of climate change (FAO, 2011; Fujisaka et al., 2009; Ramirez-Villegas et al., 2010). The international collections conserved and hosted by the CGIAR centres have been built up over decades. They currently include 746,611 accessions of crops and forages collected from over 100 countries (see the CGIAR's SINGER database from 2011). Most countries in the world do not have the resources to assemble and maintain collections of similar size and diversity, and therefore they need to rely on access to these international collections. Apart from transfers of genetic materials between the centres themselves, approximately 81 per cent of the materials distributed from the international collections have gone to developing countries and countries with economies in transition, and 19 per cent have gone to developed countries (CGIAR, 2006). The vast majority of these transfers – approximately 94 per cent – have gone to public research organizations, universities, regional organizations, germplasm networks and gene banks (SGRP, 2011).

It is widely recognized that there are increasing threats of erosion to crop, forage and agroforestry plant diversity in *in situ* conditions, including wild relatives of domesticated crops (FAO, 2011). On the one hand, it is critically important to support the *in situ* conservation of this material. On the other hand, it is equally important that potentially useful materials that cannot be preserved *in situ* are collected and conserved in *ex situ* collections. In the absence of such interventions, it is possible that a significant range of crop genetic diversity will become extinct.

Despite the value of the international crop and forage collections conserved and hosted by the CGIAR centres and the threats to the existence of a range of potentially useful diversity 'in the field', there has been a significant decline in the last 15 years in the centres' ability to acquire and conserve additional plant genetic resources for food and agriculture (PGRFA).[1] The managers of these collections, who were surveyed first in 2006 and many of them again in 2011, cite a range of contributing factors, including the lack of funds for collecting missions, a backlog in the characterization and evaluation of materials they already hold and the fact that, as the collections grow larger, it becomes more challenging to ensure that new deposits/acquisitions actually add new genetic diversity to the collection. However, the gene bank managers most consistently attribute the lower rates of new deposits during this period to the highly politicized nature of access and benefit-sharing issues at the international, national and local levels, combined with low levels of legal certainty. More particularly, they attribute their decreasing ability to acquire new materials to: (1) wide-spread distrust among many groups of actors engaged in the use and/or conservation of genetic resources; (2) uncertainty on the part of would-be suppliers (and the centres as recipients) about who can authorize transfers of materials and under what conditions; and (3) complex application and decision-making procedures in the relatively few countries that have access and benefit-sharing laws in place. These factors, in combination, have contributed to reluctance on the part of numerous countries to deposit new materials in the international collections hosted by the CGIAR centres. It has also contributed to reluctance on the part of the centres to attempt to initiate new collecting missions or request copies of materials already in other *ex situ* collections.[2]

The analysis in this chapter is divided conceptually into two time periods. The first time period is from the mid-1990s, when the rates of new deposits to the gene banks began to decline, to the end of 2004, the year that the International Treaty on Plant Genetic Resources for Food and Agriculture (ITPGRFA) came into force.[3] On the one hand, this terminal date seems appropriate for the first period of examination because the multilateral system of access and benefit sharing (multilateral system) created by the ITPGRFA was designed to address many of the political controversies and legal uncertainties that have affected countries' willingness to deposit new genetic resources in the centre-hosted international collections. One indicator of the ITPGRFA's success should be that countries make more previously unavailable materials available for international distribution, either by way of international gene banks or from their own collections. On the other hand, as we highlight later in this chapter, it is clearly not fair to expect the Treaty and the multilateral system to have had a significant impact immediately after it came into force. A number of things were necessary before it could begin to function, such as the adoption of the Standard Material Transfer Agreement (SMTA), which took two years to achieve.[4] Nonetheless, our first exercise in data collection covers the period up to 2004, and this period coincides with the date of the entry into force of the ITPGRFA – those two factors define the first time period under examination.

The second period of time examined in this chapter covers the first six years that elapsed after the ITPGRFA came into force, to the end of 2010. While, on the one hand, it is clear that the Treaty has not created, thus far, a sea-change in most actors' willingness to make new materials internationally available through the CGIAR centres' gene banks, there are some encouraging signs that the situation is improving. In 2010, for the first time since the mid-1990s, there was a significant increase in materials sent to the centres from developing countries to be included in the international collections. This chapter will examine the factors contributing to these phenomena, including the ITPGRFA and an international programme co-ordinated by the Global Crop Diversity Trust (GCDT) to support the regeneration and safety back up of at-risk materials in collections in numerous countries.

The last section of the chapter addresses the question: in the absence of the impediments highlighted in this chapter, what kinds of material would the gene banks try to obtain? It then presents a synthesis of the gene bank managers' accounts of what materials should be collected on a priority basis and made internationally available, either directly by countries or through the mechanism of the collections hosted by the CGIAR centres.

Methodology

The data for the acquisitions and distributions of PGRFA, from 1980 to 2004, for each of the international crop and forage collections hosted by the CGIAR centres was acquired from the System-Wide Information Network on Genetic Resources (SINGER) and the CGIAR centres individually. The centres' gene bank managers were also asked to provide additional accession-level data that is not available on SINGER concerning the sources of the materials they acquired and the recipients to whom they have distributed materials as well as to complete a survey/questionnaire addressing the causes of significant shifts in the rates of acquisition or the distribution identified. The generic questionnaire is included in Appendix 2.

In light of the fact that the responses to the questionnaire indicated that policy- and law-related problems were the most significant variables affecting the decreasing rates of acquisition, it was decided that the study should be deepened by collecting additional information from the CGIAR centres concerning (1) their experiences and perceptions of political and legal obstacles to their acquisition of new materials; (2) the potential future effect of those obstacles on their ability to acquire priority materials; and (3) the criteria that the managers would use to decide what outstanding diversity should be collected. Accordingly, semi-structured follow-up interviews were conducted with the questionnaire respondents. A list of respondents is included in Appendix 3. A paper summarizing the main findings was circulated back to the respondents for comment and subsequently revised.

The analysis of the initial impact of the ITPGRFA, which is presented later in this chapter, draws on data supplied directly from the CGIAR centres for the years 2005–10

(SGRP, 2007, 2009, 2011). These reports include data from each centre's gene bank and breeding programmes concerning the acquisition and distribution of PGRFA using the ITPGRFA's SMTA from 2007 to 2009 inclusive.

A second source of information concerning the post-Treaty period was the respondents to the questionnaire referred to earlier, who were re-contacted in 2011 and asked to comment on their perceptions of how things had evolved (or not) in the six years since they responded to the initial questionnaire. In those cases where the original respondent was no longer the manager of a gene bank, the current gene bank manager, as of March 2011, was asked to provide feedback. A third source of information was the GCDT, which provided information about the materials safety duplicated in international collections hosted by a CGIAR centre with support from its regeneration project during the years 2008, 2009 and 2010.

For the most part, the data provided by the CGIAR centres covering the period up to 2006 refer to the dates when samples previously received by gene banks were actually registered as accessions conserved in the gene bank and made available for distribution (see Tables 5.1 and 5.2 and Figure 5.1). These data also generally refer to accessions – not to the number of samples received. This distinction is potentially important because not

TABLE 5.1 Distributions and acquisitions by all CGIAR centres' gene banks

Year	Samples distributed	Accessions acquired
1980	44,668	16,984
1981	42,165	17,091
1982	40,494	11,497
1983	47,455	17,055
1984	49,691	37,720
1985	65,014	26,414
1986	93,211	20,679
1987	115,597	14,539
1988	143,479	13,024
1989	121,547	17,220
1990	114,309	10,294
1991	120,926	23,932
1992	108,090	13,173
1993	121,585	8,004
1994	118,527	30,029
1995	83,915	10,460
1996	112,055	11,514
1997	82,278	17,884
1998	97,896	14,978
1999	74,115	7,566
2000	63,390	11,742
2001	75,237	8,840
2002	95,611	7,832
2003	81,449	7,335
2004	90,594	5,037

Total Accessions Acquired-All Centers

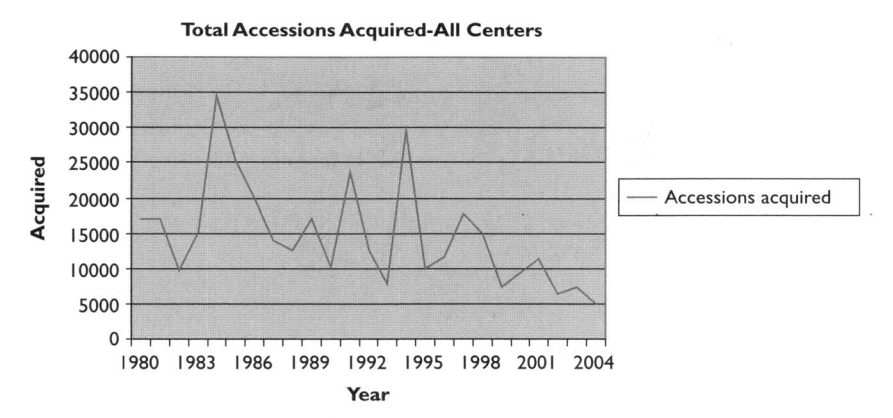

FIGURE 5.1 Total accessions acquired by all CGIAR centres

all samples received by a gene bank necessarily end up being included as an accession, and, if they do, there is sometimes a long lag time between receipt and registration by a centre, sometimes years. A sample received may be rejected in quarantine; it may be rejected because it was found after examination to duplicate other accessions; it may require one or more generations of seed multiplication to produce enough healthy seed to qualify for registration; it may require experimental determination of dormancy-breaking needs or special techniques such as embryo rescue in order to generate plants from which healthy fresh seed can be harvested; or the seed may be in such poor condition that no plants can be grown. For all of these reasons, not all samples received necessarily become accessions. The International Rice Research Institute (IRRI), for example, reports that, on average, 75 per cent of samples received between 1974 and 2010 became accessions. Alternatively, the International Center for Agricultural Research in the Dry Areas (ICARDA), the International Center for Tropical Agriculture (CIAT), the International Livestock Research Institute (ILRI) and Bioversity International report that virtually 100 per cent of materials received are eventually accessioned.

Data provided for 2007 onwards in this chapter (as summarized in Table 5.3) refer to materials transferred to the CGIAR centres using the SMTA under the ITPGRFA. The data refer to the years that such materials were actually transferred to the centre concerned, regardless of if or when those materials were subsequently registered as accessions. As a result, comparisons between data up to 2006 and from 2007 onward must be treated with some caution because of the potential difference between (1) the receipt of a sample by a gene bank and (2) the registration of that sample as an accession conserved in the gene bank and available for distribution, with the result that the numbers reported for some centres after 2006 will be somewhat higher, as they would also include whatever materials the centre received but which did not eventually become accessions. Unfortunately, for the purposes of this study, it was not possible to harmonize data for the two periods up to the end of 2006 and beginning in January 2007.

This study does not include consideration of the rates of acquisition and distribution by the CGIAR centres' breeding programmes. The breeding programmes actually acquire and distribute approximately four times more samples of PGRFA every year than the gene banks (SGRP, 2011). However, there are two reasons for not including the breeding programmes in the scope of this study. First, before they started making reports to the Governing Body of the ITPGRFA, the centres did not collectively pool all of their data across the centres concerning breeding programmes' acquisition and distributions of PGRFA. Thus, there are no ready-to-analyse data sets for these breeding programmes that are comparable to those that exist for the gene banks. Second, the gene banks' experiences are, arguably, a clearer indicator of the state of countries' willingness to make new materials internationally available and, by extension, of the impact of the ITPGRFA. This is the case partly because of the nature of the materials that the CGIAR gene banks hold and seek to obtain and partly because of the way in which the gene banks use these materials. The gene banks make long-term commitments to conserving PGRFA in the form in which they are deposited. The gene banks are interested in conserving the overall inter-specific and intra-specific diversity of the gene pools of crops and forages that make up their collections. As a result, the gene banks are generally much more interested (compared to the breeders, for instance) in collecting or receiving materials that have developed '*in situ*' – that is, farmers' varieties and forages that have been developed over long periods of time in local conditions as well as wild relatives of crops. In most countries, these kinds of material are more closely associated with national patrimony than are the improved lines developed by formal sector breeders. It is within the mandate of the gene banks of the CGIAR centres to facilitate international distribution of these conserved materials. Depositors of materials to the CGIAR gene banks know that, through the gene bank, they are making their material available to the world. The mandate of the CGIAR centres' breeding programmes, on the other hand, is to develop new, improved germplasm and to facilitate the diffusion of improved material to end users. Generally, CGIAR breeders exchange materials that are already under development through networks of other breeders, and they are generally not seeking access to *in situ* landraces or wild relatives.

Our survey efforts did not extend beyond the CGIAR gene banks. Future efforts to monitor the impact of the multilateral system will need to consider the experiences of a broader range of users. We hope that this study will be a useful starting place for such studies.

Rates of CGIAR gene banks' distribution and acquisition up to 2004: the overall picture

The international *ex situ* collections of crops and forages hosted by 11 CGIAR centres were formally initiated and strengthened at different times throughout the 1960s and 1980s. The rates of acquisition and distribution of materials by the collections varies considerably from year to year. Between 1980 and 2004, the centres' gene banks distributed approximately 2.2 million samples of PGRFA and acquired approximately 370,000 accessions.

The numbers of samples distributed was highest from the mid-1980s to the mid-1990s. Thereafter, the number of samples distributed dropped somewhat but remained substantially higher than in the early 1980s. Since the mid-1990s, the rate of annual distribution has remained relatively constant. The surveyed gene bank managers attribute this relatively small decrease since the mid-1990s in the number of samples distributed to the fact that responses to requests from the gene banks are more targeted than they used to be since more characterization and evaluation data are being collected, with more emphasis on determining which materials would be particularly well-suited for the requestor. Furthermore, some requests for materials are also more targeted than they used to be.

The rate of new acquisition of materials by the same gene banks, considered all together, was highest and steadiest from year to year throughout the 1980s. Rates of acquisition overall started to fluctuate dramatically in 1993. In the three years from 2002 to 2004, the number of new acquisitions dropped each successive year to levels that had not been recorded since the CGIAR gene banks were created. The figures in Table 5.1 and Figure 5.1 present data for all of the centres' acquisition of germplasm (and registration as accessions) for all crops conserved in the international collections. Appendix 1 provides a centre-by-centre breakdown of the same data. In this context, it is important to note that the acquisition data referred to here, and throughout this chapter, refer to PGRFA that the centres are free to distribute internationally. Details about the agreements pursuant to which the CGIAR centres are able to make these international distributions are included in Box 5.1.

In summary, a highly charged political environment, countries' unwillingness to share their materials and restrictive laws and policies are the reasons most often cited by the CGIAR centres' gene bank managers for decreased rates of acquisition of additional materials to conserve in, and distribute from, their gene banks. It is not surprising, therefore, that some of the largest 'batches' of genetic resources acquired by the centres' gene banks since the mid-1990s have come from sources, or involved materials, that are less affected by political and legal uncertainties and tensions. Many of the new acquisitions have come from other international gene banks or national gene banks situated in developed countries. Or they are materials that were originally collected and deposited in collections from which they were made internationally available before 1993. In other words, much of the PGRFA that the centres' gene banks have received since the mid-1990s has not come from new collecting missions or directly from *ex situ* collections in developing countries. For example, the approximately 5,500 forages received by the ILRI in 2000, and registered as 'accessions' from 2000 to 2002 – the single largest acquisition in the gene bank's history – came from Australia's Commonwealth Scientific Industrial Research Organization (CSIRO). These accessions had been collected prior to 1993 and were received under a germplasm acquisition agreement that allowed them to be treated and distributed as in-trust materials. They were subsequently designated as such. The approximately 4,000 accessions of tropical forages acquired by the CIAT in 2003 also came from the CSIRO. In 2004–5, the International Crop Research Institute for the Semi-Arid Tropics (ICRISAT) received 2,000 accessions of chickpeas and 800 accessions of sorghum – both the largest annual acquisitions in over 15 years – from the US Department of Agriculture (USDA). It also received another 700 accessions of chickpeas from ICARDA (another CGIAR centre). The record high number of wheat accessions acquired by the International

Centre for Maize and Wheat Improvement (CIMMYT) since 1980 – approximately 16,000 – came from the USDA in 1994. The CIMMYT places, on average, 2,000 accessions of breeding lines developed by its own wheat-breeding programme in the CIMMYT gene bank every year. The CIMMYT acquired approximately 10,000 accessions of maize between 1994 and 2001 through the operation of the USDA-funded maize regeneration programme, in partnership with the United States, the CIMMYT and national partners. These accessions were originally collected in missions supported by the International Board for Plant Genetic Resources (IBPGR) in the 1970s and 1980s. Duplicates were sent for storage in the United States because at that time the CIMMYT did not have space in its gene bank. Since then, through the regeneration programme, the countries concerned and the CIMMYT, supported by the USDA, have been regenerating the seeds. When they are regenerated, copies are sent to the CIMMYT to hold as 'designated' germplasm in its gene bank.

If the PGRFA acquired from these sources (mostly international gene banks and developed countries) were not included in the data presented in Table 5.1 and Figure 5.1, the centres' collective rate of acquisition of materials that they can make globally, publically available would drop even more precipitously.[5] The fact that CGIAR centres have acquired these materials from sources that are less affected by the politicization of genetic resource issues and related legal uncertainties does not detract from their value as additions to the international crop and forage collections. However, this information does underscore the challenges that the centres are facing *vis-à-vis* a range of alternative sources in obtaining materials to add to the international collections.

Box 5.1 From the 1994 in-trust agreements with the Food and Agriculture Organization (FAO) to the 2006 agreements with the Governing Body of the ITPGRFA

In 1994, 12 CGIAR centres signed agreements with the UN FAO, undertaking to hold 'designated germplasm in trust for the benefit of the international community' and to 'make samples of the designated germplasm and related information available directly to users or through FAO, for the purpose of scientific research, plant breeding or genetic resources conservation, without restriction' (SGRP, 2010). The centres recognized the overall policy guidance of the Commission on Genetic Resources for Food and Agriculture (CGRFA) *vis-à-vis* the management of the in-trust materials. The centres 'designated' materials to which they had a long-term conservation commitment and that were not subject to any restrictions on distributions. In 1998, the centres adopted a material transfer agreement (MTA) to be used when distributing designated in-trust germplasm. The MTA did not include mandatory monetary benefit-sharing provisions. The in-trust agreements were created on the understanding that they were temporary and would be replaced when the negotiations that led to the ITPGRFA were finished.

In 2006, the 11 centres with *ex situ* crop and forage collections signed agreements with the Governing Body of the Treaty, placing the 'in-trust' materials under the Treaty's framework, to be made available using the SMTA (which includes mandatory benefit-sharing provisions). Pursuant to these agreements, the centres recognized the authority of the Governing Body of the ITPGRFA to provide overall policy guidance relating to the collections, subject to the provisions of the Treaty. They also agreed to distribute 'in-trust' Annex 1 materials using the SMTA and to use a MTA amended by the Governing Body during its second session in 2007 when distributing non-Annex 1 materials. The Governing Body decided that the centres should use the same SMTA when distributing non-Annex 1 materials. The SMTA includes a mandatory monetary benefit-sharing provision.

The content of the international collections hosted by the CGIAR centres is not static – the centres add new materials to these collections. As a rule, the centres only accept or add new materials to the collection when they constitute previously unrepresented diversity in the collection; when the centres think it is worth conserving that diversity over time and when the centres have the right to include the material in the collection and to distribute it using the SMTA.[1] Appendix 1 includes a breakdown of the numbers of accessions in the international collections hosted by the CGIAR centres.

Note 1: The CGIAR gene banks may, on occasion, accept deposits of non-multilateral system materials that the depositor does not permit the gene bank to redistribute. While the CGIAR gene banks generally prefer not to accept materials under these conditions, they will do so if it seems likely that important genetic diversity will otherwise be lost forever.

Centres' experiences of political and legal obstacles in acquiring new materials

Challenges faced

The kinds of political and legal obstacles that the CGIAR centres have encountered in their attempts to acquire new materials are numerous, and the situations in which they have arisen are wide in scope. Many of these situations involve combined elements of mistrust on geopolitical levels; intentions or promises to share materials on the part of technical-level partners that have been thwarted by political higher-ups; insecurity on the part of national partners because of unclear lines of authority and fear of being accused, *post-facto*, of selling out the country's patrimony; insecurities on the part of the centres to even ask, given the vagaries of national procedures and the possibilities of backlashes and a considerable degree of uncertainty throughout the entire system.

This section attempts to cluster together the most frequently repeated scenarios that have been related by the gene bank managers between the mid-1990s and 2004. It is not possible at this time to quantify how many times these situations have arisen or which scenarios are the most common across the 11 relevant centres currently hosting international collections.

- CGIAR centres have undertaken joint collecting missions with national partners on the understanding that duplicates of the materials collected would eventually be sent to the centre (after being registered in the national collection), 'designated' and then made internationally available by the centre. In one case, a centre waited for over ten years for the materials to be sent; in another case, seven years. Despite sending numerous reminders, the materials never arrived. In many cases (possibly all), the agreements to send the materials to the centres for 'designation' was explicit in the contracts between the centres and the national partners. The centres concerned have not brought, or threatened to bring, legal actions for breach of contract. As far as the authors are aware, there are no instances of the donors who supported the original collecting missions intervening to speed up the delivery of the duplicates.

- CGIAR centres have made direct requests to national gene banks for materials. These requests were rarely explicitly or directly rejected. More often, the requesting centre experienced extremely long delays in responses or received no responses at all. Sometimes the responses were vague and non-committal. Often, requests originated informally in the context of a conversation between a representative of a centre's gene bank and a national partner, at a conference or through joint research.
- Sometimes it is informally communicated to the centre's gene bank representative that the political climate is such that it would not be possible to transfer materials at that time. Alternatively, the national representative may be enthusiastic to share material, but it becomes evident later on when the material is ultimately not sent that higher-level political authorities have become involved.
- Country representatives have made public statements that CGIAR centres would not be allowed to collect or otherwise acquire materials from their countries.
- In the mid-1990s, national agricultural research organization (NARO) partners in the International Network for the Genetic Evaluation of Rice (INGER) stated that they would only continue to contribute materials through the network on the condition that they were not included in the IRRI gene bank for international distribution under the centre's MTA pursuant to the FAO-CGIAR agreements. Instead, the material could only be circulated to other members of the network using the network's own MTA (even so, the NARS contributions to INGER declined to near zero because of the absence of benefit-sharing provisions in the MTA used).
- Countries have made materials available for use in a centre's breeding programme or for research purposes only on the condition that they are not included in the same centre's gene bank as 'designated' germplasm to be distributed internationally.
- National authorities have allowed CGIAR centres to participate in collecting missions in their own countries but have specified that duplicates of the collected material can only be held in the relevant CGIAR centre's gene bank if it was not made publicly available. In other similar situations, it was agreed that the CGIAR centre would not receive any duplicates and that all collected materials would remain in the country concerned.
- CGIAR centres have come into receipt of materials collected after 1993 for which the paper work was not clear or was non-existent. When the centre contacted the authorities of the countries concerned to request permission to designate the

materials, they were asked not to designate the material and to send it back to the country concerned.
- National authorities have agreed to allow CGIAR centres' staff to conduct research within their borders, but they have not allowed the concerned material to be taken out of the country for the purposes of that research.
- National technical-level partners have said they would like to share materials, but they do not know how to get permission given the political/legal situation in their country.
- National representatives have delayed authorization for joint collecting missions, citing the need to develop national policies and legislations to implement the ITPGRFA and/or the need to have more clarity concerning the kinds of benefit that would flow as a result of membership in the Treaty.
- Countries have delayed authorization until the best time for collecting is past (that is, they have authorized it only after physiological maturity of the targeted species is past).
- Even when a country's materials have been available on a restricted basis to other members of the research consortia, these conditions have not proven to be restrictive enough for some national partners. Consequently, the national partners in the international research consortia have not been able to get permission from their own national competent authorities to make materials from their countries available as part of the joint research projects.
- There have been a few high profile cases of allegations of recipients not complying with the terms of the MTA used by the CGIAR centres pursuant to their 1994 in-trust agreements with the FAO (RAFI, 1998).

It is important to bear in mind that while rates of acquisition from developing countries have slowed down considerably, they have not stopped completely. Despite this general account of declining rates of deposits, a number of developing countries have made material available for global distribution through the CGIAR centres' gene banks during the period analysed.[6]

Causal forces

In the responses of the gene bank managers, the event most closely associated with the drop in rates of acquisition was the coming into force of the Convention on Biological Diversity (CBD) in 1993.[7] Some country representatives have cited the CBD as providing a pretext for refusing to provide access to materials within their borders. Of course, countries have always had the right to refuse such access. These responses are therefore better understood as evidence of the social function that the CBD has served to simultaneously raise awareness about the value of genetic resources and to cast the issue of 'access' to them in the highly political/legal context of exercises of national sovereignty. Uncertainty about how the CBD was being interpreted and implemented by would-be access-providing countries was another issue raised by the gene bank managers. Vagaries about who could authorize the provision of material to the gene bank and under what circumstances appears to account for a number of thwarted efforts to provide/obtain materials.[8]

Another potentially contributing factor was that some countries did not feel entirely satisfied with the 1994 FAO-CGIAR in-trust agreements, at least as an interim measure pending the negotiation of the ITPGRFA. On the one hand, the development and conclusion of these agreements were greeted with considerable enthusiasm by both developing and developed countries in the Commission on Plant Genetic Resources (CPGR, 1993, 1994) and the Intergovernmental Committee on the CBD (UNEP, 1994). On the other hand, in the years that followed, the CGIAR centres' staff have occasionally heard reports that some countries were dissatisfied with the process leading to the formalization of those agreements. Even more frequently, they have heard complaints that the in-trust agreements were partly deficient because there were no geographical restrictions on where the materials could be sent and because there were no mandatory benefit-sharing provisions in the MTA that the centres used to distribute materials under the in-trust agreements. There were also a few relatively high profile cases of alleged misuse by recipients of in-trust materials obtained from the centres. As stated earlier, shortly after the in-trust agreements were signed, the INGER network opted for using its own MTA with a limited number of countries. In this context, it is relevant that some respondents noted that after 2001 when the Treaty negotiations were completed, it became even more difficult to obtain non-Annex 1 crops and forages. They attributed the increased difficulty to the fact that being left out of the multilateral system had the effect of highlighting the political sensitivity of providing access to such materials.

Centres' reactions

The responses of the CGIAR centre gene banks to these situations varied considerably. Some gene banks lapsed into the practice relatively quickly after the CBD came into force – and in some centres it developed into an unofficial policy – of ceasing to make overtures to countries requesting materials, either through new collecting missions or through transfers of duplicates of *ex situ* materials already in national collections.[9] They assumed, in the mid-1990s, that it would be best to wait until countries clarified their own positions and developed their own request-considering procedures through national laws. However, they also assumed that this process would take just a few years. They had no idea that legal uncertainties and political sensitivities would continue to characterize the atmosphere in which they worked up until the present time. One centre's gene bank, in anticipation of the relief that the ITPGRFA would provide once it came into force and was made operational, curtailed its efforts to obtain new materials more recently, in approximately 2001. In some cases, the unfortunate result was a lack of communication between the CGIAR's gene bank and would-be national partners/counterparts for a number of years.

Most CGIAR centres continued to seek materials from other international gene banks as well as national gene banks with strong track records of sharing materials, as highlighted earlier. Some centres participated in collecting activities under conditions that forbade them from keeping duplicates in their gene banks and/or designating and making them publicly available. Other centres have declined such opportunities based on the principle that there is no point collecting and conserving material if it cannot be made internationally, publicly available.

Some light at the end of the tunnel: a preliminary analysis of the impact of the Treaty's multilateral system

The final text of the ITPGRFA was adopted by the FAO Council in 2001, and it came into force in 2004. However, the multilateral system could not start to function until 2006 when the Governing Body adopted the SMTA, which is used for every transfer of material in this system. Other important elements of the multilateral system have taken still longer to finalize, such as the procedural rules for the third party beneficiary to monitor transfers and bring legal actions in cases of suspected violations of the SMTA or the Funding Strategy (both adopted in 2009), or the global information system, which still does not exist. The schedule for reporting transfers under Article 5(e) of the SMTA was not established by the Governing Body until 2009, and the reporting system is currently in the process of being established by the Secretariat. Therefore, the majority of transfers within the multilateral system are still largely unreported. Most of the information about what is being transferred, and where, arrives via voluntarily submitted reports from the CGIAR centres. Very importantly, Article 12 (3)(h) of the ITPGRFA states that if countries do not have legislation in place to make *in situ* materials available, 'Contracting Parties agree that access found in *in situ* conditions will be provided … in accordance with such standards as may be set by the Governing Body.' As already noted, to the best of the authors' knowledge, no countries have specially designed legislation in place. The impact of this legal lacuna has been multiplied by the fact that, to date, there have been no efforts by the Governing Body to develop the default standards anticipated in Article 12(3)(h). While the multilateral system could at least start to formally operate without final agreement by the Governing Body about these latter elements, the absence of these elements has clearly been a disincentive for many would-be participants in the multilateral system (see Chapter 11 by Isabel López Noriega, Peterson Wambagu and Alejandro Mejías in this volume; Chapter 12 by Godfrey Mwila in this volume and Chapter 14 by Bert Visser in this volume).

Similarly, there have been delays in the ability of the CGIAR centres (and other international organizations) to participate in the multilateral system. Article 15 of the ITPGRFA 'calls upon' the CGIAR centres to sign agreements with the Governing Body of the Treaty, placing the in-trust materials under the Treaty's framework. However, the text of the generic agreements had to be approved by the ITPGRFA's Governing Body, which met for the first time in 2006. Thereafter, the boards of trustees of each of the CGIAR centres had to approve the agreements. Following this decision, it took an additional six months until those agreements came into force, with the effect, reported earlier, that the centres' international collections were not operating under the Treaty's framework until January 2007. (Meanwhile, the collections continued to be managed pursuant to the 1994 in-trust agreements.)

Perhaps most importantly for the purposes of this chapter, it is continuing to take a long time for member states to put in place the requisite mechanisms to be able to provide materials to and through the multilateral system using the SMTA. As of January 2011, only 22 countries had published on the official ITPGRFA website lists of the *ex situ* collections within their borders that are part of the multilateral system.[10] And, so far, no country has shared information with the Governing Body about what regulations or administrative rules apply for processing requests for *in situ* multilateral system materials within their borders. Many countries have reported on the need for additional time to be able to address technical challenges associated with developing such systems. However, as Godfrey Mwila (in Chapter 12 of this volume) and Isabel López, Peterson Wambagu and Alejandro Mejías (in Chapter 11 of this volume) write, other countries appear to be holding back implementing the multilateral system until other benefits under the ITPGRFA – such as support for national level capacity building and sustainable use – are manifest.

All of these factors have contributed to the fact that it has taken several years for the ITPGRFA's multilateral system to start to be operational and that it will likely take several more years before it is operating at anything like its full potential. An assessment of the impact of the ITPGRFA to date with respect to countries' willingness to make new materials internationally available – an issue that lies at the very core of the operation of the multilateral system – has to be approached cautiously, with this caveat in mind.

Despite this caveat, the data on CGIAR centres' experiences operating under the ITPGRFA's framework provides some room for optimism. With the imminent coming into force of their own agreements with the Governing Body, the CGIAR centres contacted some countries where there was uncertainty as to whether the centre could make material internationally available pursuant to the 1994 in-trust agreements. In some of these cases, the countries responded that they would be more willing to consider allowing the centre to make the material internationally available in light of the fact that they would soon be using the SMTA, with its mandatory benefit-sharing provisions. As a result, in some instances, the competent authorities directed the centres to make previously jointly collected, but restricted access, materials publicly available using the SMTA. For example, Indonesia asked the International Potato Center (CIP) to designate as 'in trust' sweet potato accessions that were originally collected pursuant to a mid-1990s agreement that prohibited duplicates to be held by the CIP.

From 1996 to 2000, the IRRI had collected landraces and wild rice from 22 countries, with approximately 24,000 accessions being held in the gene bank. These accessions were held on restrictive conditions, which did not allow the IRRI to further distribute this material. After the ITPGRFA came into force, over the course of 2005 and 2006, national authorities from most of the relevant countries authorized the IRRI to make the materials internationally available using the SMTA. Only two (non-party) countries have refused permission for all of their material, and one country has selectively refused permission for a subset of its materials.

After the Governing Body adopted the SMTA in 2006, and after the IRRI signed its agreement with the Governing Body, the INGER network, welcoming the benefit-sharing provisions of the SMTA, decided to reverse its decision from the mid-1990s to stop depositing countries' materials in the IRRI gene bank. Since 2006, INGER

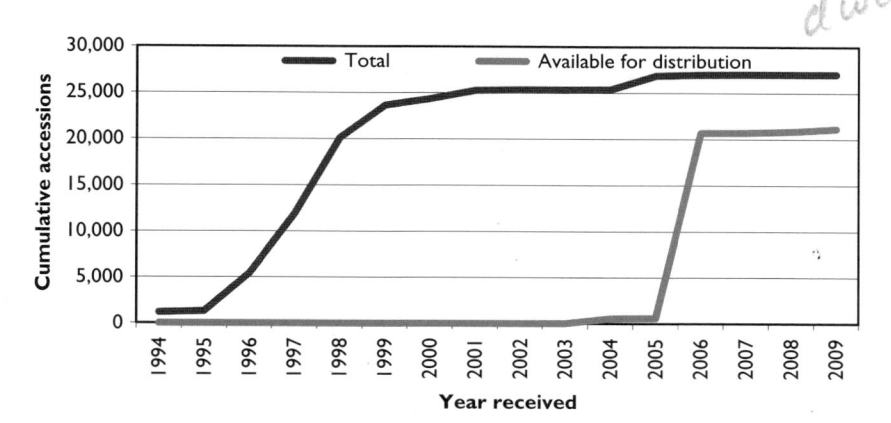

FIGURE 5.2 Acquisitions by the International Rice Collection hosted by the IRRI, 1994–2009

countries have started making contributions of national germplasm to the network again, allowing the IRRI to deposit such material in the gene bank and to distribute it to countries that are not actually in the network, using the SMTA.

The GCDT is supporting collection holders around the world to regenerate unique Annex 1 crop and forage germplasm that is at risk of being lost. As one of the conditions of support, the organizations involved have agreed to send sub-samples of the regenerated materials to gene banks in other countries for safety back-up. They are also giving permission to the gene banks holding those back-up duplicates to distribute them internationally, using the SMTA. Not all of the countries that are participating in the project are actually ITPGRFA members, but they are agreeing, nonetheless, that the regenerated, backed-up material should be distributed using the SMTA. (Five of the 20 countries that had deposited regenerated materials with the CGIAR centres by the end of 2010 under the programme were not Treaty members.) The centres' gene banks have been selected as the safety back-up location by most of the participating organizations that are regenerating materials under the project. The first such regenerated materials – 91 accessions – were deposited by the national gene bank of Azerbaijan in the CIMMYT and ICARDA-hosted international collections in 2008. Another 2,394 accessions were deposited in 2009, in centre-hosted international collections, by organizations from eight countries. In 2010, 12,128 accessions were deposited in centre-hosted international collections by organizations from 12 countries. The total number of deposits will rise still further over the next few years, as approximately 90,000 accessions of unique, at-risk materials are regenerated, backed up, and made internationally available using the SMTA through the GCDT (Jane Toll, personal communication). Thereafter, the programme will come to a close. While the financial support and international stimulation and co-ordination provided by the GCDT have clearly been critical contributing factors, it is also fair to assume that the presence of the ITPGRFA's multilateral system, including the SMTA with an internationally endorsed benefit-sharing formula, has also contributed to countries' willingness to make these materials available (Fowler and Hawtin, 2011).

114 M. Halewood et al.

TABLE 5.2 CGIAR centres' acquisitions, 2005–10

Dates acquired[a]	Number of new acquisitions
2005[b]	4,254
2006	4,499
1 January 2007–31 July 2007	2,098
1 August 2007–31 July 2008	6,358
1 August 2008–31 December 2008	2,202[c]
2009	4,888[d]
2010	12,128 (at least)[e]

Notes:
a Unfortunately, comparative analysis is a little complicated given the time periods covered by the periodic reports from the CGIAR centres to the Governing Body. It was not possible for the purposes of this chapter to obtain the relevant data from each centre for calendar years.
b As stated in the methodology section earlier, 2005 and 2006 data refer to years when accessions were registered in the gene banks. For 2007 onwards, the data refers to years when materials were received by the centres.
c Of the 2,002 accessions that were acquired by the CGIAR centres between August and December 2008, 1,750 were from another CGIAR gene bank. As a result, the number of accessions acquired from sources outside the CGIAR during this period is 252.
d Of these accessions, 1,555 were acquired from another CGIAR gene bank. As a result, the number of accessions acquired from sources outside the CGIAR during this period is 3,333. Of those, 2,394 were deposited by countries supported through the GCTD regeneration programme.
e This figure was provided by the GCDT for materials for which they have records being sent to the CGIAR centres as part of the regeneration programme.

As Table 5.2 shows, these positive developments have not contributed to an overall increase in the total number of acquisitions by the CGIAR centres from 2005–9.[11] In 2010, however, there was a significant upswing in materials being sent to the centres' gene banks, which is almost entirely accounted for by deposits made by organizations regenerating at-risk *ex situ* materials with support from the trust.

In this context, it is important to differentiate between newly deposited materials that come from *ex situ* and *in situ* sources. The GCDT regeneration project is focused on unique *ex situ* collections (that is, materials that have been previously collected – in some cases decades before – that are already in gene banks) that have been at risk of being lost as a result of the stored seeds dying before they were planted to produce new seeds (thereby replenishing the collections with viable seed).

In the meantime, new deposits to the CGIAR gene banks from materials newly collected from *in situ* conditions – from recent collecting missions – are still very rare. Table 5.3 provides details from 2007 to 2010 (inclusive) of countries that (1) gave permission to the CGIAR centres to engage in joint missions to collect materials from *in situ* materials and/or (2) transferred materials from their national collections to the CGIAR gene banks, with the intention, in both cases, of making those materials internationally available. In this context, it is also interesting to note that several countries that are not Treaty members are among those countries that have organized joint collecting missions in recent years.

On the one hand, it was predictable that the earliest impacts of the multilateral system would involve *ex situ* plant genetic resources, particularly those held by national public organizations. Pursuant to Article 11.2 of the ITPGRFA, plant genetic resources that

TABLE 5.3 Materials sent to international collections hosted by CGIAR centres, 2007–2010 (inclusive), with permission to re-distribute using the SMTA

CGIAR gene bank[1]	From new collecting missions Country (genus, species or common name)	Transferred from ex situ collections Country (genus, species or common name)
ICARDA	Syria (Aegilops, barley, wild barley, bread wheat, wild lentils, Medicago, Pisum, Trifolium, Vicia, other forages and ranges)	Azerbaijan (barley) Ukraine (chickpeas, grasspeas and lentils)
	Jordan (Aegilops, barley, wild barley, wheat, wild wheat, chickpea and lentils)	Mongolia (seeds did not pass quarantine)
	Libya (Aegilops, barley, wild barley, wheat, chickpeas, faba beans, Medicago, Trifolium, Vicia, other forages and ranges)	
	Tajikistan (Aegilops, barley, wild barley, wheat, chickpeas, wild cicer, lentil, faba beans, Medicago, Lathyrus, Pisum and Vicia)	
	Georgia (Aegilops, barley, wild barley, wheat, chickpeas, lentils, faba beans, Medicago, Lathyrus, Pisum, Vicia and other forages and ranges)	
IITA	Angola (*Manihot esculenta* – cassava) Guinea (*Manihot esculenta* – cassava) Nigeria (*Vigna* species – wild relative of cowpeas)	Benin (yams)[2] Ghana (cocoyams, yams)[2] Azerbaijan (cowpeas)[2] Togo (yams)
ICRISAT	Kenya (pigeonpeas) Uganda (pigeonpeas) Tanzania (pigeonpeas) Niger (pearl millet)	Mali (sorghum)[2] Senegal (pearl millet)[2] China (small millet) India (small millet) United States (chickpeas, groundnuts, sorghum) Ukraine (chickpeas)
IRRI	None	Madagascar (rice)[2] Pakistan (rice)[2] South Korea (rice)[2]
Africa Rice	Togo (rice) Benin (rice)	Guinea (rice)[2]
ILRI	None	None
Bioversity International	Congo DR (bananas) Tanzania (bananas)	Cameroon (bananas) Cuba (bananas) Finland (bananas) France (bananas) Guadeloupe (bananas) India (bananas) Taiwan (bananas) Uganda (bananas) Vietnam (bananas)

TABLE 5.3 (continued)

CGIAR gene bank[1]	From new collecting missions Country (genus, species or common name)	Transferred from ex situ collections Country (genus, species or common name)
CIAT	Costa Rica (wild species of *Phaseolus* beans)	Indonesia (*Manihot esculenta* – cassava) Myanmar (*Phaseolus beans*) Puerto Rico (*Phaseolus beans*) Dominican Republic (*Phaseolus beans*) Haiti (*Phaseolus beans*) United States (*Phaseolus beans*)
CIMMYT	Mexico (*teosinte, maize*)	Azerbaijan (wheat)[2] Albania (wheat)[2] Ukraine (wheat)[2] Georgia (wheat)[2] Israel (wheat)[2] Peru (maize)[2] Azerbaijan (maize)[2] South Korea (maize)[2] Zambia (maize)[2] Georgia (maize)[2]
CIP	None	Rwanda (sweet potatoes)[2] Costa Rica (wild potatoes) Costa Rica (yams, beans)

Source: CGAR Centres and The Global Crop Diversity Trust.
Notes:
[1] Data from the World Agroforestry Centre and from CIMMYT concerning wheat germplasm acquisitions was not available to include in this table.
[2] Indicates that the material was generated and sent to the centre as part of the GCDT co-ordinated regeneration project.
[a] Of the 42 countries in this table that have transferred materials to the CGIAR centres gene banks in the last four years with the intention of making those materials internationally available through the centres' gene banks using the SMTA, 12 are not Treaty members. Out of the 22 countries reported in this table that have organized missions to collect and provide materials from in situ conditions, six are not Treaty member states.

are in the management and control of contracting parties and in the public domain are automatically included in the multilateral system when a country ratifies the ITPGRFA. The plant genetic resources that are most clearly included in the multilateral system by virtue of this formula are *ex situ* materials in national public collections. It is beyond the scope of this chapter to determine which of the materials that were recently regenerated with support from the GCDT and deposited in the CGIAR centres fit this description, but it would appear, at first blush, that a sizeable proportion probably does.

By comparison, there are considerably less well-defined means of identifying and supplying *in situ* materials that are in the management and control of the contracting parties and in the public domain. Presumably Annex 1 materials that can be found in national government-controlled protected areas and parks are automatically included, but the existence and location of such materials is still unknown in many countries, and additional resources and systems are required to actually make them available.

More to the point, however, is that most *in situ* PGRFA located in Treaty member states – that which can be found in most farmers' fields – are likely not automatically included in the multilateral system. Such *in situ* materials would need to be voluntarily included in conformity with whatever laws apply in the particular country. It is the authors' hope that once actors become more familiar and comfortable with the multilateral system, they will be willing to voluntarily place *in situ* materials in the system. Perhaps the simplest way to do so would be to deposit collected samples from *in situ* conditions into an *ex situ* (national or international) collection that is part of the multilateral system. The ongoing injection of new and evolving diversity from *in situ* conditions into *ex situ* collections (and *vice versa*) is essential to the long-term vitality and viability of globally co-ordinated systems of conservation and use. Incorporating much of the existing (and developing future) *in situ* diversity in the multilateral system is not a legal requirement under the ITPGRFA, but the inclusion of such material is, nonetheless, an important indicator of the ITPGRFA's long-term success in addressing the political and equity concerns and legal uncertainties that have been affecting actors' willingness to make PGRFA internationally available.

In this context, it is interesting to note that the second *Report on the State of the World's Plant Genetic Resources for Food and Agriculture* states that several countries have indicated that they have conducted significant new collecting missions (FAO, 2010). In total, between 1996 and 2007, it is reported that up to 240,000 new accessions were collected and stored in gene banks. On the other hand, so far, most of the countries reported to have been actively collecting during this period have not yet voluntarily sent messages to the secretary of the ITPGRFA clarifying which *ex situ* collections within their borders are within the multilateral system. Nor have most of those countries shared information at the level of the ITPGRFA about transfers they are (or are not) making using the SMTA. So, while there may be significant levels of collecting, there is very little direct evidence that the newly collected material is making its way *de jure* or *de facto* into the multilateral system.

The gene bank managers are encouraged by the evidence of the positive influence of the ITPGRFA, highlighted earlier in this chapter, particularly the high rates of deposits of safety duplicated materials with support from the GCDT. Overall, however, they report that they have not yet witnessed widespread confidence or enthusiasm about the multilateral system from those actors who were hesitant or unwilling to deposit new materials in the centres' gene banks before the Treaty came into force. As a result, it is quite possible that, in the absence of a globally co-ordinated effort such as the regeneration project supported by the GCDT, the rates of new deposits will drop down again in the future.

Some of the respondents feel that until national partners more clearly understand what is actually meant by an 'equitable share of the benefits derived' from the use of PGRFA, they will be reluctant to fully participate in the multilateral system as providers of germplasm. The CGIAR centres' gene bank managers are more generally pessimistic about their collective ability to access additional non-Annex 1 materials – that is, those crops and their wild relatives and forages that are not included in the list of materials under the multilateral system of access and benefit sharing created by the ITPGRFA. The relative importance that the centres place on their work with non-Annex 1 materials varies enormously. None of the species that the World Agroforestry Centre (ICRAF) works with are included in Annex 1. At the other end of the spectrum, the Africa Rice Center (WARDA) is the only centre that reports working exclusively with Annex 1 materials. The remaining nine centres with *ex situ* collections fall in between these two extremes. Some of these non-Annex 1 materials are among the priorities for collection and *ex situ* conservation identified by the centres – for example, Andean roots and tubers, groundnuts, tropical forages and wild relatives of cassava.

Establishing priorities for future acquisitions

All of the CGIAR centres confirm that there is considerably more diversity that needs to be collected and made globally publicly available to meet major global problems of food security, environmental degradation and climate change. They confirm that in the absence of political and legal obstacles, and with sufficient funds, they would be working to acquire significantly more materials. In general, the managers are cautious about estimating the completeness of their collections' 'coverage' of global diversity. They are most cautious, of course, with respect to predicting the coverage of wild relatives, forages and tree species since, in most cases, they do not know what diversity still exists. Furthermore, most of the managers stated that they need considerably more information about what percentage of coverage was held in other organizations' *ex situ* collections. On the other hand, the CIAT gene bank manager states that, with respect to *Phaseolus* beans, the situation is relatively well understood, namely: (1) what the gaps are in the *ex situ* collections (measured against what exists *in situ*); (2) where the 'missing' materials are located *in situ*; and (3) which of those resources are at risk of being extinct soon. Ramirez-Villegas and his colleagues (2010) propose a methodology for analysing gaps in *ex situ* collections of crop wild relatives and summarize the results of their test of the methodology with respect to *Phaseolus*.

For some of the 'global' crops, a few of the CGIAR centres' collections – for example, the IRRI's *O. sativa* collection (over 106,000 accessions) and the CIMMYT's maize, as far as Latin America is concerned (approximately 28,000 accessions) – may perhaps be considered close to complete.[12] However, there is substantially more collecting of wild relatives of both *O. sativa* and maize that needs to be undertaken. For example, the IRRI has accessions of 22 wild species in the genus *Oryza*, most of which have been shown to be genetically more diverse than *O. sativa* itself. To date, the IRRI has only 3,880 accessions for all of these wild relatives. It is worth noting in this context that there is still no agreed metric for determining the percentage of diversity of a crop that may be held in an *ex situ* collection. Among the issues that need to be considered is whether or not the unit of measurement is genes, alleles, genotypes or varieties/populations. It has been estimated that ICRISAT's collections – excluding groundnuts – are between 70–75 per cent complete based on some experts' general knowledge of the crop. Bioversity International may hold 80 per cent of the global diversity of *Musa* landraces and cultivars, at the level of genotypes, but it holds very little diversity of its wild relatives. ICRAF's coverage of agroforestry tree diversity in its network of field gene banks with national partners is hard to estimate and probably very thin. Again, many of the gene bank managers expressed caveats when making such estimates. Some of them reported to have been surprised in recent years by learning about new populations of crops that they specialized in growing in *in situ* conditions.

Most of the CGIAR centres are now placing particular emphasis on the need to collect wild relatives of their mandate crops. This focus on wild relatives can be accounted for by the combined facts that most centres already have substantial collections of landraces/cultivars. (Again, it must be kept in mind that this is not the case for all crops and that in many cases crop coverage is still far from complete. This is particularly true for clonal crops such as yam and may be partly due to the difficulty in maintaining germplasm in *ex situ* conditions, in comparison to seed crops.) Another reason for the attractiveness of collecting wild relatives is that marker-assisted breeding now makes wild relatives potentially more valuable as sources of genes not present in cultivated gene pools, not only for resistance to pests/diseases but also for a range of other factors such as adaptation to climate change. Furthermore, since they are not actively managed by farmers, wild relatives are particularly vulnerable to being lost due to climate change and other land-use changes (Hunter and Heywood, 2011). In 2006, the CGIAR Science Council noted that 'wild relatives are greatly underrepresented in most of the CGIAR collections' and states that 'a concerted effort is needed to study the remaining distribution of such species and ensure that the accessions collected adequately represent the broad range of diversity within them' (CGIAR Science Council, 2005). In an initiative led by the GCDT and the United Kingdom's Millennium Seed Bank, attempts are now in progress to systematically identify gaps in collections of wild relatives of 23 crops.

It is important to note in this context that the CGIAR centres' gene banks' motivation is not to fill gaps in collections as an end in itself. The gene bank managers stressed the importance of prioritizing acquisition strategies in order to: (1) target materials that are undergoing erosion at accelerated rates as a result of factors such as mining, urban expansion and grazing lands being converted for cultivation and (2) target materials that are particularly important to assemble as part of strategies to respond to biotic and abiotic stresses and diseases or other specific traits (for example, Ug 99, stem rust virulence).

The CGIAR centres are clear that outstanding diversity does not have to be housed in their own collections, but it does need to be held by organizations with the capacity and willingness to play an active role in making this diversity available in the long term for research and conservation purposes. This spirit was reflected by the CGIAR Science Council (2005, 24), which noted that while there is a continuing need to collect, conserve and characterize diversity, 'the CGIAR should not act alone, but must become a well-defined component of a rational, coordinated, forward-looking global system with clearly described areas of responsibility.' It also noted that as far as underutilized PGRFA are concerned, many of them are not included in Annex 1 of the ITPGRFA, and, therefore, working closely with national partners concerning these materials will be important. Finally, the Science Council noted that 'it is not anticipated that the CGIAR would not necessarily establish new germplasm collections [for underutilized plant genetic resources]; rather it would work by assisting partners (who have a comparative advantage in developing collections with international access for these species) to do so' (*ibid.*, 27).

The gene bank managers are clear that a 'rational, coordinated, forward-looking global system' requires global level co-ordination to ensure complementary responsibility sharing when it comes to collecting, conserving and making PGRFA available for use (CGIAR Science Council, 2005, 24). The gene bank managers stressed in their responses, in both 2005 and 2011, that the ITPGRFA sets the scene for a significant shift in paradigm, whereby a much broader range of countries and international organizations will be partners in a global conservation and use system. Ultimately, therefore, the focus for future evaluation of the impact of the Treaty should not be exclusively on whether countries have made more materials available to the CGIAR centres for further global distribution but, rather, whether they have made such materials globally available, either on their own or through other organizations that have the capacities to assist them under the Treaty's framework.

Conclusion

A number of factors contributed to the decline in rates of acquisition of PGRFA by the CGIAR centres for inclusion in the international crop and forage collections from the mid-1990s until 2010: decreased levels of international support for collecting missions, overstretched staff, an inability to characterize and evaluate the materials already collected and challenges associated with targeting gaps in existing collections. According to the managers of these collections, however, the most consistent, all embracing contributing factors have been, in combination, high levels of political controversy concerning the control, use and sharing of benefits associated with genetic resources and low levels of legal certainty about who has the authority within countries to decide to provide access to those resources.

In 2005, the CGIAR centres undertook a study to discover and synthesize the trends in acquisitions by their own gene banks and to document the experiences of the gene bank managers, particularly with respect to the challenges they were facing. The study covered the period from 1980 to 2004, which is the year the ITPGRFA came into force. Since the Treaty was negotiated in the hopes that it would both 'bring closure' to many of the highly politicized debates and provide badly needed legal certainty concerning the appropriate conditions of access and benefit sharing, the data and analysis provided in this chapter constitute a useful baseline study against which to measure at least one area of the Treaty's impact.

For a variety of reasons that are beyond the scope of this chapter to analyse, the ITPGRFA's multilateral system is still not fully operational. Most notably, countries have been slow to put into place structures and processes to be able to routinely make materials available through the multilateral system. Nonetheless, over the course of the last six years, there has been a slowly accumulating body of evidence that the Treaty, and the multilateral system in particular, is contributing to some actors being more willing to deposit materials in the centres' gene banks and thereby make them internationally available. The most notable circumstantial evidence is that organizations from 20 developing countries have elected to deposit, in the centre-hosted international collections, up to 90,000 accessions as part of an internationally co-ordinated regeneration project. While the multilateral system on its own did not lead to these new deposits, it was clearly an important factor contributing to the depositors' comfort in making these materials available internationally.

It seems likely that it will take several more years before the requisite systems are in place (including national systems of implementation) to support the hoped-for density of providers and recipients participating in the multilateral system. It is our hope that this chapter will provide a basis for evaluating future implementation of the multilateral system, with respect to at least one indicator of success – that is, the number of deposits of PGRFA to the CGIAR centres' gene banks that can be made internationally available under the ITPGRFA's framework.

Appendix 1: gene bank acquisitions

TABLE 5.A1 Gene bank acquisitions: CIAT, CIMMYT and CIP, 1980–2004

Year	CIAT Percentage against total	CIAT Accessions acquired	CIMMYT Percentage against total	CIMMYT Accessions acquired	CIP Percentage against total	CIP Accessions acquired
1980	6.27	1,427	3.01	2,969	0.00	0
1981	7.44	1,693	1.19	1,171	0.00	0
1982	5.48	1,248	0.82	812	0.00	0
1983	5.29	1,204	2.01	1,985	8.83	1,005
1984	14.36	3,268	15.58	15,350	1.39	158
1985	10.85	2,470	4.42	4,359	5.90	671
1986	9.97	2,268	2.46	2,428	12.68	1,443
1987	4.56	1,037	0.36	359	11.03	1,255
1988	3.95	900	1.89	1,863	4.96	564
1989	2.80	637	3.21	3,165	4.77	543
1990	2.32	529	2.46	2,422	2.06	234
1991	1.51	344	8.32	8,196	3.39	386
1992	0.47	107	0.54	532	16.37	1,863
1993	2.32	529	1.84	1,817	5.81	661
1994	0.33	76	19.63	19,347	3.64	414
1995	0.25	56	3.75	3,698	2.50	284
1996	0.87	198	5.73	5,650	0.64	73
1997	0.22	51	6.05	5,965	7.22	822
1998	0.36	83	2.46	2,420	0.46	52
1999	0.34	78	2.65	2,614	1.28	146
2000	0.26	60	2.09	2,058	2.40	273
2001	0.59	135	1.61	1,586	0.48	55
2002	1.47	335	4.35	4,282	2.00	228
2003	17.68	4,024	0.41	402	0.10	11
2004	0.00	0	3.13	3,087	2.10	239
Total	100.00	22,757	100.00	98,537	100.00	11,380

Source: SINGER and C.G. Coutres.

TABLE 5.A2 Gene bank acquisitions: ILRI, WARDA and ICRAF, 1980–2004

Year	ILRI Percentage against total	ILRI Accessions acquired	WARDA Percentage against total	WARDA Accessions acquired	ICRAF Percentage against total	ICRAF Accessions acquired
1980	0.00		0.00	0	0.00	0
1981	0.00	0	0.00	0	0.00	0
1982	9.10	1,701	0.00	0	0.00	0
1983	9.92	1,853	0.00	0	0.00	0
1984	16.90	3,157	0.00	0	0.00	0
1985	6.29	1,175	78.99	6016	0.00	0
1986	3.86	721	0.00	0	0.00	0
1987	3.47	649	0.00	0	0.00	0
1988	2.92	545	0.00	0	0.00	0
1989	0.71	133	0.00	0	0.00	0
1990	0.19	35	0.00	0	0.00	0
1991	1.25	233	0.00	0	0.00	0
1992	2.75	514	0.00	0	0.00	0
1993	0.70	130	0.00	0	1.40	25
1994	2.69	503	0.00	0	2.91	52
1995	2.57	480	0.00	0	51.99	928
1996	0.06	11	0.00	0	0.00	0
1997	0.08	15	0.00	0	43.70	780
1998	0.07	13	0.00	0	0.00	0
1999	1.58	295	0.00	0	0.00	0
2000	12.73	2,379	21.01	1,600	0.00	0
2001	14.10	2,635	0.00	0	0.00	0
2002	8.05	1,504	0.00	0	0.00	0
2003	0.00	0	0.00	0	0.00	0
2004	0.02	4	0.00	0	0.00	0
Total	100.00	18,685	100.00	7,616	100.00	1,785

TABLE 5.A3 Gene bank acquisitions: IPGRI, ICRISAT and IITA, 1980–2004

Year	IPGRI Percentage against total	IPGRI Accessions acquired	ICRISAT Percentage against total	ICRISAT Accessions acquired	IITA Percentage against total	IITA Accessions acquired
1980	0.09	1	6.48	3,791	12.15	1,431
1981	0.00	0	7.88	4,615	0.71	84
1982	0.00	0	5.19	3,036	3.65	430
1983	0.51	6	9.09	5,324	0.67	79
1984	1.28	15	8.11	4,750	2.20	259
1985	2.05	24	6.93	4,059	4.28	504

TABLE 5.A3 (continued)

Year	IPGRI Percentage against total	IPGRI Accessions acquired	ICRISAT Percentage against total	ICRISAT Accessions acquired	IITA Percentage against total	IITA Accessions acquired
1986	5.20	61	9.76	5,711	5.87	691
1987	13.14	154	4.97	2,911	12.38	1,458
1988	12.54	147	5.05	2,956	23.61	2,780
1989	15.19	178	6.33	3,704	4.30	506
1990	26.45	310	4.21	2,465	1.49	175
1991	5.29	62	3.31	1,935	0.05	6
1992	1.19	14	6.84	4,007	3.88	457
1993	3.50	41	3.96	2,320	0.14	16
1994	1.54	18	4.19	2,454	12.87	1,516
1995	1.54	18	2.89	1,689	1.49	176
1996	2.56	30	0.86	503	0.20	24
1997	2.22	26	0.97	566	0.20	24
1998	2.22	26	2.61	1,529	4.23	498
1999	0.43	5	0.00	1	2.62	308
2000	0.26	3	0.06	33	0.20	23
2001	0.00	0	0.30	177	1.96	231
2002	1.79	21	0.00	0	0.85	100
2003	1.02	12	0.01	5	0.00	0
2004	0.00	0	0.00	0	0.00	0
Total	100.00	1,172	100.00	58,541	100.00	11,776

TABLE 5.A4 Gene bank acquisitions: ICARDA and IRRI, 1980–2004

Year	ICARDA Percentage against total	ICARDA Accessions acquired	IRRI Percentage against total	IRRI Accessions acquired
1980	5.19	4,823	4.60	2,542
1981	6.58	6,115	6.17	3,413
1982	1.97	1,830	4.41	2,440
1983	3.63	3,373	4.02	2,226
1984	7.42	6,897	6.94	3,839
1985	3.72	3,462	6.64	3,674
1986	5.23	4,865	4.50	2,491
1987	5.30	4,922	3.24	1,794
1988	2.19	2,033	2.23	1,236
1989	6.94	6,453	3.44	1,901
1990	3.20	2,976	2.08	1,148
1991	11.86	11,025	3.14	1,736
1992	4.15	3,860	3.29	1,819
1993	2.36	2,195	0.49	270

TABLE 5.A4 (continued)

Year	ICARDA Percentage against total	ICARDA Accessions acquired	IRRI Percentage against total	IRRI Accessions acquired
1994	4.81	4,471	2.13	1,178
1995	3.23	3,004	0.23	127
1996	1.02	946	7.37	4,079
1997	3.60	3,347	11.37	6,288
1998	2.54	2,357	14.46	8,000
1999	0.72	672	6.23	3,447
2000	4.97	4,624	1.25	689
2001	3.07	2,858	1.56	863
2002	1.40	1,298	0.12	64
2003	3.09	2,870	0.02	11
2004	1.80	1,669	0.07	38
Total	100.00	92,945	100.00	55,313

Appendix 2: gene bank acquisition and distribution trend survey

Gene bank survey, July 2005

1. Name of gene bank:
2. Name of individual respondent(s):
3. Date:

Part 1: regarding general trends in rates of acquisition and distribution

4. In the last 15 years (1990–2005), the rates of acquisition of new accessions by your gene bank have:

   - declined?
   - increased?
   - remained steady?

5. If there is a decline in your gene banks' overall rate of acquisition, please identify which of the following factors have contributed to that decline:

   - most relevant diversity is already collected
   - no space left in the gene bank for additional accessions
   - no money for additional collecting missions
   - no capacity/money to evaluate or characterize what is already in the gene bank, so there is no point adding more
   - restrictive policies or laws make it difficult to obtain new material

- unwillingness of countries to allow collecting missions or to share their *ex situ* materials
- other

Please add additional information to further clarify your answer

6. In the last 15 years (1990–2005), the rates of distribution of material from your gene bank have:
   - declined?
   - increased?
   - remained steady?

7. If there has been a decline in the rate of distributions from your gene bank, please identify which of the following factors have contributed to that decline:
   - decline in requests for materials?
   - requests are more targeted (i.e., requestors are more specific about the range of materials they are seeking, so you end up sending less material per request)?
   - your responses to requests are more targeted (i.e., you spend more time in the past, with more information about your collection, determining which particular materials are best suited to the needs of the requestor, so you end up sending less material per request)
   - you do not have sufficient resources to regenerate enough material to send materials in response to all requests?
   - restrictive policies or laws or conditions upon specific materials that you are holding make it difficult to distribute materials?
   - other?

Please add additional information to clarify your answer

Part 2: Regarding requests by CGIAR Centres to states for materials

(i.e., excluding CGIAR centres' collections and other collections held by international organizations)

8. How many requests has your gene bank made to countries (i.e., not to international institutions and their gene banks) for materials in the last five years?
   - 0
   - 1–5
   - 6–10
   - 11–15
   - other (please indicate number)

9. Regarding Question 8 above
    a. how many were rejected?
    b. how many were ignored (i.e., simply no answer)?
    c. how many agreed-to requests were subject to the same conditions as set out in the FAO-In Trust Agreements of 1994?
    d. how many agreed-to requests were subject to conditions different from the standard conditions of the FAO-In Trust Agreements of 1994 (i.e., were more restrictive in terms of your gene bank's rights to use and distribute that material)?

10. Regarding Question 9d above, what were the details of those conditions?

11. What percentage of the rejected or ignored requests in 9a or 9b involved proposals for new collecting missions for *in situ* materials?
    - If you do know, please indicate the percentage
    - If you do not know, please mark with an 'X'

12. What percentage of the rejected or ignored requests in 9a or 9b were for *ex situ* material originally collected in the country to which the request was made?
    - If you do know, please indicate the percentage
    - If you do not know, please mark with an 'X'

13. What percentage of the rejected or ignored requests in 9a or 9b were for *ex situ* material originally collected in another country (and subsequently transferred to the country to which the request was made)?
    - If you do know, please indicate the percentage
    - If you do not know, please mark with an 'X'

14. What per centage of the rejected or ignored requests in 9a or 9b were for improved material?
    - If you do know, please indicate the percentage
    - If you do not know, please mark with an 'X'

Appendix 3: List of respondents to the 2005–6 questionnaire and semi-structured interviews

Bioversity/INIBAP	Ines Van den Houwe
	Nicolas Roux
CIAT	Daniel G. Debouck
CIMMYT	Thomas Payne
	Suketoshi Taba
CIP	Enrique Chujoy
	Willy Roca
ICARDA*	Jan Konopka
ICRAF	Tony Simons

ICRISAT   Hari D. Upadhyaya
IITA   V. Mahalakshmi
           Dominique Dumet
ILRI   Jean Hanson
IRRI GRC   Ruaraidh Sackville Hamilton
WARDA   Eklou A. Somado
           Ines Sanchez

* no follow-up interview conducted

Appendix 4: Holdings that the centres may distribute using the Standard Material Transfer Agreement (SMTA)[13]

TABLE 5.A5 Holdings that the centres may distribute using the Standard Material Transfer Agreement

CGIAR centre	Crop	Accessions designated in trust (current)
CIAT	cassava	6,592
	forages	23,140
	beans	36,249
	Total	**65,981**
CIMMYT	maize	
	teosinte	
	Tripsacum	
	Sub-total	**27,856**
	Triticum aestivum aestivum	92,011
	Triticum turgidum durum	22,051
	Triticosecale spp	20,390
	Hordeum vulgare	15,133
	Triticum turgidum dicoccon	2,632
	Triticum aestivum spelta	1,271
	Triticum monococcum	943
	Triticum hybrid	616
	Secale cereal	587
	Triticum and *Aegilops* spp	5,000
	Sub-total	*160,634*
	Total	**188,490**
CIP	Andean roots and tubers	1,183
	sweet potatoes	7,777
	potatoes	6,620
	Total	**15,240**
ICARDA	barley	26,020
	chickpeas	12,179
	faba beans	10,375
	wheat	37,277

TABLE 5.A5 (continued)

CGIAR centre	Crop	Accessions designated in trust (current)
	forages	30,626
	lentils	10,041
	Total	**126,518**
ICRAF	N/A	N/A
ICRISAT	chickpeas	17,124
	groundnuts	14,803
	pearl millet	21,563
	pigeonpeas	13,389
	sorghum	36,771
	small millet	10,180
	Total	**113,830**
IITA	Bambara groundnuts	1,815
	cassava	2,078
	cowpeas	15,003
	soybeans	1,909
	wild vigna	1,507
	yams	3,087
	maize	
	Musa	
	African yam beans	66
	soybeans	1,742
	Total	**27,207**
ILRI	forages	17,032
	Total	**17,032**
Bioversity International	*Musa*	1,203
	Total	**1,203**
IRRI	*Oryza sativa* (Asian cultivated rice)	97,917
	Oryza glaberrima (African cultivated rice)	1,657
	Annex 1 wild relatives and interspecific hybrids	4,235
	non-Annex 1 wild relatives	18
	Total	**103,827**
WARDA	Rice	14,751
	Total	**14,751**
Overall total		**674,079**

Source: Centres and SINGER

Notes: This table includes materials that the centres have designated as in–trust under the earlier 1994 in–trust agreements with FAO, materials that they have received under an SMTA and conserved and materials that they have received under other agreements that allow the centres to conserve and redistribute those materials using the SMTA.

Notes

The authors wish to acknowledge contributions to the development of this chapter from a range of people. We recognize the members of the CGIAR Genetic Resources Policy Committee, which first suggested the system-wide examination of rates of acquisition by the centres' genebanks. The members and/or observers of the GRPC during the time that it reviewed the early drafts of the chapter were: Carlos Correa, Emile Frison, Mike Gale, Tony Gregson, Masa Iwanaga, Leonardo Montemayor, Maria José Sampaio, Carl-Gustaf Thornström, Juan Lucas Restrepo, Orlando de Ponti, Clive Stannard, José T. Esquinas-Alcázar and Victoria Henson-Appolonio. We also want to recognize the original respondents to the 2006 survey listed in Appendix 1 of the chapter (some of whom are co-authors). In addition, we want to acknowledge the following people who commented on more recent drafts of the chapter: Isabel López Noriega, Gerald Moore, Ronnie Vernooy, Bert Visser, Luigi Guarino and Daniel Debouck. Finally, we thank the CGIAR Inter-centre Working Group – Genetic Resources (ICWG-GR) for setting aside time during its most recent annual meeting (in April 2011) to review the main findings of the chapter and allow centre representatives an opportunity to provide additional information.

1 For more details, see Appendix 1, which provides details about how many accessions each centre acquired each year, up to 2004, and the percentage of that year's acquisitions, which represents the centre's total holdings.
2 It has been suggested – though not by any of the respondents surveyed for this study – that another reason why a number of countries do not transfer materials to the CGIAR centres is simply because they have established facilities themselves and no longer need to transfer materials out of their country for conservation purposes (Bert Visser, personal communication).
3 International Treaty on Plant Genetic Resources for Food and Agriculture, 29 June 2004, www.planttreaty.org/texts_en.htm.
4 Standard Material Transfer Agreement, 16 June 2006, ftp://ftp.fao.org/ag/agp/planttreaty/agreements/smta/SMTAe.pdf.
5 For the purposes of this chapter, it was not possible to identify (and subtract) all acquisitions of this nature for all of the centres' gene banks over the decades analysed. That is why instead, we have chosen to provide just a few examples of the sources of some of the biggest acquisitions (in terms of numbers of accessions).
6 For the purposes of this chapter, we have limited the lists of names of provider countries to those countries that made deposits during the period 2007–10 (see Table 5.3).
7 Convention on Biological Diversity, 31 I.L.M. 818 (1992).
8 This finding is consistent with a number of recently released papers (see, for example, Correa, 2005; Brazil, n.d.). The latter is a report of a meeting hosted by the Brazilian government in the lead up to the eighth Conference of the Parties, which involved participation of approximately 200 scientists from across Latin America. The report includes the statements that '[b]asic biological research is seriously hampered by many of the current national ABS regimes', and '[d]istrust, rather than trust, is presently dominating the situation in many countries, hampering biological research. This holds for national as well as international research.' The participants included a recommendation that 'all countries are encouraged to review their processes for permits on research, collection, import, and export of specimens to rationalize and streamline the ABS process. In addition, rules and regulations need to be practicable.'
9 The ILRI and the International Potato Center, for example, informally adopted this approach in 1994. More recently, in approximately 2001, the CIAT considered that it would be best to postpone most requests until the ITPGRFA came into force.
10 Note that publication of such information is not a condition precedent for material being included in the multilateral system. Pursuant to Article 11.2 of the ITPGRFA, all PGRFA that are 'under the management and control of Contracting Parties and in the public domain' are automatically included in the multilateral system. The publication of

lists of such materials is merely a practical (perhaps the only) way of informing people what is actually included in the multilateral system.

11 Again, it is important to note that not all acquisitions 'count' in the same way when considering whether the ITPGRFA is working to overcome the political and legal tensions that were affecting rates of acquisition in the pre-Treaty period. For example, almost half of the materials acquired by the gene banks from August 2007 to July 2008 came from other CGIAR gene banks (see Table 5.1).

12 On the other hand, it will probably always be easy to find evidence of further substantial missing diversity. The IRRI, for example, reports that in an Indian trial of the suitability of landraces of *O. sativa* for recovery in tsunami-affected soils, none of the best six varieties were in the IRRI gene bank. This may be evidence that important Indian diversity is absent from the collection. Or is it just that the same genetic entities are in the gene bank but known by different names or that all of the genes are present in other varieties but in different combinations? And, based only on sampling density, other potential gaps may always be identified fairly readily, albeit without evidence that they correspond to real genetic gaps – for example, Myanmar is within the centre of the diversity of rice and still is home to much traditional farming and many varieties, but it has been sampled relatively little compared with its neighbouring countries and so arguably may have novel diversity that is not conserved in the IRRI. Similarly, at the northern limits of rice agriculture in central Asia and northern China, there may possibly be genes for cold tolerance that are not available in the gene bank at the IRRI.

13 This includes materials that the centres have designated as in-trust under the earlier 1994 in-trust agreements with FAO, materials that they have been received under an SMTA and conserved, and materials they have been received under other agreements that allow the centres to conserve and redistribute those materials using the SMTA.

References

Brazil (n.d.) *Outcomes and Recommendations of the Meeting of Biodiversity: The Megascience in Focus*, Doc. UNEP/CBD/COP/8/INF/46, online: http://www.cbd.int/doc/?meeting=cop-08 (last accessed 30 May 2012).

Commission on Plant Genetic Resources (CPGR) (1993) *Progress Report on the Global System for the Conservation and Utilization of Plant Genetic Resources*, Doc. CPGR/93/5.

——(1994) *Report of the Commission on Plant Genetic Resources*, First extraordinary session, Rome, Doc. CPGR-Ex1/94/REP.

Correa, C. (2005) 'Do National Access Regimes Promote Use of Genetic Resources and Benefit Sharing?' *International Journal of Environment and Sustainable Development* 4(4): 443–463.

CGIAR (2005) *System Priorities for CGIAR Research*, Doc. 2005–20015.

——(2006) *Safeguarding the Worlds Agricultural Legacy*, online: www.cgiar.org/pdf/cg_gene banks_2006.pdf (last accessed 1 April 2011).

Food and Agriculture Organization (FAO) (2010) *Second Report on the State of the World's Plant Genetic Resources for Food and Agriculture*, FAO, Rome.

——(2011) *Draft Updated Global Plan of Action for the Conservation and Sustainable Utilization of Plant Genetic Resources for Food and Agriculture*, Doc. CGRFA/WG-PGR-5/11/2 Rev. 1.

Fowler, C., and G. Hawtin (2011) 'The Global Crop Diversity Trust: An Essential Element of the Treaty's Funding Strategy', in C. Frison and F. Lopez Esquinas, *Plant Genetic Resources and Food Security: Stakeholders' Perspectives on the International Treaty on Plant Genetic Resources for Food and Agriculture*, Earthscan, London.

Fujisaka, S., D. Williams and M. Halewood (2009) *The Impact of Climate Changes on Countries' Interdependence on Genetic Resources for Food and Agriculture*, Background Study Paper no. 48, online: ftp://ftp.fao.org/docrep/fao/meeting/017/ak532e.pdf (last accessed February 2012).

Hunter, D., and V. Heywood (eds) (2011) *Crop Wild Relatives: A Manual of In Situ Conservation*, Earthscan, London.

Ramirez-Villegas, J., C. Khoury, A. Jarvis, D.G. DeBouck and L. Guarino (2010) 'A Gap Analysis Methodology fo Collecting Crop Genepools: A Case Study with *Phaseolus* Beans', *PLos ONE* 5(1): e13497, doi: 10.1371/journal.pone.0013497.

Rural Advancement Foundation International (RAFI) and Heritage Seed Curators Australia (HSCA) (1998) 'Plant Breeders' Wrongs: An Inquiry into the Potential for Plant Piracy through International Intellectual Property Conventions', Rural Advancement Foundation International, Ottawa, online: www.etcgroup.org/upload/publication/400/01/occ_plant.pdf (last accessed February 2012).

System-Wide Genetic Resources Programme (SGRP) (2007) *Experiences of the IARCS of the CGIAR Implementing Their Agreements with the Governing Body of the Treaty, with Particular Focus on Their Use of the Standard Material Transfer Agreement*, online: ftp://ftp.fao.org/ag/agp/planttreaty/gb2/gb2i11e.pdf (last accessed February 2012).

——(2009) *Experiences of the IARCS of the CGIAR Implementing Their Agreements with the Governing Body of the Treaty, with Particular Focus on Their Use of the Standard Material Transfer Agreement*, online: ftp://ftp.fao.org/ag/agp/planttreaty/gb3/gb3i15e.pdf (last accessed Feb 2012).

——(2010) *Booklet of CGIAR Centre Policy Instruments, Guidelines and Statements on Genetic Resources, Biotechnology and Intellectual Property Rights*, online: www.sgrp.cgiar.org/sites/default/files/Policy_Booklet_Version3.pdf (last accessed February 2012).

——(2011) *CGIAR Centres' Experience with the Implementation of Their Agreements with the Treaty's Governing Body, with Particular Reference to the Use of the SMTA for Annex 1 and non-Annex 1 Materials*, online: www.itpgrfa.net/International/sites/default/files/gb4i05e.pdf (last accessed February 2012).

United Nations Environment Programme (UNEP) (1994) *Report of the Intergovernmental Committee on the Convention on Biological Diversity*, Doc. UNEP/CBD/COP/1/4.

——(2006) *Outcomes and Recommendations of the Meeting of 'Biodiversity: The Megascience in Focus'*, Doc. UNEP/CBD/COP/8/INF/46.

Part II
The history and design of the International Treaty's multilateral system of access and benefit-sharing

6

A BRIEF HISTORY OF THE NEGOTIATIONS ON THE INTERNATIONAL TREATY ON PLANT GENETIC RESOURCES FOR FOOD AND AGRICULTURE

José Esquinas-Alcázar, Angela Hilmi and Isabel López Noriega

> Caminante no hay camino, se hace camino al andar
> There is no road, you make your own path as you go along
> Antonio Machado, *Proverbs and Songs*

Introduction

To understand the negotiations that resulted in the International Treaty on Plant Genetic Resources for Food and Agriculture (ITPGRFA), it is necessary to understand the growing economic and social importance of these resources throughout the twentieth century, the historical framework of the negotiations, and the scientific and political climate in which they developed.[1] Consequently, this chapter will discuss these factors over the various phases of the political debate that began in 1979.

In a certain sense, the history of the exchange of genetic resources *is* the history of mankind. The struggle for access to useful plants for agriculture and food originating from other places has been one of mankind's main motivations to travel since early times and has not only frequently led to encounters and alliances but also to conflicts and war between different cultures (Esquinas-Alcázar, 2005; Harlan, 1992; Parry, 1978; Vasey, 1992; Zohari, 2000).

The history of the Treaty reflects these efforts to access and control genetic resources as well as concern for the future of mankind. The Treaty is a result of a long historical negotiation process that underwent technical, financial, political, institutional and economic phases (Cooper, 2002, 2; Kate and Lasen, 1997, 284–86; Mekour, 2002, 3–5; Rose, 2003). This chapter will illustrate the most salient features of this long process.

First phase: international technical and scientific discussions up to the start of the negotiations

Ever since the 1950s, some international organizations, above all, the Food and Agriculture Organization (FAO) of the United Nations, have had serious concerns over the loss of the diversity of genetic resources in the world. In the early 1960s, the FAO (1961) held a technical meeting that led to the creation, in 1965, of a Panel of Experts in Plant Exploration and Introduction (FAO, 1969a, 1969b, 1970, 1973a, 1975a). From 1965 to 1974, this group met regularly to advise the FAO on this issue and to set international guidelines for germplasm collection, conservation and exchange (FAO, 1975b). Technical issues gradually emerged related to the assessment of biodiversity and genetic erosion, the identification of collecting sites, sampling techniques, germplasm conservation methods and assessment and documentation methods. In 1967, 1973 and 1981, the FAO held international technical conferences on plant genetic resources (FAO 1973b; FAO/IBP, 1967), each resulting in a publication (Frankel and Bennet, 1970; Frankel and Hawkes, 1975; Holden and Williams, 1984).

Arising from the need to organize and finance new programmes to conserve these plant genetic resources, in 1967, the Crop Ecology and Genetic Resources Unit was created and a fund established. In 1972, the Consultative Group on International Agricultural Research (CGIAR), following up on the recommendations of the UN Conference on the Human Environment, which was held in Stockholm, Sweden, in June of the same year, and of its own Technical Advisory Committee, decided to create the International Board of Plant Genetic Resources (IBPGR, which is now known as Bioversity International) (CGIAR, 1972) – an organization with its own budget. The IBPGR would become part of the international programme of the CGIAR, and its Secretariat would be provided by the FAO's Plant Genetic Resources Unit. The IBPGR was established in 1974, with its headquarters at the FAO in Rome, and, since then, it has promoted and carried out a number of activities for the collection, conservation, assessment, documentation and use of plant germplasm.

In parallel with the FAO and IBPGR activities, and in some cases due to their role as catalyser, in the 1970s, international, regional, national and private organizations created or strengthened programmes aimed at safeguarding and using plant resources, especially *ex situ* resources. Among these initiatives, those of the Tropical Agricultural Research and Higher Education Centre (CATIE) deserve mention.

Second phase: political discussions and initial negotiations

In 1979, at the FAO conference (the highest decision-making body in the organization, where all member countries are represented), the first political discussions began, which in only a few years led to the adoption of the non-binding International Undertaking on Plant Genetic Resources for Food and Agriculture (International Undertaking) and much later, negotiations and approval of the legally biding ITPGRFA.[2] The questions raised by the developing countries during the 1979 conference reflect the basic outline of the difficult negotiations the following years and are the basis of the Treaty and its multilateral system of access and benefit sharing. These questions may be summarized as follows:

- Plant genetic resources are found throughout the world, but the greatest diversity is in tropical and subtropical areas, where most developing countries are situated. When seeds are collected and deposited in germplasm banks, often in developed countries, to whom do these stored samples belong? The country where they are collected? The country where they are being stored? Humanity at large?
- If the new varieties obtained are the result of applying technology to raw material or genetic resources, why are the rights of the providers of the technology recognized (plant breeders' rights, patents and so on) and not the rights of the providers of genetic resources?

The answers to these questions were neither clear nor convincing. At times, they opened the floor to strong, dialectic confrontations.

During the 1979 FAO conference, the Spanish delegation proposed that, in order to solve these problems, an international agreement should be developed and a germplasm bank established under the jurisdiction of the FAO. Despite the strong support for the proposal during the conference, it did not result in a draft resolution (FAO, 1979).

In the autumn of that same year, in the months prior to the FAO conference, Mexico, with the support of the Group of Latin America and Caribbean Countries (GRULAC) and the Group of 77,[3] promoted a draft resolution, which included the two essential elements of the 1979 Spanish proposal (FAO, 1981a, 1981b):

- a call for an international agreement on genetic resources and
- a request for a germplasm bank or network of germplasm banks under the jurisdiction of the FAO.

During the FAO conference in November 1981, the Mexican draft resolution, presented by the Group of 77, provoked intense discussions among the countries (FAO, 1981a, 1981b). A debate that had been planned to last two or three hours, actually lasted several days. Although some developed countries joined the developing countries' proposal, others strongly opposed it. A working group, with the exceptional participation of the FAO's director-general, was created to revise the draft resolution so that it would be acceptable to all. A consensus was reached on the basis of a diluted resolution (FAO, 1981c), in which the director-general was requested to carry out feasibility studies on an international agreement and a germplasm bank network under the jurisdiction of the FAO.

The requested studies, presented in the spring of 1983 at the meeting of the FAO Committee on Agriculture (COAG) (FAO, 1983a, 1983b), reached the conclusion that the agreement was not necessary and the germplasm bank network was not technically feasible. The controversy raised from these conclusions ended with an offer by the Spanish government to place its national germplasm bank under the jurisdiction of the FAO, thereby showing that the problem was not a question of technical feasibility but, rather, of political will. As a result, the COAG requested the director-general to prepare a new document on the basis of the Spanish proposal, which would be presented at the FAO conference that same year (FAO, 1983c, 1983d).

In November 1983, the twenty-second session of the conference of the FAO witnessed long and difficult discussions in a strained atmosphere, in which political tensions were felt. Finally, on the last day, and after various voting sessions, and amid cries, applause, tears and a great ovation, both the International Undertaking and the permanent intergovernmental commission responsible for its monitoring were created. However, they were approved without consensus – eight countries expressed their reservations.[4]

> **Box 6.1 International Undertaking on Plant Genetic Resources**
>
> The International Undertaking is the first comprehensive international agreement on plant genetic resources for food and agriculture. It was approved at the FAO Conference in 1983 (Resolution 8/83) as a non-binding instrument to promote international harmony and cooperation in matters related to plant genetic resources for food and agriculture. According to the approved text, the objective of the International Undertaking is to ensure that plant genetic resources of economic and/or social interest, particularly for agriculture, will be explored, preserved, evaluated and made available for plant breeding and for scientific purposes. The International Undertaking, with 11 articles, formally recognizes plant genetic resources, including improved and commercial varieties, as a common heritage of mankind, and seeks to guarantee its freedom of exchange without restrictions through a network of germplasm banks under the auspices and/or jurisdiction of the FAO. The International Undertaking is supervised by the participating countries through the FAO Commission on Genetic Resources for Food and Agriculture (CGRFA).

The conference then requested the FAO Council to develop the statutes of the intergovernmental Commission on Plant Genetic Resources (renamed in 1995 as the Commission on Genetic Resources for Food and Agriculture). In this commission, between 1983 and 1991, negotiations were held between the countries, which made it possible to reach agreed interpretations of the International Undertaking to make it acceptable to all, and later, from 1993 and 2001, renegotiations of the International Undertaking in order to transform it into a binding agreement, the ITPGRFA.

From 1983, the newly established commission served as an intergovernmental forum where countries continued to debate on agreed interpretations of some of the provisions of the International Undertaking, which would allow the countries that had not joined to lift their reservations. This is how three resolutions were negotiated, which were integrated in the International Undertaking as annexes. In these resolutions, the concept of national sovereignty was introduced, and plant breeders' rights and farmers' rights were recognized in parallel and simultaneously. In this process, it was also agreed that farmers' rights would be developed through an international fund. Some countries felt that this fund should comprise a percentage of the benefits derived from the use of the genetic resources, whereas most countries felt that it should be tied to the needs of the countries to ensure their conservation and sustainable use.

A process began with the aim of quantifying these needs, which led to the fourth International Technical Conference on Plant Genetic Resources, which was held in Leipzig in 1996. During the preparation of this conference, 155 countries drafted national reports on the situation of their genetic resources, needs and priorities. Moreover, over the course of 12 regional meetings, the respective regional reports were developed. This process culminated in the Leipzig Technical Conference, where the Declaration of Leipzig on the conservation and sustainable use of plant genetic resources for food and agriculture was adopted, the first *State of the World's Plant Genetic Resources for Food and Agriculture* was launched (FAO, 1996a), and the first Global Plan of Action on Plant Genetic Resources for Food and Agriculture was approved (FAO, 1996b). In addition, a document was prepared by the Secretariat, which quantified the necessary funds for projects, programmes and activities in line with the priorities defined in the Global Plan of Action. This plan later became the basis for Article 14 of the ITPGRFA, which recognizes the importance of the Global Plan of Action for the Treaty and encourages the contracting parties to promote its effective implementation.

During the years following the adoption of the International Undertaking, major emphasis was put on the implementation of its Article 7 on the development of an 'international network of base collections in gene banks under the auspices or the jurisdiction of FAO'.[5] In compliance with this Article, on 26 October 1994, 12 International Agricultural Research Centres (IARCs) of the CGIAR signed agreements with the FAO, placing their *ex situ* collections in this international network under the auspices of the FAO.[6] The agreements also recognized the intergovernmental authority of the CGIAR to set policies for these *ex situ* collections, while the commission and the FAO, on behalf of the international community, recognized the IARCs as 'trustees' of these international collections. Under these agreements, the IARCs made a commitment to provide the international community with the designated germplasm for the benefit of developing countries, in particular, in accordance with the International Undertaking. Through this agreement, the IARCs reaffirmed the principles of not claiming ownership over the designated germplasm and not seeking any intellectual property rights over the said germplasm or over related information. Since then, the agreements have been renewed every four years (SGRP, 2010).

Following the approval of the ITPGRFA, these agreements were replaced by those signed by the centres on 16 October 2006, placing their germplasm collections under Article 15 of the Treaty. A similar agreement was signed by CATIE on the same day. Since then, agreements have also been signed by the International Coconut Genebank for Africa and the Indian Ocean, the International Coconut Genebank for the South Pacific, the Mutant Germplasm Repository of the FAO/IAEA Joint

Division, the International Cocoa Genebank and the Centre for Pacific Crops and Trees of the South Pacific Community.

> **Box 6.2 The agreements between the Governing Body of the Treaty and the International Agriculture Research Centres (IARCs) of the Consultative Group on International Agricultural Research (CGIAR)**
>
> In a ceremony held on World Food Day in 2006, the IARCs holding germplasm collections signed an agreement with the FAO acting on behalf of the Governing Body of the Treaty. The text of the agreement had previously been discussed and agreed on by the Governing Body at its first session.
>
> These agreements substituted the in-trust agreements signed between the IARCs and the FAO in 2004; reaffirmed the status of the *ex situ* collections held by the centres as global public goods; and put the collections under the auspices of the Treaty. Furthermore, through these agreements, the IARCs recognized the authority of the Governing Body to provide policy guidance relating to *ex situ* collections held by them and subject to the provisions of the Treaty.
>
> As of January 2007, in line with the conditions defined in the agreements, in order to transfer germplasm the IARCs began to use the Standard Material Transfer Agreement (SMTA) adopted by the Governing Body in its first session for the crops and forages included in Annex I of the Treaty. As of January 2008, following the decision adopted by the Governing Body at its second session, the IARCs have been using this same SMTA to also distribute plant germplasm held in the centres' collections and not included in Annex 1 of the Treaty. For non-Annex 1 crops and forages, the IARCs agreed to inform the Governing Body of the transfer agreements that they enter into, and to deliver to the contracting parties to the Treaty, samples of plant genetic resources for agriculture and food that have been collected in *in situ* conditions.

It is appropriate here to emphasize the fundamental role played by the non-governmental organizations (NGOs) in the whole process. An important initiative of the NGOs was the International Keystone Dialogue Series on Plant Genetic Resources (1988–91), which brought together representatives from civil society, farmers, indigenous organizations, the private sector, governments of developed and developing countries and international research centres. This initiative served to identify the main constraints and possible areas of agreement that would then permit intergovernmental meetings of the CGRFA to reach consensus on many of the earlier-described agreements. The most important meetings of this initiative took place in Keystone, Colorado, in the United States in 1988, Madras, India, in 1990 and Uppsala, Sweden, and Oslo, Norway, in 1991. The meetings, which were chaired by M.S. Swaminathan, were vital in resolving some of the most engaging issues, such as farmers' rights and the need to finance plant genetic resources conservation and use in support of poor farmers in developing countries. The needed funds were estimated at US $300–500 million.

Another similar initiative was that of the Crucible Group. The Crucible Group met twice in 1993 in Uppsala, Sweden, and in Berne, Switzerland. The results of the conversations of the group were published in the book *People, Plants and Patents: The Impact of Intellectual Property on Trade, Plant Biodiversity, and Rural Society*.[7] Due to its wide representativeness, the agreed recommendations of this group were very well received in the negotiation fora. Among the many resulting recommendations, the recommendation to place the *ex situ* collections of the CGIAR under the auspices of the FAO should be highlighted.

Third phase: a binding agreement for the agrarian sector – development of the Treaty

In parallel with the processes described earlier, from 1988 to 1992, the first international binding agreement on biological diversity in general was negotiated in the United Nations Environment Programmeme (UNEP). The Convention on Biological Diversity (CBD) was submitted for signature at the UN Conference on Environment and Development in June 1992.[8] This agreement, which also includes agricultural biological diversity, does not sufficiently take into account the specific needs of the agricultural sector because this sector was weakly represented during the negotiation process. It was only in May 1992, at the last moment and at the last negotiation meeting in Nairobi, that it was possible to bring together the few representatives of countries who were directly or indirectly linked to the agriculture sector. This group managed to draft a resolution on agricultural biodiversity to be incorporated into the Nairobi Final Act, at which point the convention was approved. It became Resolution 3 of the Nairobi Final Act, which concerned the relationship between the CBD and the promotion of sustainable agriculture. It highlighted the importance of the agreements reached at the FAO headquarters and requested a revision to the International Undertaking in line with the CBD.

Box 6.3 Resolution 3 of the Nairobi Final Act and Resolution 7/93 of the FAO Conference

In adopting the agreed text of the Convention on Biological Diversity (CBD) in May 1992, countries also adopted Resolution 3 of the Nairobi Final Act, which recognized the need to seek solutions to outstanding matters concerning plant genetic resources, in particular:

(1) access to *ex situ* collections not addressed by the convention and
(2) the question of farmers' rights.

It was requested that these matters be addressed within the FAO's forum.

> In 1993, the FAO Conference accordingly adopted Resolution 7/93 for the revision of the International Undertaking and requested the FAO to provide a forum in the CGRFA for the negotiation among governments, for:
>
> (1) the adaptation of the International Undertaking on Plant Genetic Resources, in harmony with the CBD;
> (2) consideration of the issue of access on mutually agreed terms to plant genetic resources, including *ex situ* collections not addressed by the CBD; and
> (3) the issue of the realization of farmers' rights.

Soon after, in 1994, during the Uruguay Round (also with minimal participation from the agricultural sector), trade agreements were developed and approved in Marrakesh, Morocco. These agreements culminated in the establishment of the World Trade Organization (WTO) in 1994 and also concerned genetic resources for agriculture and food. They included the Agreement on Trade-Related Aspects of Intellectual Property Rights (TRIPS Agreement), in particular, Article 27.3(b) of the TRIPS Agreement, which required the contracting parties to provide for 'the protection of plant varieties either by patents or by an effective *sui generis* system or by any combination thereof'.

The approval of both the CBD and the TRIPS Agreement as binding agreements raised an alarm bell for the agricultural sector, which was locked between two binding agreements, without its specific needs being adequately taken into account.

The International Undertaking, with its merely voluntary nature, lacked sufficient weight to be able to protect agricultural interests. Growing pressure from other sectors – and, in particular, the commercial and environmental sectors – on the agricultural sector, made it possible to achieve what seemed unimaginable just a little while ago – namely a united front of developed and developing countries, the seed industries and NGOs, working towards a common political objective. The objective was to turn the International Undertaking into a binding agreement, which would permit dialogue on an equal footing with the commercial and environmental sectors. It would also legally guarantee the conservation of important crop genetic resources for future generations and fair and equitable access to them for research and genetic improvement. Here began the last negotiation phase of what is today known as the ITPGRFA, in a highly constructive atmosphere.

In 1995, the Conference of the Parties to the CBD provided an important support towards progress on this negotiating process, through Decision no. II/15, which was taken in Jakarta in 1995: 'Recognizing the special nature of agricultural biodiversity, its distinctive features and problems needing distinctive solutions' (UNEP/CBD, 1995). The formal negotiations, which started with the International Undertaking and its annexes and culminated in the adoption of the ITPGRFA, lasted for seven years (IISI, 1996, 1997a, 1997b, 1997c, 1998, 2000a, 2000b, 2001a, 2001b, 2001c, 2001d, 2002a, 2002b, 2006). In this period, from 1994 to 2001, the CGRFA was held in three regular sessions and six extraordinary sessions. A Contact Group chaired by Fernando Gerbasi, ambassador of Venezuela and president of the commission at that time, held six intermediate sessions to discuss controversial issues – in particular, the list of plants to include in the multilateral system of access and benefit sharing; the form and manner for sharing the monetary benefits derived from commercialization; intellectual property rights to the materials from the multilateral system; financial resources; genetic materials conserved in the IARCs; and the definition of key terms.

The sixth Extraordinary Session of the CGRFA, which was held in Rome in June–July 2001, sought to conclude the negotiations. However, its delegates did not reach an agreement on various points: the definitions of 'plant genetic resources for food and agriculture' and 'genetic material'; the application of intellectual property rights to the materials of the multilateral system; the relationship between the International Undertaking and other international agreements; and the list of plants to be included in the multilateral system. The pending issues were transmitted to the FAO Council, which finalized the negotiations at its twenty-first session in Rome from 30 October to 1 November 2001, making it possible to resolve them (FAO, 2001a).[9]

The negotiations concluded with the adoption of the ITPGRFA by consensus, with only two abstentions from Japan and the United States (FAO, 2001b), in a climate of widespread euphoria in the thirty-first session of the FAO Conference on 3 November 2001 (FAO, 2001c).

Box 6.4 Annex 1 of the Treaty

One of the most complex and controversial issues in the formal negotiation process was the selection of the genera or crops to be included in the multilateral system and that appear in Annex 1 of the Treaty. With the aim of providing a solid technical and scientific basis for the negotiators, the following selection criteria were adopted:

(1) the importance of the crop for world food security and
(2) the interdependence of the countries with respect to the genetic resources of the crop of interest.

The support of the International Plant Genetic Resources Institute (IPGRI) in the preparation of the technical documents was essential in this phase.

In compliance with Article 25, the ITPGRFA was open for signature at the FAO headquarters from 3 November 2001 to 4 November 2002, for all members of the FAO and any states that were not members of the FAO but were members of the United Nations, or any of its specialized agencies, or of the International Atomic Energy Agency. Under Article 26, the Treaty is subject to ratification, acceptance or approval, and, under Article 27, it has been open to accession, once closed to signature. The Treaty entered into force on 29 June 2004, 90 days following its ratification by 40 governments, and it became operative with the first session of its Governing Body in Madrid in June 2006.

As part of the interim arrangements, the CGRFA (acting as the Interim Committee of the Treaty) met in order to: prepare the draft rules of procedure; draft the financial rules for the Governing Body of the Treaty and a budget proposal; propose procedures for its compliance; prepare draft agreements to be signed by the IARCs and the Governing Body; draft a material transfer agreement in order to facilitate access, including the terms for sharing commercial benefits; and initiate cooperation arrangements with the Conference of the Parties of the CBD.

The first meeting of the Governing Body of the Treaty, which took place in Madrid, Spain, in June 2006, brought together 105 countries and the European Union, which up until then had ratified it. This meeting resolved important issues and approved a Standard Material Transfer Agreement (ITPGRFA, 2006, Appendix G), which determined the level, form and manner of monetary payments on commercialization under the multilateral system of the Treaty.[10] During the meeting, the Relationship Agreement between the Governing Body of the Treaty and the Global Crop Diversity Trust, 'an essential element of the Treaty funding strategy', was signed (ITPGRFA, 2006, Appendix M), and the text of the agreements between the Governing Body and the IARCs and CATIE was adopted. It also achieved great advances towards resolving other issues, such as the mechanisms to promote compliance with the Treaty and its funding strategy. At its second (Rome, Italy, October 2007) and third sessions (Tunis, Tunisia, June 2009), the Governing Body made progress in the implementation of the funding strategy and the establishment of the principles and rules of the third party beneficiary (see Chapter 9 in this volume). During these sessions, the Governing Body also adopted important resolutions on farmers' rights and on the functioning of the multilateral system (ITPGRFA, 2007, 2009). The fourth session took place in Bali, Indonesia, in March 2011. To date, the Treaty has been ratified by 126 countries and the European Union.

Box 6.5 The Standard Material Transfer Agreement

The Standard Material Transfer Agreement (SMTA) was adopted at the first session of the Governing Body pursuant to Article 12.4 of the Treaty, which calls for the adoption of such an agreement to ensure facilitated access and benefit sharing in accordance with the provisions of the Treaty.

The SMTA is a contract between two parties (the provider and the recipient of the material), which sets out the conditions for access to, and use of, all plant genetic resources for food and agriculture in the multilateral system, as well as the conditions for benefit sharing in the case that the recipient commercializes a product that incorporates material received through the SMTA. In this way, the SMTA is the legal instrument through which the multilateral system of access and benefit sharing under the Treaty will operate.

The negotiations of the text of the SMTA were long and difficult. They began with the establishment of an expert group by the Commission on

> Treaty done → implementation & the challenge now

Genetic Resources for Food and Agriculture (CGRFA) at its first meeting, acting as the Interim Committee for the Treaty. This expert group met in Brussels, Belgium, on 4–8 October 2004, with the objective of developing and proposing recommendations on the terms of the SMTA. The outputs of the expert group were considered at the second meeting of the CGRFA, which decided to establish a Contact Group for the drafting of the SMTA. This Contact Group met twice, in Hamammet, Tunisia, on 18–22 July 2005, and in Alnarp, Sweden, on 24–28 April 2006. A Friends of the Chair Group, created at this last meeting, facilitated the resolution of the outstanding issues at the first session of the Governing Body, where the final text of the SMTA was approved.

Conclusion

The ITPGRFA marks a turning point in the history of international cooperation and is an important instrument for combating hunger and poverty. Society will benefit from the Treaty in various ways: consumers will benefit from a greater variety of food and agricultural products as well as increased food security; farmers will benefit from the provisions of the Treaty on farmers' rights; the scientific community will have access to crucial plant genetic resources for the improvement of plants and research; the IARCs will benefit because the Treaty recognizes their role and provides a legal framework for plant breeding, conservation and access to the collections they hold; and the public and private sectors will also benefit because they will be guaranteed access to a wider range of genetic diversity for agricultural development.

The Treaty is the last step of a long journey that allowed humanity to develop a binding agreement that guarantees and regulates the conservation, sustainable use and access to genetic resources for agriculture and food as well as a fair and equitable sharing of the benefits derived from its use. The speed with which countries are ratifying the Treaty in their parliaments is very encouraging.

The entry into force of the ITPGRFA is also the first step of a new path – its compliance. The Treaty is legally binding for countries that have ratified it. However, this is not enough – each country should develop the necessary regulations to put its provisions into practice and, in some cases, legislate or modify its national legislations in harmony with the Treaty.

The new phase that we are now facing is characterized by a communications challenge – through the media, information and awareness raising. The laws and regulations will only be a paper exercise if the message of the importance of these resources and the need for their conservation and sustainable use does not reach all members of society. To achieve this awareness, it is necessary that children in school learn to recognize the value of, and to develop respect for, natural resources as an integral part of their home, the Earth, and that adults maintain this respect throughout their lives.

Notes

1 International Treaty on Plant Genetic Resources for Food and Agriculture, 29 June 2004, www.planttreaty.org/texts_en.htm (last accessed 30 March 2011) [ITPGRFA].
2 International Undertaking on Plant Genetic Resources for Food and Agriculture, 1983, www.fao.org/waicent/faoinfo/agricult/cgrfa/IU.htm (last accessed 30 March 2011) [International Undertaking].
3 The Group of 77 is the largest intergovernmental organization of developing states in the United Nations. It currently has 131 member states. See www.g77.org (last accessed December 2010).
4 The delegation of New Zealand reserved its position on the text of the International Undertaking on Plant Genetic Resources because there was no provision which took account of plant breeders' rights. The delegations of Canada, France, Germany (Federal Republic of), Japan, Switzerland, United Kingdom and the United States of America reserved their positions with respect to the Resolution and the International Undertaking on Plant Genetic Resources. These same seven countries and the Netherlands reserved their positions with respect to the establishment of a Plant Genetic Resources Commission.
5 International Undertaking, *supra* note 2, Article 7.1.
6 International Centre for Research in Agroforestry (ICRAF), International Centre for Tropical Agriculture (CIAT), International Maize and Wheat Improvement Center (CIMMYT), International Centre for Agricultural Research in the Dry Areas (ICARDA), International Crops Research Institute for the Semi-Arid Tropics (ICRISAT), International Institute of Tropical Agriculture (IITA), International Network for the Improvement of Banana and Plantain (IPGRI/INIBAP), International Livestock Research Institute (ILRI), International Potato Center (CIP), International Rice Research Institute (IRRI), West Africa Rice Development Association (WARDA) and World Agroforestry Centre.
7 This publication is available in English, French and Spanish at www.idrc.ca/en/ev-93177-201-1-DO_TOPIC.html (last accessed December 2010).
8 Convention on Biological Diversity, 31 I.L.M. 818 (1992).
9 The documents of the Open-ended Working Group are available at www.fao.org/waicent/faoinfo/agricult/cgrfa/docswg.htm (last accessed December 2010).
10 Standard Material Transfer Agreement, 16 June 2006, ftp://ftp.fao.org/ag/agp/planttreaty/agreements/smta/SMTAe.pdf (last accessed 10 November 2010).

References

Consultative Group on International Agricultural Research (CGIAR) (1972) *Fourth Meeting of the Technical Advisory Committee of the Consultative Group on International Agricultural Research*, Report, 2–4 August 1972, Washington, DC, online: www.cgiar.org/corecollection/docs/cg7211c.pdf (last accessed November 2010).
——(2003a) 'Joint Statement of FAO and the CGIAR Centres on the Agreement Placing CGIAR Germplasm Collections under the Auspices of FAO', in *Centre Policy Instruments, Guidelines and Statements on Genetic Resources, Biotechnology and Intellectual Property Rights*, Booklet, version III, System-wide Genetic Resources Programme, Rome, Italy, online: www.sgrp.cgiar.org/sites/default/files/Policy_Booklet_Version3.pdf (last accessed November 2010).
——(2003b) 'Second Joint Statement of FAO and the CGIAR Centres on the Agreement Placing CGIAR Germplasm Collections under the Auspices of FAO', in *Centre Policy Instruments, Guidelines and Statements on Genetic Resources, Biotechnology and Intellectual Property Rights*, Booklet, version III, System-wide Genetic Resources Programme, Rome, Italy, online: www.sgrp.cgiar.org/sites/default/files/Policy_Booklet_Version3.pdf (last accessed November 2010).
Cooper, H.D. (2002) 'The International Treaty on Plan Genetic Resources for Food and Agriculture', *Review of European Community and International Environmental Law* 11(1): 1–16.

Esquinas-Alcázar, J. (2005) 'Protecting crop genetic diversity for food security: political, ethical and technical challenges', *Nature Reviews Genetics* 6: 946–53.

Food and Agriculture Organization (FAO) (1961) *Report of the Technical Meeting on Plant Exploration and Introduction* (10–20 July 1961), FAO, Rome.

—— (1969a) *Report of the Third Session of the FAO Panel of Experts on Plant Exploration and Introduction, Rome, Italy, 25–28 March 1969*, Plant Production and Protection Division, FAO, Rome.

—— (1969b) *Report of the FAO Panel of Experts on Plant Exploration and Introduction, 25–28 March 1969, third session*, Plant Production and Protection Division, FAO, Rome.

—— (1970) *Report of the Fourth Session of the FAO Panel of Experts on Plant Exploration and Introduction, Rome, Italy, 14–17 April 1970*, Plant Production and Protection Division, FAO, Rome.

—— (1973a) *Report of the Fifth Session of the FAO Panel of Experts on Plant Exploration and Introduction, Fifth Session, Rome, 8–10 March 1973*, Plant Production and Protection Division, FAO, Rome.

—— (1973b) *Report of the Technical Conference on Crop Genetic Resources. Rome, 12–16 March 1973*, Plant Production and Protection Division, FAO, Rome.

—— (1975a) *Report of the Sixth Session of the FAO Panel of Experts on Plant Exploration and Introduction, Rome, 3–5 December 1974*, Plant Production and Protection Division, FAO, Rome.

—— (1975b) 'Proposed Standards and Procedures for Seed Storage Installations Used for Long-term Conservation of Base Collections', in *Report of FAO Panel of Experts on Plant Exploration and Introduction, Sixth Session, Rome, 3–5 December 1974*, Plant Production and Protection Division, FAO, Rome.

—— (1979) *Verbatim Records of Commission II of the Twentieth Session of the FAO Conference, Rome, 10–28 November 1979*, Meeting 12 of Commission II, 21 November 1979, Doc. C 79/11/PV/12, FAO, Rome.

—— (1981a) *Report of the Twenty-first Session of the FAO Conference, Rome, 7–26 November 1981*, Doc. C 81/LIM/29, FAO, Rome.

—— (1981b) *Verbatim Records of Commission II of the Twenty-first Session of the FAO Conference*, 16 November 1981, Doc. C 81/II/PV/8, FAO, Rome.

—— (1981c) *Resolution 6/81 (Plant Genetic Resources)*, adopted at the Twenty-First Session of the FAO Conference, November 1981, FAO, Rome.

—— (1983a) *Report of the Eighty-third Session of the FAO Council, Rome, 13–24 June 1983*, FAO, Rome.

—— (1983b) *Report of the Seventh Session of the Committee on Agriculture (COAG), Rome, 21–30 March 1983*, Doc. CL 83/9, FAO, Rome.

—— (1983c) *Plant Genetic Resources: Report of the Director-General*, presented at the Twenty-Second Session of the FAO Conference, Doc. C83/25, FAO, Rome.

—— (1983d) *Report of the Twenty-second Session of the FAO Conference, Rome, 5–23 November 1983*, FAO, Rome, online: www.fao.org/docrep/x5563E/X5563e0a.htm#e.%20plant%20genetic%20resources%20(follow%20up%20of%20conference%20resolution%2068> 1 (last accessed December 2010)

—— (1996a) *Report on the State of the World's Plant Genetic Resources*, prepared for the International Technical Conference on Plant Genetic Resources, Leipzig, Germany, 17–23 June 1996, online: www.fao.org/AG/aGp/agps/pgrfa/pdf/SWRSHR_S.PDF (last accessed December 2010)

—— (1996b) *Global Plan of Action for the Conservation and Sustainable Use of Plant Genetic Resources for Food and Agriculture*, adopted by the International Technical Conference on Plant Genetic Resources, Leipzig, Germany, 17–23 June 1996, FAO, Rome.

—— (2001a) *Document CL 121/5-Sup. of the Hundred and Twenty-first Session of the FAO Council. Rome, 30 October – November 2001*, entitled *The International Undertaking on Plant Genetic Resources*, Information provided according to Article XXI.1 of the General Rules of the Organization, FAO, Rome.

—— (2001b) *Verbatim Records of the Plenary Sessions of the Thirty-first Session of the FAO Conference, Rome, 2–13 November 2001*, FAO, Rome.

——(2001c) *Resolution 3/2001 (Approval of the International Treaty of Plant Genetic Resources for Food and Agriculture and interim arrangements for its application) adopted at the Thirty-first Session of the FAO Conference, Rome,* November 2001, FAO, Rome, online: www.fao.org/ DOCREP/MEETING/004/Y2650e/Y2650e01.htm#3 (last accessed January 2010).
FAO/IBP (1967) *Recommendations of the Technical Conference on the Exploration, Utilization and Conservation of Plant Genetic Resources. Rome,* 18–26 September 1967, FAO, Rome, online: www.fao.org/docrep/006/85704E/85704E02.htm#ch21 (last accessed November 2010).
FAO / UN Environment Programme / International Board for Plant Genetic Resources (1981) *Report of the Technical Conference on Crop Genetic Resources. Rome,* 6–10 April 1981, FAO, Rome.
Frankel, O.H., and E. Bennet (1970) *Genetic Resources in Plants: Their Exploration and Conservation.* Blackwell Scientific, Oxford.
Frankel, O.H. and J.G. Hawkes (1975) *Crops Genetic Resources for Today and Tomorrow.* Cambridge University Press, Cambridge, UK.
Harlan, J.R. (1992) *Crops and Man.* Crop Science Society of America, University of Illinois, Urbana, IL.
Holden, J.H.W., and J.T. Williams (1984) *Crop Genetic Resources: Conservation and Evaluation.* Allen and Unwin, London.
International Institute for Sustainable Development (IISI) (1996) *Earth Negotiations Bulletin* 9 (40), online: www.iisd.ca/vol09/ (last accessed 15 April 2011).
——(1997a) *Earth Negotiations Bulletin* 9(66), online: www.iisd.ca/vol09/ (last accessed 15 April 2011).
——(1997b) *Earth Negotiations Bulletin* 9(68), online: www.iisd.ca/vol09/ (last accessed 15 April 2011).
——(1997c) *Earth Negotiations Bulletin* 9(76), online: www.iisd.ca/vol09/ (last accessed 15 April 2011).
——(1998) *Earth Negotiations Bulletin* 9(97), online: www.iisd.ca/vol09/ (last accessed 15 April 2011).
——(2000a) *Earth Negotiations Bulletin* 9(161), online: www.iisd.ca/vol09/ (last accessed 15 April 2011).
——(2000b) *Earth Negotiations Bulletin* 9(167), online: www.iisd.ca/vol09/ (last accessed 15 April 2011).
——(2001a) *Earth Negotiations Bulletin* 9(180), online: www.iisd.ca/vol09/ (last accessed 15 April 2011).
——(2001b) *Earth Negotiations Bulletin* 9(191), online: www.iisd.ca/vol09/ (last accessed 15 April 2011).
——(2001c) *Earth Negotiations Bulletin* 9(197), online: www.iisd.ca/vol09/ (last accessed 15 April 2011).
——(2001d) *Earth Negotiations Bulletin* 9(213), online: www.iisd.ca/vol09/ (last accessed 15 April 2011).
——(2002a) *Earth Negotiations Bulletin* 9(245), online: www.iisd.ca/vol09/ (last accessed 15 April 2011).
——(2002b) *Earth Negotiations Bulletin* 9(246), online: www.iisd.ca/vol09/ (last accessed 15 April 2011).
——(2006) *Earth Negotiations Bulletin* 9(369), online: www.iisd.ca/vol09/ (last accessed 15 April 2011).
International Treaty on Plant Genetic Resources for Food and Agriculture (ITPGRFA) (2006) *Report of the First Session of the Governing Body of the Treaty on Plant Genetic Resources for Food and Agriculture, Madrid, Spain,* 12–16 June 2006, Doc. IT/GB-1/06/Report, FAO, Rome, online: ftp://ftp.fao.org/ag/agp/planttreaty/gb1/gb1repe.pdf (last accessed January 2011).
——(2007) *Report of the Second Session of the Governing Body of the Treaty on Plant Genetic Resources for Food and Agriculture, Rome,* 29 October–2 November 2007, Doc. IT/GB-2/ 07/Report, FAO, Rome.
——(2009) *Report of the Third Session of the Governing Body of the Treaty on Plant Genetic Resources for Food and Agriculture, Tunis,* 1–5 June 2009, Doc. IT/GB-3/09/Report, FAO, Rome.

Kate, K., and C. Lasen (1997) 'The Undertaking Revisited: A Commentary on the Revision of the International Undertaking on Plant Genetic Resources for Food and Agriculture', *Review of European Community and International Environmental Law* 6: 284–86.

Mekour, M.A. (2002) *A Global Instrument on Agrobiodiversity: The International Treaty on Plant Genetic Resources for Food and Agriculture*. FAO Legal Papers Online no. 24, January 2002, Rome, Italy.

Parry, M.L. (1978) *Climatic Change, Agriculture and Settlement*. Archon Books, Hamden, CT.

Rose, G. (2003) 'International Law of Sustainable Agriculture in the 21st Century: The International Treaty on Plant Genetic Resources for Food and Agriculture', *Georgetown International Environmental Law Review* 15: 583–632.

System-wide Genetic Resources Programme (SGRP) (2010) *Booklet of CGIAR Centre Policy Instruments, Guidelines and Statements on Genetic Resources, Biotechnology and Intellectual Property Rights*, Version III, August, Rome, online: www.sgrp.cgiar.org/sites/default/files/Policy_Booklet_Version3.pdf (last accessed January 2011).

UN Environment Programme (UNEP) / Convention on Biological Diversity (CBD) / Conference of the Parties 2/19 (1995) Decision II/15: FAO Global System for Conservation and Utilization of Plant Genetic Resources for Food and Agriculture. UNEP/CBD, Montreal, Canada, online: www.cbd.int/doc/meetings/cop/cop-02/official/cop-02-19-en.pdf (last accessed December 2010).

Vasey, D.E. (1992) *An Ecological History of Agriculture*. Iowa State University Press, Ames, IA.

World Bank (1992) *World Development Report 1992: Development and the Environment*. Oxford University Press, New York.

World Trade Organization (WTO) (1994) *Agreement Establishing the World Trade Organization*, WTO, Geneva, online: www.wto.org/english/docs_e/legal_e/04-wto.pdf (last accessed January 2011).

Zhu, Y., H. Chen, J. Fan, Y. Wang, N. Li, J. Chen, J. Fan, S. Yang, L. Hu, H. Leung, W. Mew, P.S. Teng, Z. Wang and C.C. Mundt (2000) 'Genetic Diversity and Disease Control in Rice', *Nature* 406: 718–22.

Zohari, D. (2000) *Domestication of Plants in the Old World: The Origin and Spread of Cultivated Plants in West Asia, Europe and the Nile Valley*, Oxford University Press, Oxford.

7

THE DESIGN AND MECHANICS OF THE MULTILATERAL SYSTEM OF ACCESS AND BENEFIT SHARING

Daniele Manzella

Introduction

The Multilateral System of Access and Benefit-Sharing (multilateral system) was created under the International Treaty on Plant Genetic Resources for Food and Agriculture (ITPGRFA).[1] The multilateral system embodies the key policy and legal concepts that shaped the negotiations of the Treaty, such as food security, sovereign rights of States over their own plant genetic resources for food and agriculture (PGRFA), intellectual property rights, and applies those concepts to an operational instrument of access and benefit-sharing that aims to serve the community of farmers and breeders worldwide.

In this chapter, we will try to describe the multilateral system 'as is', meaning as it was established in Part IV of the ITPGRFA. We will take readers through the nuts and bolts of the multilateral system, including its scope, operational solutions for access and benefit sharing, and governance mechanisms.

A (short) description of the multilateral system

The multilateral system is a global gene pool of crops and forages. It is global because it was established by state governments in a binding agreement of international law. To date, the ITPGRFA has a membership of 127 contracting parties, from all regions of the world. The multilateral system pools samples of genetic material from a set of crops, which are listed in Annex 1 to the ITPGRFA. These crops provide about 80 per cent of our food from plants.[2]

Samples are included in the gene pool by the state governments and the institutions that they control. Samples also come into the gene pool from international institutions as well as from natural and legal persons – anyone, that is – within the jurisdiction of the contracting parties. These samples are pooled in that they are administered under a common set of rules. These rules are contained in the ITPGRFA and further specified in a contractual instrument, namely the Standard Material Transfer Agreement (SMTA).[3] The rules apply to individual transfers of these samples (for example, from a gene bank to a breeder) for certain purposes, namely the utilization and conservation for research, breeding and training for food and agriculture. The rules specify not only how to obtain access to the plant genetic material but also how to share the results of research and breeding on that material.

Why was the multilateral system created?

The multilateral system was created to address the specific features and needs of PGRFA in relation to access and benefit-sharing. The multilateral system is enshrined in the ITPGRFA, the objectives of which are the conservation and sustainable management of PGRFA as well as the fair and equitable sharing of the benefits arising from their use.[4] These objectives are in harmony with the Convention on Biological Diversity (CBD).[5] In fact, the ITPGRFA applies the general principles established in the CBD in ways that are particularly well suited to PGRFA.[6]

The multilateral system constitutes a special regime in the context of the international principles of access to genetic resources and benefit-sharing dictated by the CBD. The CBD established a regime of access and benefit sharing based on the two principles of prior informed consent and mutually agreed terms. Prior informed consent is to be given by the country of origin of the resources or by the country that has acquired the resources under the Convention. The terms of such access shall be mutually agreed. The implementation of these two principles in the realm of the CBD oriented access and benefit sharing a bilateral perspective – that is, distinct agreements between individual providers and recipients with terms negotiated on a case-by-case basis.[7]

Agriculture has characteristics and needs that do not fully fit into this logic. Agriculture has always been based on seed exchange. Farmers and farming communities have been swapping their crops, and the genes within their crops, since the beginnings of agriculture. Through 12,000 years of cultivation and exchange, many plant varieties for food and agriculture were, and continue to be, developed (Rose, 2003, 586). As a result, countries have become interdependent as they all depend very largely for food and agriculture on crops that have originated elsewhere. Today, drawing on the widest possible range of genetic resources to maintain and adapt those crops remains necessary for the sustainable production of food from plants (Moore and Tymowski, 2005, 2–6).

Agriculture needs an enabling access and benefit-sharing system that recognizes interdependence, triggers the sharing of genetic material of plant origin on a multilateral and facilitated basis and, most importantly, instills fairness into such a system and recognizes that the global pool to which access is facilitated is continuously enriched by the contributions of farmers worldwide. A practical and fair access and benefit-sharing system for PGRFA is to ensure that genetic resources continue to flow worldwide, while those individuals who conserve and develop those resources are adequately rewarded.

After the CBD, governments, in the exercise of their sovereign rights over their own biological resources, established a special system of access and benefit-sharing for PGRFA in order to devise and operate distinctive solutions to facilitate access to PGRFA and share the benefits produced by the use of these resources.

With the recent adoption of the Nagoya Protocol on Access and Benefit Sharing and the Fair and Equitable Sharing of Benefits Arising from Their Utilization by the CBD Tenth Conference of the Parties in October 2010, the underpinning reasons for a multilateral system of facilitated access and benefit sharing for PGRFA have been reaffirmed.[8] The special nature of agricultural biodiversity, its distinctive features and the problems needing distinctive solutions are recognized. The interdependence of all countries with regard to genetic resources for food and agriculture as well as their special nature and importance for achieving food security worldwide and for sustainable development of agriculture in the context of poverty alleviation and climate change, are also reaffirmed. In this regard, the fundamental role of the ITPGRFA is acknowledged, with a specific recognition of the multilateral system developed in harmony with the CBD.[9]

What is the legal basis of the multilateral system?

The multilateral system was established under the ITPGRFA, which sets forth certain obligations for states that are contracting parties. The contracting parties have agreed that facilitated access to PGRFA under the multilateral system will be regulated by the conditions established in the ITPGRFA itself and have also agreed to take the necessary legal or other appropriate measures to provide such facilitated access to other contracting parties as well as to legal and natural persons under their jurisdiction. These obligations are the legal cornerstone of the multilateral system.[10]

However, the legal basis of the multilateral system is not limited to states' obligations under international law. Under the ITPGRFA, the contracting parties have agreed that the modality for providing facilitated access is a standard contract, namely the SMTA, which reproduces and defines in detail the relevant provisions of the ITPGRFA and projects them into the sphere of private law (meaning, the law that two parties to a contract create to regulate their transaction).[11] The SMTA is the legal instrument that allows for the conditions of facilitated access pursuant to the ITPGRFA to be passed from states to natural and legal entities under their jurisdiction – that is, from a provider of PGRFA to a recipient. Hence, the multilateral system practically functions through a standard contract between two individuals or legal entities.

International law dictates the basic principles and rules of a system to administer an international public good, such as PGRFA, and the system is practically effected through a contractual instrument of private law, which projects a global system and its principles and rules into the individual transactions that occur under the system. An example of this intersection is in the SMTA provisions on applicable law. The applicable law to the SMTA includes the objectives and the relevant provisions of the ITPGRFA and, when necessary for interpretation, the decisions of the Governing Body, which is the forum grouping states that are contracting parties to the ITPGRFA.[12]

What is in the multilateral system?

The scope of the ITPGRFA covers all PGRFA.[13] However, the obligations that are in Part IV of the Treaty, which are those concerning the multilateral system, apply only to the PGRFA of certain crops and forages, which are listed in Annex 1 to the ITPGRFA.[14] This list was developed according to the criteria of food security and interdependence.[15] Thus, the multilateral system applies to the PGRFA of the crops and forages that are listed in Annex I to the ITPGRFA, and it 'contains' – albeit not physically in one place – these resources.

The multilateral system does not apply to all of the Annex I PGRFA that are in the territory of the contracting parties. It applies only to some of them – namely those that are 'under the management and control' of the Contracting Party and those that are 'in the public domain'.[16] When state governments negotiated the ITPGRFA, they were not in a position, politically, to commit to including resources that, for example, were subject to the rights of individuals and legal entities under national law (rights of any sort – for instance, the rights of local farming communities over resources on their territories). Thus, they limited the application of the multilateral system to resources that they could manage and control directly. In addition, they avoided using the ITPGRFA as a means of rewriting intellectual property rights laws (Halewood and Nnadozie, 2008). Materials in the public domain should be considered as those which are not the subject of intellectual property rights. As a result, the state governments have not committed to include resources in the multilateral system that are the subject matter of protection under intellectual property rights (see Chapter 9 in this volume).

The fact that governments could not commit to include PGRFA that they do not manage and control does not mean that the multilateral system's gene pool is precluded for those resources. The ITPGRFA foresees the possibility for natural and legal entities holding Annex I PGRFA to voluntarily place them in the system. Furthermore, governments have agreed to take measures to encourage those holders to include those resources.[17] As an incentive to do so, the ITPGRFA also provides that, within a certain timeline, the Governing Body should assess the progress of including these PGRFA in the multilateral system and that, following this assessment, the Governing Body should decide whether access shall continue to be facilitated to those natural and legal persons that have not included them.[18]

The multilateral system is not only constituted by PGRFA from national collections and territories, but also 'contains' PGRFA that are in the international gene banks of the International Agricultural Research Centres (IARCs) of the Consultative Group on International Agricultural Research (CGIAR).[19] The IARCs have been collecting PGRFA from farmers fields for many decades, and have been holding and distributing those PGRFA in trust for the benefit of the international community since 1994. With the advent of the ITPGRFA, the contracting parties have recognized the importance of these collections for the objectives of the Treaty and have called upon the IARCs to make them available under the terms and conditions of the multilateral system. Accordingly, the IARCs, and a number of other international institutions holding plant germplasm, signed agreements with the Governing Body of the Treaty. As a result, Annex I PGRFA that the IARCs maintain in their gene banks are accessible through the SMTA. The Governing Body authorized the IARCs to also apply the SMTA for distributing non-Annex I PGRFA that were collected before the entry into force of the ITPGRFA (FAO, 2007, para. 28).[20]

Not all the uses of PGRFA that are in the multilateral system are covered by the rules of the system. Would-be recipients cannot, as a right, request materials through the multilateral system for a limitless range of potential uses. Facilitated access through the multilateral system is for the purposes of 'utilization and conservation for research, breeding and training for food and agriculture' (Article 12.3(a) of the ITPGRFA). Such purposes do not include chemical, pharmaceutical and/or other non-food/feed industrial uses.[21] In substance, the contracting parties are obliged to grant facilitated access under the multilateral system only in cases where access is requested for certain purposes. If access is requested for other purposes, the contracting parties can apply different conditions from those of the multilateral system.

How does access work?

In practice, the multilateral system works as a common pooling, distributing and benefit-sharing system for the PGRFA that it covers. Access to such resources is facilitated in the sense that those who want to access the genetic material in the system do not need to negotiate access agreements on a case-by-case basis with national competent authorities. Instead, the resources are available to anyone who wants them under a standard contract, i.e. the SMTA. Contracting parties must ensure that the standard contract is applied to all transfers of PGRFA in the multilateral system by the natural and legal entities under their jurisdiction. The use of the SMTA cuts out all of the costs involved in the bilateral process for the benefit of farmers and gene bank managers who typically provide the genetic material and for the plant breeders and researchers who typically seek access to this material to improve it.

The text of the SMTA was negotiated over the course of two years through an intergovernmental process (Lim and Halewood, 2008) and was approved by the Governing Body of the ITPGRFA at its first session in 2006 (FAO, 2006). The terms and conditions of the SMTA have four basic objectives: (1) speeding up access; (2) creating a chain of SMTAs when the accessed material is transferred by the original recipient to someone else; (3) ensuring that access to another sample of the same type and kind remains possible to others under the same terms and conditions; and (4) recognizing that, once the accessed material is changed by the recipient (for instance, by plant breeding), such a recipient can decide not to make it available for facilitated access.

Thus, access is to be accorded expeditiously, without the need to track individual accessions (that is, individual samples of PGRFA that are transferred through the SMTA) and free of charge.[22] If the recipient conserves PGRFA accessed under the multilateral system, he or she shall assure the continued access to this PGRFA by others. Whoever obtains resources from the system, and conserves them, must continue to make them available to other people. He or she must do so by passing on the same obligations to the subsequent recipients who, in turn, will be bound to the same obligation. In practice, this is achieved by requiring the original recipient who wants to transfer the accessed material to use the same contract under which he or she has received the material, which will create a sequence of SMTAs.[23]

The recipient is not allowed to appropriate the received material by claiming intellectual property rights, or other rights that limit the facilitated access to the same material, in the form that it is received from the system.[24] If the recipient changes the material or works on it, he or she will have the discretion of either making or not making the changed material ('PGRFA under development') available to others, and, if the recipient decides to transfer the changed material to someone else, he will be able to request additional terms and conditions, separate from the SMTA, to the next recipient.[25]

In substance, access under the multilateral system is centred on a contractual instrument that allows the legal obligations provided for in the ITPGRFA to be passed on to recipients, and from them to subsequent recipients, by means of a contractual agreement.

How does benefit sharing work?

The logic underpinning benefit sharing in the multilateral system revolves around the following points. Since PGRFA in the multilateral system are treated as pooled goods, there is no individual owner with whom individual contracts for access and benefit sharing must be negotiated. As such, benefits resulting from their use do not go back to the provider. Rather, they must be shared in multilateral ways. The ITPGRFA recognizes that facilitated access to PGRFA is in itself a major benefit, making it possible for farmers, plant breeders and researchers, in both the public and private sectors, to have access to the widest possible range of PGRFA.[26]

The ITPGRFA identifies and makes a provision for a wide range of non-monetary benefit sharing, including the exchange of information, the access to, and transfer of, technology, and capacity building. These forms of benefit-sharing are largely based on general obligations of state governments under the ITPGRFA.[27] However, monetary benefit sharing remains central in the multilateral system, and innovative solutions have been introduced in the ITPGRFA and translated into contractual obligations in the SMTA. The recipients of material under the SMTA are subject to the following benefit-sharing scheme:

> A recipient who commercializes a product that is a plant genetic resource for food and agriculture and that incorporates material accessed from the Multilateral System, shall pay ... an equitable share of the benefits arising from the commercialization of that product, except whenever such a product is available without

restriction to others for further research and breeding, in which case the recipient who commercializes shall be encouraged to make such payment.[28]

In a nutshell, the mechanism is devised to work in these terms: providers make genetic material available; recipients work on this material through research and breeding; once the material reaches the product stage as PGRFA (for example, a new plant variety is generated and this variety contains genetic material accessed under the multilateral system) and if such a product is commercialized, part of the generated revenue is shared. However, the revenue is shared mandatorily only in cases where the research and breeding on the commercialized product is restricted. If research and breeding are still free (typically because there is no restriction on the product due to intellectual property right or because the consent of the right holder for acts of research and breeding is not required), the monetary benefits are shared voluntarily.[29]

The ITPGRFA sets forth the basic structure of monetary benefit sharing under the multilateral system, but it is the SMTA that defines how much is to be shared. The recipient has two alternative options for monetary benefit sharing: either he or she pays 0.77 per cent on the net sales of the commercialized (and restricted) product for a period corresponding to the duration of such restriction (for instance, 20 years in the case of intellectual property rights-based restrictions), or he pays 0.5 per cent on the sales of all PGRFA products of the same crop to which the accessed material belongs.[30] In the latter case, the recipient pays regardless of the restrictions for further research and breeding on the products, and for a period of ten years, which is renewable. He or she obtains, in return for this greater payment obligation, access to all of the genetic material of that crop in the multilateral system – that is, he can obtain more material of the same crop from others under separate SMTAs but will have to pay only once.

SMTA-generated monetary benefits flow into a multilateral fund – namely the Benefit-Sharing Fund.[31] This fund is also open to direct contributions from the contracting parties, the private sector, non-governmental organizations and other sources, such as institutional donors (FAO, 2006, Appendix F, paras 2.1(d)–(e)). Under the ITPGRFA, the state governments agreed that benefits arising from the use of PGRFA that are shared under the multilateral system would flow primarily to farmers, especially in developing countries, who conserve and use PGRFA in a sustainable manner. Accordingly, the Governing Body, in the context of a general funding strategy for the ITPGRFA, has determined that financial resources in the benefit-sharing fund be allocated to three main priorities, namely: (1) information exchange, technology transfer and capacity building; (2) managing and conserving PGRFA on-farm; and (3) the sustainable use of PGRFA (FAO, 2007, Appendix D.1).

Resources in the Benefit-Sharing Fund are under the direct control of the Governing Body, which means that the contracting parties decide how much to allocate, to whom and for what, and also how much the fund is expected to capitalize within a certain period of time. In practice, calls for proposals under the Benefit-Sharing Fund are open on a regular basis, so that any governmental or non-governmental organization, including gene banks and research institutions, farmers and farmers' organizations and regional and international organizations, based in countries that are eligible contracting parties, may apply for grants (FAO, 2007, Appendix D.2). The results of those projects funded through these grants (typically, PGRFA that are managed and conserved on-farm and information generated from these projects) go back into the multilateral system, thus creating a complete loop whereby PGRFA are accessed and improved, the benefits of those improvements that are shared go to conserve and use in a sustainable manner more PGRFA, which, in turn, are made available for facilitated access (FAO, 2009, Appendix A.3).

Who maintains the multilateral system?

The Governing Body, which is comprised of the states that are contracting parties to the ITPGRFA, is the main entity responsible for the policy guidance of the multilateral system.[32] Intergovernmental technical committees that are established by the Governing Body play a fundamental supporting role. They work intersessionally (that is, in the period between one session of the Governing Body and the following one) to deal with technical and operational matters and report to the Governing Body. They also provide users of the multilateral system, including providers and recipients under the SMTA, with guidance and assistance. The Ad Hoc Advisory Committee on the SMTA and the multilateral system, which was created by the Governing Body at its third session in 2009 (FAO, 2009, Annex), advises on implementation questions raised by users of the SMTA.

The Ad Hoc Committee on the Funding Strategy, among other matters, advises on the disbursing and reporting procedures for recipients of financial resources under the benefit-sharing fund (FAO, 2009).

Although it creates a direct line of communication with users of the system, the work of these technical committees does not diminish the power and prerogatives of the Governing Body, which remains central in terms of guidance of the multilateral system. However, it proves the recognition by contracting parties that in order for the machinery of the multilateral system to serve properly the community of users, practical and detailed guidance is needed on a constant basis.

The Food and Agriculture Organization of the United Nations (FAO) also plays a role in maintaining the multilateral system. As the SMTA is the trigger of the access and benefit-sharing mechanisms of the multilateral system, some of the obligations in the SMTA are set forth in favour of the multilateral system as a whole and are not in the interest of either the individual provider or recipient. As a result, neither the provider nor the recipient would be willing to enforce these obligations. The monetary payment to be made by the recipients is an example of an obligation that the provider has no interest in enforcing, since the payment goes into a multilateral trust fund and not to the individual provider. It is for this reason that, in the context of the bilateral contractual relationship, the SMTA recognizes certain roles and responsibilities for the Third Party Beneficiary, which is an entity other than the two parties to the SMTA that represents certain interests in the execution of the contractual obligations.[33] The FAO, on behalf of the Governing Body, is the Third Party Beneficiary under the ITPGRFA. The Governing Body, at its first session in 2006, invited the FAO to act as the Third Party Beneficiary to carry out the roles and responsibilities, as identified in the SMTA, under the direction of the Governing Body and in accordance with the procedures established by the Governing Body. The FAO accepted this invitation, and the Governing Body, at its third session, approved the Third Party Beneficiary procedures to guide the FAO in the exercise of its facilitating functions for dispute settlement under the SMTA.[34]

How do we know what happens within the multilateral system?

The tracking of material is not required under the multilateral system.[35] However, simple reporting obligations are established in the SMTA in order to transmit information to the Governing Body, which, through this information, can effectively monitor the operation of the multilateral system and, in particular, the operation of the SMTA, while maintaining standards of confidentiality for each individual transaction. Much of the information to be reported by parties to the SMTA is required specifically in order to enable the Third Party Beneficiary to initiate dispute settlement under the SMTA. The SMTA requires the provider to inform the Governing Body about the individual material transfer agreements that it has entered into. This information shall be made available by the Governing Body to the Third Party Beneficiary, which keeps it secure and confidential.[36] An initial recipient, who transfers the material originally received and thus acts as provider to a subsequent recipient, has the exact same obligation.[37] If the recipient has modified the material originally received from the multilateral system and decides to transfer it (as 'PGRFA under development'), he or she must identify the original material in the SMTA and notify the Governing Body.[38]

A further set of reporting obligations relates to monetary benefit sharing for products developed from PGRFA obtained from the multilateral system. The SMTA requires a recipient who is obliged to make payments to submit to the Governing Body an annual report on the sales of the product, the amount of the payment due and the information that allows for the identification of any restrictions that have given rise to the benefit-sharing obligation.[39] If the recipient opts for the crop-based alternative payment scheme, he or she must notify the Governing Body of such choice and submit an annual report to the Governing Body on the sales of all of the products belonging to the crop and the amount of the payment that is due.[40]

Where does the multilateral system stand at present?

The multilateral system has started functioning. PGRFA are being distributed under the SMTA worldwide by the IARCs. A number of contracting parties have also started using the SMTA to grant access to PGRFA that they consider as incorporated into the multilateral system, and so have some private PGRFA holders.[41] Financial contributions from various entities have been made into the Benefit-Sharing Fund and, thanks to those contributions, two calls for proposals have been opened to date. Through these calls, financial resources have been allocated by the Governing Body to fund projects in developing countries. Governmental and non-governmental organizations, including gene banks and research institutions, farmers and farmers' organizations and regional and international organizations have recieved grants under the Benefit-Sharing Fund.[42]

The governance mechanisms of the multilateral system are also progressing, through assessments of the current status of the system and policy guidance. As the multilateral system has established a functional system that needs to process numerous transactions involving PGRFA by institutions and individuals, in a practical and coherent manner, and as the Governing Body is to impart policy guidance not only at the macro level but also at the micro level of the operations under the system, the Governing Body needs to know how the system is working and how the system instruments are being applied. This is why the efforts of the various ITPGRFA stakeholders are concentrating on ensuring that an adequate flow of information reaches the Governing Body.

There are three detectable categories of information through which the Governing Body is articulating its guidance (FAO, 2009). The first category relates to the implementation of the multilateral system in terms of scope and actual availability of PGRFA. Since the multilateral system is a virtual gene pool, it can only work insofar as its potential users have the ability to know what is in it, what genetic material they can access and where this genetic material comes from. As a result, the Governing Body has stressed the importance of documenting all of the PGRFA within the multilateral system, so that they can be accessed. The Governing Body has also requested all contracting parties to report on their PGRFA that are in the multilateral system, and to take measures to make information on these resources available to the potential users of the multilateral system.

The second category relates to the inclusion of PGRFA in the multilateral system by natural and legal persons within the jurisdiction of the contracting parties, which is essential in order for the Governing Body to make an assessment on the progress of such inclusions and to decide whether access will be granted to those who have not included their material in the system.

The third category relates to the implementation and operation of the SMTA. The Governing Body has realized that, for the system to work on a large scale, all contracting parties are to take the policy, legal and administrative measures necessary for their national plant genetic resource systems, and for the natural and legal persons within their jurisdictions, to be able to use the SMTA in order to provide facilitated access to PGRFA.

A number of stakeholders are already providing the Governing Body with such information, from contracting parties that have issued a list of the materials that they consider to be available under the SMTA, often by linking this information to publicly available databases – to private entities that have voluntarily placed their own PGRFA in the multilateral system, and to the IARCs that have published accurate and informative statistics on transfers of PGRFA under the SMTA (SGRP, 2009).

Conclusion

The multilateral system is not static, rather, continually evolving. It evolves because it expands in size, as holders agree to make plant genetic material available under the terms and conditions of the SMTA and new gene bank accessions of crops and forages listed in Annex 1 are incorporated into the system. It also evolves because its functioning is to adapt to users' need. Indeed, the evolution of the multilateral system takes place at four levels: at the international level under the aegis of the Governing Body, which groups the state contracting parties to the ITPGRFA; at the level of international organizations that manage and provide access to international collections under the policy guidance of the Governing Body; at the national level, where governments manage national public collections of plant genetic material and implement their obligations though domestic legal and administrative measures; and at the users' level, where providers and recipients exchange and use material.

The interactions among these four levels pose some challenges to the future growth system. These challenges revolve around the governance mechanism of the multilateral system. An intergovernmental forum is required to make complex technical decisions that must be informed, based on data coming not only from the governments but also from other entities (for example, the IARCs and natural and legal persons). The governance mechanisms of the multilateral system are such that the micro level (that is, the users) and the macro level (that is, the Governing Body) are directly connected. As seen above, this connection is particularly apparent in the case of the SMTA. It reproduces some provisions of the ITPGRFA (an instrument of international law that establishes obligations for states) and was approved by the Governing Body, but it is used as a means of creating contractual obligations between legal persons within the jurisdiction of these states. The Governing Body manages such contractual instrument with the advice of a technical committee, on the basis of the experience and the needs of the users. To what extent will state governments accept that the decisions of an intergovernmental body (decisions that, according to the ITPGRFA, are taken by consensus) directly reach users within their jurisdiction and, thereby, bypass the regulatory 'filters' of national authorities and reduce the margin of discretion that government authorities normally enjoy? If national authorities continue accepting, and, if so, under what limitations, that the gene pool of the multilateral system is governed by an agreed set of internationally agreed rules, and that these rules may be adjusted, or even partially redefined, collectively at the international level and based on the experiences that come directly from the users of the system, the overall design of the multilateral system will prove to be right. If the mechanics of the system are administered efficiently by the Governing Body collectively, with concrete and practical responses to the needs that users have, access to

PGRFA will be facilitated on a global scale, and benefit sharing for farmers in developing countries will be substantial.

Otherwise, the multilateral system runs the risk of losing its multilateral and functional dimensions and dissolving into the experimental dimension of a first global effort to redress the shortcomings of bilateralism in access and benefit-sharing for genetic resources for food and agriculture.

Through an examination of its policy and legal infrastructure, we have highlighted the innovative solutions for access and benefit-sharing that the multilateral system proposes. These innovations are based on the management of an international public good through a contractual instrument and the enshrining of such a contractual instrument into an international treaty that is governed by state contracting parties. These innovations inevitably bear the risk of failure if they are not managed progressively. The governance of the multilateral system, through the Governing Body and its technical committees, is complex but dynamic enough to accommodate the adaptations and evolutions of the system that the users for which the system has been designed may require. If it stands the challenges of adaptation and evolution, the multilateral system will have a long and prolific life.

Notes

1 International Treaty on Plant Genetic Resources for Food and Agriculture, 29 June 2004, www.planttreaty.org/texts_en.htm (last accessed 15 April 2015) [ITPGRFA].
2 For information on the ITPGRFA, see www.planttreaty.org/mls_en.htm (last accessed 15 May 2011).
3 Standard Material Transfer Agreement, 16 June 2006, ftp://ftp.fao.org/ag/agp/planttreaty/agreements/smta/SMTAe.pdf (last accessed 15 April 2011).
4 *Ibid.*, Article 1.1.
5 Convention on Biological Diversity, 31 ILM 818 (1992) [CBD]. The close relationship between the CBD and the ITPGRFA has its origin in the mandate for the negotiation of the Treaty, given by the Conference of the Food and Agriculture Organization (FAO) in 1993 following an invitation made at the Nairobi conference in 1992, which included the issues of access to plant genetic resources for food and agriculture on mutually agreed terms. The Treaty provides that its objectives are to be pursued by closely linking the Treaty to the FAO and the CBD as well as by the cooperation of the Governing Body with the Conference of the Parties to the CBD. ITPGRFA, *supra* note 1, Articles 1.2, 19.3(f), (g) and (l).
6 Indeed, the Conference of the Parties to the CBD recognizes 'the special nature of agricultural biodiversity, its distinctive features and problems needing distinctive solutions'. CBD, *supra* note 3, Decision II/15 of the Conference of the Parties, www.biodiv.org/decisions/default.aspx?m=COP-02& id = 7088& lg = 0 (last accessed 15 April 2011).
7 The following are the basic provisions of the CBD on access to genetic resources and benefit sharing:

Article 15.1: Authority to determine access to genetic resources rests with the national governments and is subject to national legislation.
Article 15.4: Access, where granted, shall be on mutually agreed terms.
Article 15.5: Access to genetic resources shall be subject to prior informed consent of the contracting party providing such resources.
Article 15.7: Contracting parties shall take legislative, administrative or policy measures with the aim of sharing in a fair and equitable way the results of research

and development and the benefits arising from the commercial and other utilization of genetic resources with the contracting party providing such resources.

8 Nagoya Protocol on Access and Benefit Sharing and the Fair and Equitable Sharing of Benefits Arising from Their Utilization, 29 October 2012, www.cbd.int/abs/text/ (last accessed 15 April 2011).
9 *Ibid.*, preamble, paras 14, 15 and 18.
10 ITPGRFA, *supra* note 1, Articles 12.1 and 12.2.
11 *Ibid.*, Article 12.4.
12 SMTA, *supra* note 3, Article 7.
13 ITPGRFA, *supra* note 1, Article 3.
14 *Ibid.*, Article 11.1.
15 The criteria and processes that led to the list of crops and forages in Annex I to the ITPGRFA are explained in Chapter 16 of this volume.
16 ITPGRFA, *supra* note 1, Article 11.2.
17 *Ibid.*, *supra* note 1, Article 11.3.
18 *Ibid.*, Article 11.4.
19 *Ibid.*, Article 11.5.
20 *Ibid.*, Article 15.1(b).
21 *Ibid.*, *supra* note 1, Article 12.3(a). Correspondingly, Article 6.1 of the SMTA provides that '[t]he Recipient undertakes that the Material shall be used or conserved only for the purposes of research, breeding and training for food and agriculture. Such purposes shall not include chemical, pharmaceutical and/or other non-food/feed industrial uses.'
22 ITPGRFA, *supra* note 1, Article 12.3(b); SMTA, *supra* note 3, Article 5(a).
23 *Ibid.*, *supra* note 1, Article 12.3(g); SMTA, *supra* note 10, Article 6.4.
24 *Ibid.*, *supra* note 1, Article 12.3(d); SMTA, *supra* note 10, Article 6.2.
25 *Ibid.*, *supra* note 1, Article 12.3(e); SMTA, *supra* note 10, Article 6.5.
26 *Ibid.*, *supra* note 1, Article 13.1.
27 *Ibid.*, Article 13.1(a)-(c).
28 *Ibid.*, Article 13.2(d)(ii); SMTA, *supra* note 10, Article 6.7.
29 SMTA, *supra* note 10, Article 6.8.
30 The ITPGRFA, *supra* note 1, prescribed that, at its first meeting, the Governing Body of the Treaty would determine the level, form and manner of the payment, in line with commercial practice (Article 13.2(d)(ii)). Accordingly, the Governing Body, at its first meeting, approved the text of the SMTA, whose provisions define the amount and modalities of payments. See Annexes 2 and 3, and Article 6.11 of the SMTA for the rate and modalities of payment.
31 ITPGRFA, *supra* note 1, Article 19.3(f).
32 *Ibid.*, *supra* note 1, Article 19.3(a), lists as a function of the Governing Body to: '(a) provide policy direction and guidance to monitor, and adopt such recommendations as necessary for the implementation of this Treaty and, in particular, for the operation of the Multilateral System.'
33 The 'third party beneficiary' is a legal concept derived from the English law of contracts based on which a person, who is not a party to a contract but for whose benefit the contract was concluded, has legal rights to enforce the contract.
34 The third party beneficiary procedures specify the two-fold role of the FAO acting as third party beneficiary by establishing a system of progressive escalation, which consists of: (1) the initial gathering of information with regard to disputes; (2) facilitation of amicable dispute settlement; (3) initiation of mediation; and (4) initiation of arbitration. The third party beneficiary is dealt with more extensively in Chapter 9 of this volume.
35 ITPGRFA, *supra* note 1, Article 12.3(b); SMTA, *supra* note 3, Article 5(a).
36 SMTA, *supra* note 3, Article 5(e). The Governing Body determined that providers have two options for fulfilling their reporting obligations, as follows: (1) by transmitting a copy of the completed SMTA or (2) by ensuring that the completed SMTA is at the

disposal of the third party beneficiary, stating where the SMTA in question is stored and how it may be obtained and providing a series of data contained in the SMTA itself. In the same resolution, the Governing Body also decided that the information required from the provider shall be communicated at least once every two calendar years (FAO, 2009).
37 SMTA, *supra* note 3, Article 6.4.
38 *Ibid.*, Article 6.5.
39 *Ibid.*, Article 6.7.
40 *Ibid.*, Article 6.11.
41 A list of materials that are considered as being included in the multilateral system and available under the SMTA is available at www.planttreaty.org/inclus_en.htm (last accessed 15 May 2011).
42 See Call for Proposals 2010, ftp://ftp.fao.org/ag/agp/planttreaty/funding/call2010/cfp10_0_en.pdf (last accessed 15 May 2011).

References

Food and Agriculture Organization (FAO) (2006) *Report of the First Session of the Governing Body of the International Treaty on Plant Genetic Resources for Food and Agriculture*, Doc. IT/GB-1/06/Report, Madrid, Spain.
——(2007) *Report of the Second Session of the Governing Body of the International Treaty on Plant Genetic Resources for Food and Agriculture*, Doc. IT/GB-2/07/Report, FAO, Rome.
——(2009) *Report of the Third Session of the Governing Body of the International Treaty on Plant Genetic Resources for Food and Agriculture*, Doc. IT/GB-3/09/Report, Tunis, Tunisia.
Halewood, M., and K. Nnadozie (2008) 'Giving Priority to the Commons: The International Treaty on Plant Genetic Resources for Food and Agriculture', in G. Tansey and T. Rajotte (eds), *The Future Control of Food, A Guide to International Negotiations and Rules on Intellectual Property, Biodiversity and Food Security*, Earthscan, London.
Lim E. and M. Halewood (2008) 'A short history of the Annex 1 list', in G. Tansey and T. Rajotte (eds), *The Future Control of Food: A Guide to International Negotiations and Rules on Intellectual Property, Biodiversity and Food Security*, Earthscan, London.
Moore, G., and W. Tymowski (2005) *Explanatory Guide to the International Treaty on Plant Genetic Resources for Food and Agriculture*. International Union and the Conservation of Nature, Gland, Switzerland.
Rose, Gregory (2003) 'International Law of Sustainable Agriculture in the Twenty-First Century: The International Treaty on Plant Genetic Resources for Food and Agriculture', *Georgetown International Environmental Law Review* 15: 583–631.
System-Wide Genetic Resources Programme (SGRP) (2009) *Experiences of the International Agricultural Research Centres of the Consultative Group on International Agricultural Research with the Implementation of the Agreements with the Governing Body, with Particular Reference to the Use of the Standard Material Transfer Agreement for Annex 1 and Non-Annex 1 Crops*, FAO Doc. IT/GB-3/09/Inf., 15, International Treaty on Plant Genetic Resources for Food and Agriculture, online: ftp://ftp.fao.org/ag/agp/planttreaty/gb3/gb3i15e.pdf (last accessed 15 May 2011).

8

PROTECTING THE INTERESTS OF THE MULTILATERAL SYSTEM UNDER THE STANDARD MATERIAL TRANSFER AGREEMENT

The third party beneficiary

Gerald Moore

Introduction

The Standard Material Transfer Agreement (SMTA), like any other material transfer agreement (MTA), is basically a contract setting out the terms and conditions under which material will be transferred between the provider and the recipient of the material.[1] While some of the parties to a particular SMTA may be state enterprises, the SMTA, like other MTAs, operates at the level of private contract law rather than international law. One of the basic principles of contract law is that a contract creates binding rights and obligations only as between the parties to the contract – this is known as the principle of privity of contract. The difference between the SMTA and a normal MTA is that plant genetic resources for food and agriculture (PGRFA) that are transferred under an SMTA are regarded as coming from the multilateral system on access and benefit sharing (multilateral system), and benefits under the SMTA flow not to the individual provider but to the multilateral system itself.[2] Once in the multilateral system, the benefits are to be passed on to farmers, particularly in developing countries, in order to finance further efforts to conserve and use PGRFA sustainably. In this sense, the multilateral system is the real beneficiary of the benefit-sharing provisions of the SMTA rather than the parties to the SMTA themselves. The SMTA recognizes this fact by providing for the appointment of a third party beneficiary to represent the rights of the multilateral system and by giving this third party beneficiary the power to initiate dispute settlement action, including arbitration, in the event of a breach of the terms and conditions of the SMTA affecting those rights. In so doing, the SMTA also solves one of the major problems affecting access and benefit-sharing agreements in general, namely the capacity of developing countries, or individual institutions in developing countries, to enforce the terms of these agreements.

This chapter describes the development of the concept of the third party beneficiary in the context of the negotiations on the SMTA and examines its relationship with the International Treaty on Plant Genetic Resources for Food and Agriculture (ITPGRFA), its status in national law and the way in which the concept is expressed in the SMTA.[3] The chapter then looks at the implications of the concept for genetic resource agreements outside the scope of the ITPGRFA.

The origins of the concept in the negotiations of the SMTA

The drafting of the SMTA was carried out in three distinct phases. The first phase involved the initial consideration by a regionally balanced group of experts of the possible elements to be included in a standard material transfer agreement. The second phase concerned the negotiation of the first draft of the SMTA by countries that were members of the Food and Agriculture Organization's (FAO) Commission on Genetic Resources for Food and Agriculture (CGRFA). The final phase comprised the negotiation of the SMTA by the contracting parties in the first session of the Governing Body of the ITPGRFA in June 2006 and its adoption during the closing plenary meeting.

Phase 1: the Expert Group on the Terms of the SMTA (Expert Group)

The concept of the third party beneficiary was first raised during the discussions of the Expert Group, which met in Brussels in October 2004 under the chairmanship of Lim Eng Siang from Malaysia. The Expert Group adopted firm recommendations only with respect to the process for taking forward the drafting of the SMTA. However, it also listed a number of options for possible elements to be included in the draft SMTA, together with general views on those options. These options were drawn up in the form of responses to a number of questions posed by the FAO's CGRFA, acting as the Interim Committee for the ITPGRFA (CGRFA, 2004a). Many of the views expressed during the meeting commented on the fact that benefits under the SMTA, including both monetary and non-monetary benefits, flowed to the multilateral system and not to the individual provider of the material (*ibid*. at para. 49 (first bullet point) and para. 50). Following this thought, one of the views expressed was that '[t]hird parties should be able to initiate dispute settlement' (*ibid*. at para. 61 (point 9 on dispute resolution)). In the same connection, the legal adviser to the meeting 'noted that, because there are third party beneficiaries under the MTA, through the Multilateral System it may be advantageous to allow for them to be represented in dispute settlement, which would be easier in international arbitration' (*ibid*.). Quite separately, concerns were also expressed by the Expert Group regarding compliance control. Under 'additional items' to be included in the SMTA, a view was expressed that '[t]here should be a guarantor, to ensure that the obligations the recipient accepts are fulfilled' (*ibid*. at point 10). It was these two strands of thought that were eventually to coalesce in the concept of the third party beneficiary.

Phase 2: the Contact Group

At its second meeting in November 2004, the CGRFA, acting as Interim Committee for the ITPGRFA, agreed on the establishment of a contact group to develop a draft

SMTA for consideration by the Governing Body (CGRFA, 2004b).[4] The Contact Group held its first meeting in July 2005.

The notion of the third party beneficiary under the SMTA and its possible role in initiating dispute settlement was presented, in a background study paper, to this first session of the Contact Group by the Secretariat (Moore, 2005). The paper noted that many of the rights created by the SMTA – in particular, those relating to benefit sharing – would in fact be third party beneficiary rights, and looked at the mechanisms for according rights to initiate dispute settlement proceedings to the multilateral system as a third party beneficiary under the SMTA. In this connection, the paper also raised a possible alternative approach of defining the provider of PGRFA under the SMTA as an agent for the multilateral system.

This latter notion was taken up by the African Group in the first meeting of the Contact Group in Tunisia in July 2005. The African Group pointed out that under the multilateral system the providers of PGRFA in developing countries would have limited capacity or incentive to monitor and/or enforce compliance by recipients with the terms of the SMTA, given that the benefits flow to the multilateral system rather than to the providers. The African Group put forward the notion of the provider acting as an agent for the multilateral system as a possible solution that would at the same time protect the interests of the multilateral system while providing a guarantor to ensure that the obligations undertaken by the recipient were enforced. The notion was referred to a Legal Expert Group established by the Contact Group. Reporting back to the Contact Group,[5] the Legal Expert Group advised that, while the concept of agency was compatible with the ITPGRFA, it foresaw some legal and practical difficulties in developing this approach.[6] It identified the option of establishing third party beneficiary rights in the SMTA and empowering a legal entity to initiate legal action to enforce these rights as an alternative way of dealing with the underlying concern expressed by the African Group. The Contact Group included this latter option in the draft SMTA that it drew up.[7]

Further work on the third party beneficiary concept was undertaken by the Secretariat of the CGRFA between the first and second meetings of the Contact Group, and a more detailed information document was submitted to the Contact Group at its second meeting in Sweden in April 2006 (CGRFA, 2006a).[8] The document analysed the basis for the concept of the third party beneficiary in the Treaty, in national law and in the draft SMTA and discussed the questions of what institution could constitute the third party beneficiary and how the SMTA might be drafted to include a role for the third party beneficiary.[9] With respect to this last question, the document looked at the definition of the third party beneficiary, its right to initiate dispute settlement, the scope of the rights to be protected, monitoring rights, *locus standi* and arbitration, and suggested appropriate wording.

The Contact Group, at its second meeting, accepted the concept of the third party beneficiary in principle, with the negotiations focusing on the scope of the rights to be protected, the degree of control to be exercised by the Governing Body over the third party beneficiary and the extent of any monitoring powers. These questions, including which organization was to be entrusted with the role of the third party beneficiary, were left to the Governing Body to resolve at its first session in Madrid in June 2006 (CGRFA, 2006b, para. 14(d)).[10]

Phase 3: the Governing Body

The negotiations on the SMTA in the Governing Body lasted the better part of one whole week. By the time this meeting was held, there was little questioning of the need for the concept of the third party beneficiary to be incorporated into the SMTA. As in the second session of the Contact Group, the negotiations focused on the scope of the rights to be protected and the extent of monitoring powers. In the end, the compromise reached was to define the rights to be protected somewhat broadly, but to limit monitoring mainly within the context of dispute settlement. The provisions of the SMTA relating to the third party beneficiary are described in more detail in the following sections.

The third party beneficiary in the ITPGRFA

The actual words 'third party beneficiary' are not to be found anywhere in the text of the Treaty. Nevertheless, the factual situation underlying this concept is clearly established in the provisions dealing with the multilateral system and the SMTA. Under Article 12.4 of the ITPGRFA, facilitated access is to be provided pursuant to a SMTA. This agreement is to be between the provider of the PGRFA and the recipient of these resources. However, the benefits under the SMTA, including the monetary benefits of commercialization, are not to flow to the individual provider but, rather, to the multilateral system itself for the ultimate benefit of farmers in all countries who conserve and sustainably use PGRFA. Essentially, then, the position of the multilateral system as a third party beneficiary under the SMTA is expressly provided for in the Treaty.

While the concept of a third party beneficiary is, therefore, implicit in the ITPGRFA, nothing is said as to how the interests of the third party beneficiary are to be protected. Indeed, Article 12.5 of the Treaty provides that:

> Contracting Parties shall ensure that an opportunity to seek recourse is available, consistent with applicable jurisdictional requirements, under their legal systems, in case of contractual disputes arising under such MTAs, recognizing that *obligations arising under such MTAs rest exclusively with the parties to those MTAs* [emphasis added].

The question thus arises as to whether the last clause of Article 12.5 would in any way preclude the bringing of an action by, or on behalf of, a third party beneficiary.

The answer to this question would appear to be no. From a literal, if somewhat narrow, interpretation of the wording of Article 12.5, the restriction would apply only to 'obligations' arising out of the SMTA and not to 'rights' arising out of these agreements. This reasoning is consonant with general principles of contract law, which would preclude absolutely the creation of obligations on the part of non-parties to a contract, while not necessarily precluding the creation of rights for third parties. It could also be argued that in a sense a third party beneficiary is in fact a party to the agreement, albeit on a different standing to the provider and the recipient, and, consequentially, with more limited rights. Both of these interpretations would be consonant with the objectives of the ITPGRFA – it would hardly be consistent for the Treaty to create third party beneficiary rights under the SMTA and, at the same time, to preclude the enforcement of these rights.

Given that the concept of a third party beneficiary is implicit in the ITPGRFA and that the right of the third party beneficiary to enforce its rights is not precluded by the wording of the Treaty, what remains to be determined is how the rights accorded to the third party beneficiary under the SMTA may be enforceable in practice.

The concept of third party beneficiary rights in national law

As noted earlier, the principle of privity of contract means that a contract binds only the direct parties to that contract and creates rights only for those parties. It is recognized, in particular, that a contract cannot ever create legal obligations that are binding on a third party without his or her consent. However, national contract law in many countries increasingly recognizes that there are some instances in which a contract may bestow rights on a third party. An example would be where the parties to a contract agree among themselves to make a gift to a third party or to create an insurance in which the beneficiary is a third party to the insurance contract.

Under English law, for example, the general rule under common law was until recently that of the privity of contract, although there were cases – notably in the area of so-called trusts of promise and cases of agency – where third party rights were recognized (UK Law Commission, 1996). In 1999, new legislation was introduced in the United Kingdom that specifically recognized third party rights (UK Office of Public Sector Information, 1999).

More generally speaking, where national systems of contract law recognize the enforceability of third party beneficiary rights, they will do so only where it is clearly the intention of the parties to create such legally enforceable rights and where these rights and the legal holder of these rights are clearly defined in the contract.

While the state of national law on the recognition of third party rights may not perhaps be clear in all countries, the possibility for a contract to provide for enforceable third party beneficiary rights is expressly and unambiguously recognized in the Principles of International Commercial Contracts, which were adopted by the International Institute for the Unification of Private Law (UNIDROIT) in 2004.[11] It was partly for this reason that reference was made explicitly to the UNIDROIT Principles in Article 7 of the final version of the SMTA in dealing with the choice of applicable law.[12]

The third party beneficiary in the SMTA

The provisions dealing with the third party beneficiary are set out in two separate places in the SMTA. Under Article 4 on General Provisions, paragraph 3 provides that 'the parties to this Agreement agree that (the entity designated by the Governing Body), acting on behalf of the Governing Body of the Treaty and its Multilateral System, is the third party beneficiary under this Agreement.' The rights to monitor the performance of the SMTA are set out in paragraph 4 of Article 4, which reads as follows: '4.4 The third party beneficiary has the right to request the appropriate information as required in Articles 5e, 6.5c, 8.3 and Annex, 2 paragraph 3, to this Agreement.' At first sight, both of these provisions appear somewhat limited and rather unexciting: for the most part, they merely allow the third party beneficiary to request appropriate information on matters that providers or recipients are already required to report to the Governing Body.[13] The third party beneficiary will already have access to the information provided to the Governing Body on these matters, and, indeed, the Governing Body is expressly required to ensure that the information is passed on to the third party beneficiary.[14] What Article 4.4 does add is the right of the third party beneficiary to request this information in its own right and to request it in cases where the provider or the recipient has neglected or refused to provide the appropriate information to the Governing Body.

Some of the powers given to the third party beneficiary are more substantive and far reaching. One article that is referred to in Article 4.4 gives rather broad powers to the third party beneficiary to monitor performance by the parties of their obligations under the SMTA in general, albeit in the context of dispute settlement. Article 8.3 gives the third party beneficiary 'the right to request that the appropriate information, including samples as necessary, be made available by the Provider and the Recipient, regarding their obligations in the context of the SMTA' and requires the provider and the recipient to provide the information or samples requested.[15] In fact, the powers given to the third party beneficiary under Articles 4.4 and 8.3 are a very real basis for monitoring compliance by the parties with their obligations under the SMTA. Indeed, some developing countries were keen to see the powers of the third party beneficiary expanded with respect to monitoring. Other countries preferred to limit the scope of these powers. The final wording reflects the compromise reached in the negotiations in the Governing Body that while the powers could be quite far-reaching they should be linked primarily to dispute settlement and exercised primarily in this context.

The main powers of the third party beneficiary are set out in Article 8, which deals with dispute settlement. Article 8.1 provides that dispute settlement may be initiated by the entity designated by the Governing Body (the third party beneficiary) acting on behalf of the Governing Body and its multilateral system as well as by the provider or the recipient. The rights that can be protected are defined very broadly in Article 8.2 as 'rights and obligations of the provider and the recipient under this Agreement' (the SMTA). While the natural focus may well be on the benefit-sharing obligations of the recipient, these rights and obligations would also relate, for example, to the obligation of the provider to give expeditious access to plant genetic resources in the multilateral system free of charge or at the minimal cost involved. It would also cover the provider's obligations with respect to the provision of available information on the plant genetic resources being provided as well as the reporting requirements on SMTAs entered into. From the recipient's point of view, it would cover, for example, obligations regarding the type of use to which the plant genetic resources are put as well as the restrictions on the claiming of intellectual property rights over the material received.

The dispute settlement procedures that are to be used are set out in Article 8.4 in an escalating scale and include negotiation, mediation through a neutral third party that has been mutually agreed upon, and arbitration. Arbitration can be under the arbitration rules of any international body agreed upon by the parties to the dispute. Where the parties are unable to agree, then the rules of arbitration of the International Chamber of Commerce are set as the default arbitration procedure.[16] The provisions on arbitration are also interesting in that they provide for the establishment of a list of experts by the Governing Body, from whom either party may appoint its arbitrator or from whom both may agree to appoint a sole arbitrator or presiding arbitrator as appropriate.

A good deal of discussion during the negotiations focused on the degree of direction that would be given by the Governing Body to the third party beneficiary in carrying out its functions. In the end, it was accepted that this was a matter that should be settled between the Governing Body and the third party beneficiary rather than between the parties to the SMTA. As such, it would be more appropriately dealt with in a resolution of the Governing Body addressed to the entity designated as the third party beneficiary and defining its mandate.

During the negotiations in the Governing Body, there was a certain amount of reticence on the part of some contracting parties when it came to actually naming the entity to undertake the role of third party beneficiary. This hesitation was probably a reflection of the parties' natural preference for leaving open options to change the entity selected should the need arise. In fact, the Governing Body had little choice in the matter, as had already been pointed out in the information document on the concept of the third party beneficiary presented to the Contact Group at its second meeting by the Legal Office of the FAO.[17] As pointed out in this chapter, the real third party beneficiary is undoubtedly the multilateral system itself. However neither the multilateral system nor the Governing Body possess legal personality – whatever entity was to be designated to represent the third party beneficiary rights of the multilateral system would have to have its own legal personality and the capacity to take legal action to protect those rights.[18] Indeed, the only two practical options open to the Governing Body would be to invite the FAO to undertake this responsibility given that the ITPGRFA was set up within the framework of the FAO Constitution or to set up a new entity with its own legal personality for this purpose. Obviously, the latter option would have presented considerable legal and institutional complications.

In the last moments of its first session, the Governing Body decided to invite:

> the Food and Agriculture Organization of the United Nations as the Third Party Beneficiary, to carry out the roles and responsibilities as identified and prescribed in the Standard Material Transfer Agreement, under the direction of the Governing Body, in accordance with the procedures to be established by the Governing Body at its next session.
>
> (ITPGRFA, 2006, Resolution 2/2006, para. 8)

The FAO made it known that it was prepared in principle to accept this responsibility, provided that proper procedures were drawn up by the Governing Body defining the roles and responsibilities that would be involved.[19] The Governing Body welcomed this decision at its second session, which was held in October/November 2007, and decided to: (1) mandate the secretary to prepare a draft text setting out the procedures to be followed by the FAO when acting as the third party beneficiary, taking into account its role as a UN specialized agency and its privileges and immunities and (2) to set up an ad hoc third party beneficiary committee composed of representatives of the contracting parties in order to consider the draft text as well as comments received from contracting parties and to finalize the draft procedures for submission to the Governing Body at its third session in 2009. The Governing Body invited the Director-General to bring these matters to the attention of the relevant bodies of the FAO, including its invitation to take on the role of third party beneficiary and the procedures endorsed by the Governing Body (ITPGRFA, 2007, paras 61–64).

Finally, at its third session in June 2009, the Governing Body formally adopted the Procedures for the Operation of the Third Party Beneficiary, which will come into force once they are approved by the competent bodies of the FAO (ITPGRFA, 2009, para. 42 and Resolution 5/2009 at Appendix A.5). The procedures provide that the FAO shall act as the third party beneficiary under the SMTA under the direction of the Governing Body. Under the procedures, it will seek to resolve disputes over possible non-compliance with the obligations under the SMTA first by negotiation and, if this fails, then by mediation. Only in the event of a failure to resolve the dispute by mediation within a period of six months, may the third party beneficiary submit the dispute to arbitration. All expenses incurred are to be charged to the third party beneficiary operational reserve to be set up by the Governing Body. To enable it to carry out its roles and responsibilities, the third party beneficiary is to have access to information on the SMTAs entered into, which is to be provided by the parties to those SMTAs. These reporting requirements have now been formally laid down by the Governing Body at its third session in June 2009 (*ibid.*, Resolution 5/2009).[20]

At this session, the Governing Body also took other major decisions to formalize the legal and institutional framework for the operations of the third party beneficiary. These decisions include the establishment of a third party operational reserve to fund its operations, to be filled from voluntary contributions from contracting parties and to be drawn upon by the ITPGRFA's Secretariat, and the establishment of a list of experts from which parties to the SMTA may appoint mediators and arbitrators. Operational guidelines for the commencement and management of amicable dispute resolution proceedings are to be drawn up by the ITPGRFA's Secretariat for consideration by the Third Party Beneficiary Committee. One of the interesting issues in this regard is the extent to which the Governing Body itself may be involved in the initiation of dispute settlement procedures and the extent to which such decisions will be left to the Treaty's Secretariat. On the one hand, there may be a need to ensure that action is not delayed by political considerations, while, on the other hand, the airing of disputes in a political forum could provide an important impetus for the resolution of some disputes regarding non-compliance.

While the legal and institutional framework is thus well on its way to being finalized, the effectiveness of the scheme will depend on how well it can be made to operate in practice. On this aspect, of course, the jury is still out. Given the need to establish credibility in the system, great care will need to be taken in the selection and management of the first cases.

Implications of the concept for other fora

In one sense, the concept of the third party beneficiary under the SMTA is peculiar to the circumstances of the SMTA and the multilateral system. Under the multilateral system, all transfers of PGRFA have to be subject to the SMTA, which is an agreement between an individual provider of PGRFA and an individual recipient, albeit containing standard terms negotiated by the contracting parties to the ITPGRFA on a multilateral basis. However, the SMTA differs from other MTAs in that the PGRFA being transferred are recognized as being pooled and, in this sense, as having their origins in, and flowing from, the multilateral system itself rather than from any individual provider. Similarly, the benefits arising from the use of PGRFA in the multilateral system flow not to the individual provider but, rather, to the multilateral system itself. Since individual providers will not be obtaining any benefits directly from the recipient under the SMTA, there is thus little or no incentive for those providers to take costly legal action to enforce the terms of the SMTA. As the PGRFA is transferred on to subsequent recipients, who themselves become providers in the chain of transfers, the incentive to enforce compliance becomes even more tenuous. The concept of the third party beneficiary stems from this very particular situation and addresses a particular need in particular circumstances.

That said, there is another aspect of the concept of the third party beneficiary that addresses a more widespread need that is not restricted to the transfer of PGRFA under the multilateral system. This is the need felt by many developing countries for some mechanism for guaranteeing compliance with MTAs. Compliance is a particular problem for the SMTA, which calls for particular solutions. But it is also a more general problem for MTAs covering other genetic resources – a problem that has been exercising the minds of the contracting parties to the Convention on Biological Diversity (CBD) for example.[21] It is hoped that the concept of the third party beneficiary and the empowerment of an international institution such as the FAO to initiate legal action through dispute settlement procedures to protect the integrity of the multilateral system will help to resolve this problem for the contracting parties to the ITPGRFA. Perhaps some aspects of this approach may prove useful to help resolve the problem in other fora, such as the negotiations on access and benefit sharing under the CBD. However, first it would be necessary to identify interests arising under MTAs that could properly be categorized as multilateral.

Conclusions

The third party beneficiary is an innovative concept introduced into the SMTA to solve the problems of compliance arising from the fact that benefits arising from the use of PGRFA transferred under the SMTA flow to the multilateral system itself rather than to the individual parties to the SMTA. The FAO has accepted in principle to represent the interests of the third party beneficiary under the SMTA and, in particular, to initiate dispute settlement procedures where necessary to protect those interests. In undertaking these responsibilities, the FAO will act under the direction of the Governing Body. The procedures for the operation of the third party beneficiary have now been established by the Governing Body at its third session, and an operational reserve is being set up to finance those operations. How well the system will operate in practice remains to be seen.

While the concept of the third party beneficiary has been designed to deal with the particular problems raised by the multilateral nature of the SMTA, it also responds to a more deep-seated need on the part of developing countries for an institutional mechanism to guarantee compliance with SMTA obligations. Similar needs have been expressed by developing countries in other fora, such as the CBD. However, whether the concept can be extended to MTAs for other genetic resources may depend on the identification of the multilateral interests to be protected under those MTAs.

Notes

1 Standard Material Transfer Agreement, 16 June 2006, ftp://ftp.fao.org/ag/agp/plant-treaty/agreements/smta/SMTAe.pdf (last accessed 10 October 2010) [SMTA].
2 The benefits include payments on the commercialization of products incorporating material accessed from the multilateral system and information resulting from research and development carried out on the material.
3 International Treaty on Plant Genetic Resources for Food and Agriculture, 29 June 2004, www.planttreaty.org/texts_en.htm (last accessed 10 October 2010) [ITPGRFA].
4 The terms of reference of the Contact Group are set out in Appendix C to the report.
5 The reports of the Legal Export Group were delivered orally by the chairman, although informal written copies were provided to the various regions.
6 The Legal Working Group pointed out that the agency proposal might lead to a principal being held responsible for all of the obligations of the provider under Article 6, while this was clearly not the intention. It also noted that the agency proposal might be perceived as interfering with the sovereignty of the contracting parties.
7 The Contact Group's text was as follows:

> 5.2 The parties to this Agreement agree that the (legal person representing the Governing Body), as a third party beneficiary, has the right to monitor the execution of this Agreement and to initiate dispute resolution procedures in accordance with Article 9.2, in the case of a breach of this Agreement.
>
> 5.3 The monitoring and *locus standi* rights, referred to in Article 5.2, include but are not limited to, the rights to:
>
> (a) request samples of any Product from the Provider and the Recipient, and information relating to the execution of their obligations under Articles 6.1 and 7.1, 7.2, 7.4, 7.5, 7.6, 7.7, 7.10, 7.11 and 7.13, including statements of account;
> (b) initiate dispute resolution procedures in conformity with Article 9 of this Agreement, in the case of a breach of the obligations referred to in paragraph (a) above.

5.4 The rights granted to the (legal person representing the Governing Body) above do not prevent the Provider and the Recipient from exercising their rights under this Agreement].

9.1 Dispute settlement may [only] be initiated by the Provider or the Recipient [or a person duly appointed to represent the interests of third party beneficiaries under this Agreement] [but acknowledging that this does not preclude the Governing Body from taking any action it deems appropriate if it considers that this agreement has been breached].

8 The information document was prepared by the Legal Office of the Food and Agriculture Organization (FAO) of the United Nations, in consultation with the Secretariat of the ITPGRFA and Bioversity International.
9 The description of the basis for the concept in the ITPGRFA and in national law presented later in this chapter draws much on the information document submitted to the Contact Group.
10 The Contact Group adopted a recommendation calling on the Governing Body to establish the operational procedures necessary to enable the third party beneficiary to carry out the role assigned to it in the Standard Material Transfer Agreement (SMTA).
11 The Principles of International Commercial Contracts 2004 were adopted by the Governing Council of the International Institute for the Unification of Private Law (UNIDROIT) in 2004. They differ from the 1994 version of the principles in a number of ways, including, in particular, the inclusion for the first time of the recognition of third party rights. UNIDROIT is an independent intergovernmental organization whose purpose is to study the needs and methods for modernizing, harmonizing and coordinating private and, in particular, commercial law as between states and groups of states. UNIDROIT was originally set up in 1926 as an auxiliary organ of the League of Nations and, following the demise of the league, was re-established in 1940 on the basis of a multilateral agreement, the UNIDROIT Statute. At present, some 60 states are party to the statute. The 2004 principles were adopted by consensus by the UNIDROIT Governing Council.
12 Article 7 provides that '[t]he applicable law shall be General Principles of Law, including the UNIDROIT Principles of International Commercial Contracts 2004, the objectives and the relevant provisions of the Treaty, and, where necessary for interpretation, the decisions of the Governing Body.' The General Principles of Law were chosen mainly to avoid the difficulties of deciding on a particular body of national law, whether this is a named national body of law, such as that of Switzerland, or a generic choice, such as the law of the provider, the law of the recipient or the law where the SMTA was entered into. The Governing Body also took into account the difficulties that the FAO, which was invited to serve as the third party beneficiary under the SMTA, would have, as an organization of the UN system, in submitting itself to any system of national law under arbitration proceedings. All agreements entered into by the FAO that have a choice of law clause provide for General Principles of Law as the applicable law.
13 Article 5e of the SMTA provides for notification by the provider of the SMTAs entered into; Article 6.5c provides for notification of transfers of plant genetic resources for food and agriculture under development; and paragraph 3 of Annex 2 provides for notifications regarding sales of products by the recipient, its affiliates, contractors, licensees and lessees as well as amounts of payments due and information allowing for the identification of restrictions that give rise to benefit-sharing payments.
14 Article 5e provides as follows: 'The Provider shall periodically inform the Governing Body about the Material Transfer Agreements entered into, according to a schedule to be established by the Governing Body. This information shall be made available by the Governing Body to the third party beneficiary.'
15 Article 8.3 provides as follows: '8.3 The third party beneficiary has the right to request that the appropriate information, including samples as necessary, be made available by the

Provider and the Recipient, regarding their obligations in the context of this Agreement. Any information or samples so requested shall be provided by the Provider and the Recipient, as the case may be.'

16 The applicable law for the arbitration is to be General Principles of Law, including the UNIDROIT Principles of International Commercial Contracts 2004, the objectives and the relevant provisions of the ITPGRFA, and, when necessary for interpretation, the decision of the Governing Body (Article 7).
17 See CGRFA (2006a) and note 7 earlier in this chapter.
18 Since the ITPGRFA is concluded under Article XIV of the FAO Constitution (www.fao.org/docrep/003/x8700e/x8700e01.htm#14 (last accessed 10 October 2010)), the usual practice would be for the Governing Body to draw on the legal personality of the FAO itself in all matters requiring the exercise of legal personality.
19 In Circular State Letter no. G/X/AGD-10, dated 22 December 2006, the director-general of the FAO communicated to the contracting parties his 'agreement in principle' for the organization to act as the third party beneficiary foreseen in the STMA. This agreement in principle is subject to formal approval, upon review of the procedures to be established by the Governing Body at its next session, defining the roles and responsibilities of the third party beneficiary.
20 The providers of material under the SMTA are required to provide the Governing Body once every two calendar years with the following:

- a copy of the completed SMTA; or
- in the event that the Provider does not transmit a copy of the SMTA:
  - ensuring that the completed SMTA is at the disposal of the third party beneficiary as and when needed;
  - stating where the SMTA in question is stored and how it may be obtained; and
- providing the following information:
  - the identifying symbol or number attributed to the SMTA by the provider;
  - the name and address of the provider;
  - the date on which the provider agreed to or accepted the SMTA and, in the case of shrink wrap, the date on which the shipment was sent;
  - the name and address of the recipient and, in the case of a shrink-wrap agreement, the name of the person to whom the shipment was made;
  - the identification of each accession in Annex I to the SMTA and of the crop to which it belongs.

21 Convention on Biological Diversity (CBD), 31 ILM 818 (1992). There have been many discussions in the Conference of the Parties to the CBD on the difficulties facing providers of genetic materials in enforcing the terms of MTAs in foreign countries. A number of developing countries engaged in the negotiations for the establishment of an international regime on access and benefit sharing have called for mechanisms of compliance and observance, including instruments for legal sanctions. See, for example, CBD (2004) and CBD (2005).

References

Commission on Genetic Resources for Food and Agriculture (CGRFA) (2004a) *Report on the Outcome of the Expert Group on the Terms of the Standard Material Transfer Agreement*. Doc. CGRFA/IC/MTA-1/04/Rep, FAO, Rome, online: ftp://ftp.fao.org/docrep/fao/meeting/014/j3477e.pdf (last accessed 10 October 2010).
——(2004b) *Report of the Commission on Genetic Resources for Food and Agriculture acting as the Interim Committee for the International Treaty on Plant Genetic Resources for Food and Agriculture, Second Meeting*, Doc. CGRFA/MIC-2/04/REP, FAO, Rome, online: ftp://ftp.fao.org/docrep/fao/meeting/014/j3477e.pdf (last accessed 10 October 2010).

——(2006a) *Third Party Beneficiary, Including in the Context of Arbitration.* Doc. CGRFA/IC/CG-SMTA-2/06/Inf.4, FAO, Rome, online: ftp://ftp.fao.org/docrep/fao/meeting/014/j7582e.pdf (last accessed 10 October 2010).

——(2006b) *Report of the Second Meeting of the Contact Group for the Drafting of the Standard Material Transfer Agreement, Alnarp, Sweden,* 24–28 April 2006, Doc. CGRFA/IC/CG-SMTA-2/06/REP, FAO, Rome, online: ftp://ftp.fao.org/docrep/fao/meeting/014/j7715e.pdf (last accessed 10 October 2010).

Convention on Biological Diversity (CBD) (2004) *Compilation of Views, Information and Analysis on the Elements of the International Regime on Access and Benefit-sharing.* Doc. NEP/CBD/WG-ABS/3/3 (30 November), online: www.cbd.int/doc/meetings/abs/abswg-03/official/abswg-03-03-en.pdf (last accessed 10 October 2010).

——(2005) *Report of the Ad Hoc Working Group on Access and Benefit-sharing on the Work of its Third Meeting, Bangkok, 3 March 2005.* Doc. UNEP/CBD/WG-ABS/3/7, online: www.cbd.int/doc/meetings/abs/abswg-03/official/abswg-03-07-en.pdf (last accessed 10 October 2010).

International Institute for the Unification of Private Law (UNIDROIT) (2004) *Principles of International Commercial Contracts,* UNIDROIT, Rome, online: www.unidroit.org/english/principles/contracts/principles2004/integralversionprinciples2004-e.pdf (last accessed 10 October 2010).

International Treaty on Plant Genetic Resources for Food and Agriculture (ITPGRFA) (2006) *Report of the First Session of the Governing Body of the International Treaty on Plant Genetic Resources for Food and Agriculture.* Doc. IT/GB-1/06/Report, FAO, Rome, online: ftp://ftp.fao.org/ag/cgrfa/gb1/gb1repe.pdf (last accessed 10 October 2010).

——(2007) *Report of the Second Session of the Governing Body.* Doc. IT/GB-2/07/Report, FAO, Rome, online: ftp://ftp.fao.org/ag/agp/planttreaty/gb2/gb2repe.pdf (last accessed 10 October 2010).

——(2009) *Report of the Third Session of the Governing Body.* Doc. IT/GB-3/09/Report, Doc. IT/GB-2/0, FAO, Rome, online: ftp://ftp.fao.org/ag/agp/planttreaty/gb3/gb3repe.pdf (last accessed 10 October 2010).

Moore, G. (2005) *International Arbitration.* Background Study Paper no. 25, Commission on Genetic Resources for Food and Agriculture Acting as Interim Committee for the International Treaty, FAO, Rome, online: ftp://ftp.fao.org/docrep/fao/meeting/014/aj344e.pdf (last accessed 10 October 2010).

UK Law Commission (1996) *Privity of Contract: Contracts for the Benefits of Third Parties.* Report no. LC242, UK Law Commission, London, online: www.lawcom.gov.uk/docs/lc242.pdf (last accessed 10 October 2010).

UK Office of Public Sector Information (1999) *Contracts (Rights of Third Parties) Act,* UK Office of Public Sector Information, London, online: www.opsi.gov.uk/acts/acts1999/19990031.htm (last accessed 10 October 2010).

9

PLANT GENETIC RESOURCES UNDER THE MANAGEMENT AND CONTROL OF THE CONTRACTING PARTIES AND IN THE PUBLIC DOMAIN

How rich is the ITPGRFA's multilateral system?

Carlos M. Correa

Introduction

Following seven years of negotiations, the text of the International Treaty on Plant Genetic Resources for Food and Agriculture (ITPGRFA) was finally adopted in November 2001 at the thirty-first session of the conference of the Food and Agriculture Organization (FAO).[1] As articulated in Article 1, the main objectives of the Treaty are the conservation and sustainable use of plant genetic resources for food and agriculture and the fair and equitable sharing of the benefits derived from their use. Although the Treaty is concerned with the access to, and conservation and use (for plant breeding, research and teaching) of *all* plant genetic resources for food and agriculture, it has established a special regime of 'facilitated access' for a group of crops that are important for food security, which are enumerated in Annex I of the Treaty. This regime, known as the multilateral system of access and benefit sharing (multilateral system), considers the materials in Annex I to be part of a common pool from which the contracting parties, and the organizations located within them, may benefit without payment or any other condition for access,[2] except the signature of a Standard Material Transfer Agreement (SMTA).[3]

The multilateral system is an important step towards the realization of the concept of 'global public goods' (Kaul et al., 2003, 8). The ITPGRFA aims to ensure that resources belonging to the multilateral system remain freely accessible, and it specifically restricts obtaining intellectual property rights over materials in the form received from the multilateral system that impede the use of such materials for plant breeding, research and teaching.[4] A key step in establishing the scope of the multilateral system is to determine which materials the contracting parties are obligated to include in the system. This chapter explores this issue, focusing on the contracting parties' obligations, as laid out in Article 11.2.[5]

Resources included in the multilateral system

In accordance with Article 11 of the ITPGRFA, the multilateral system is comprised of resources from five sources:

- all plant genetic resources for food and agriculture listed in Annex I that are under the management and control of the contracting parties and in the public domain;
- contributions from all others holding plant genetic resources for food and agriculture listed in Annex I, upon invitation from the contracting parties;
- resources included voluntarily in the multilateral system by natural or legal persons within the jurisdiction of the contracting parties, 'who hold plant genetic resources for food and agriculture listed in Annex I' (Article 11, para. 3);
- plant genetic resources for food and agriculture listed in Annex I and held in the *ex situ* collections of the International Agricultural Research Centres (IARCs) of the Consultative Group on International Agricultural Research (CGIAR), as provided for in Article 15.1(a); and
- the resources held by other international institutions, in accordance with Article 15.5.

While the first and third sources are defined relatively well, the second source creates a vague category ('all other holders'), which, by simple exclusion, includes all of those sources that possess plant genetic resources for food and agriculture that are not included in the remaining categories. This category includes resources held by non-contracting parties and natural and legal persons not under the jurisdiction of a contracting party. While it is evident that those sources possessing these resources could provide them when invited by the contracting parties, it is not clear if they could also do so based solely on their own initiative.[6] It would not be very rational, however, to exclude this possibility given the objectives of the Treaty and the multilateral system.

Regarding the third source of materials, the contracting parties agree 'to take appropriate measures to encourage' the indicated natural and legal persons to contribute resources to the multilateral system, making it clear that their inclusion is neither automatic nor mandatory (Article 11, para. 3). Article 11.4, however, foresees the possibility of implementing measures in case this does not take place. This article states:

> Within two years of the entry into force of the Treaty, the Governing Body shall assess the progress in including the plant genetic resources for food and agriculture referred to in paragraph 11.3 in the Multilateral System. Following this assessment, the Governing Body shall decide whether access shall continue to be facilitated to those natural and legal persons referred to in paragraph 11.3 that have not included these plant genetic resources for food and agriculture in the Multilateral System, or take such other measures as it deems appropriate.

Contributions of resources to the multilateral system from the fourth source, the IARCs, depend on the prior establishment of agreements with the centres. Such agreements were in fact signed by the IARCs and by the Tropical Agricultural Research and Higher Education Centre (CATIE), on 16 October 2006, following their approval at the first meeting of the Governing Body.

Finally, the inclusion of resources held by other international organizations will depend on agreements signed with the Governing Body. One of the functions of the Governing Body is to secure such agreements, with the goal of expanding the resources available in the common fund. Since 2006, when the IARCs and CATIE signed agreements with the Governing Body, another three international centres have also signed agreements: two of the coconut (*Cocos nucifera*) gene banks operated under the Coconut Genetic Resources Network (COGENT) as well as the Mutant Germplasm Repository operated by the FAO/International Atomic Energy Agency Joint Division in Vienna. The other three coconut gene banks under COGENT are currently in the process of finalizing agreements with the Governing Body, and negotiations are also underway to sign agreements with the South Pacific Community Gene Bank and the International Cocoa Centre Gene Bank of the University of the West Indies in Trinidad and Tobago.

Resources from the contracting parties included in the multilateral system

In accordance with the ITPGRFA, plant genetic resources are considered to be part of the multilateral system, when such resources 'are under the management and control of the Contracting Parties and are in the public domain' (Article 11, para. 2). Various aspects of this provision need to be further examined.

Resources and materials

Article 11 alludes to 'plant genetic resources for food and agriculture' in various paragraphs. In Article 2, they are defined as 'any genetic material'. It is clear, therefore, that the intention of the Treaty is to establish obligations regarding the materials as such, while being aware that the intangible aspect of the material (the genetic information) is what gives it its value as a 'resource'. The reference to holding these resources in Article 11 confirms this conclusion since only physical things and titles that represent particular rights (such as corporate shares or credits) can be possessed.

Anticipated consent

Article 11.2 contains the contracting parties' formal and anticipated expression of their resolve to include the resources defined in this article in the multilateral system. Consequently, no further action or declaration of consent of any kind is necessary. The resources included in the concept of Article 11, which are discussed later in this chapter, are automatically and unconditionally part of the multilateral system. In other words, a contracting party may not impose conditions for inclusion in the system of such resources that are different from those mentioned in the Treaty.

Crops included

To dispel any doubt, Article 11.2 refers to 'all' resources defined in it and does not leave room for any exceptions or reservations.[7] Consequently, a contracting party could not declare, for example, that it unilaterally excludes materials from certain crops listed in Annex I of the Treaty from the system. The list of crops is fixed. A state may decide whether or not to ratify the Treaty, but it may not subject itself to a partial application of the list of crops in Annex I.

Ex situ *and* In situ *materials*

Article 11.2 does not distinguish between materials maintained in *ex situ* and *in situ* conditions.[8] There is no possibility, therefore, to interpret that the obligation assumed applies only to one category or the other. Specifically, Article 12.3(h) of the ITPGRFA makes it clear that *in situ* resources form part of the multilateral system, but with an important difference compared to those preserved in *ex situ* conditions. Specifically, access to the resources can be subject to national legislation and, where there is an absence of national legislation on the matter, to the rules that the Governing Body may establish. However, national legislation should not impose additional limitations to accessing the *in situ* materials of Annex I that are under the management and control of the contracting parties and in the public domain.[9]

Resources under development

The only exception temporarily limiting the obligation imposed by Article 11.2 is provided by the Treaty in relation to resources 'under development, including material being developed by farmers', whose access 'shall be at the discretion of its developer, during the period of its development' (Article 12.3 (e)). Despite the seemingly mandatory language, which is suggested by phrases such as 'shall be', the expression 'at the discretion of' suggests that resources 'under development' may not necessarily be accessible under the conditions established in the multilateral system. However, strictly speaking, as long as these resources are included in Annex I, they also belong to the multilateral system. The SMTA, which was approved by the Governing Body during its first meeting, confirms this interpretation. The agreement authorizes the establishment of additional conditions for the transfer of improved material, which still belongs to the multilateral system.[10]

Management, control and the public domain

Having examined the coverage of Article 11.2, an important task still remains, which is to establish what is meant by 'under the management and control of the Contracting Parties and in the public domain'. The three components of this definition (management, control and public domain) must be considered in conjunction in order for certain materials (from crops listed in Annex I) to be regarded as automatically forming a part of the multilateral system.

The original text of the ITPGRFA, which was negotiated in English, refers to 'management'. This is a concept linked to 'administration'.[11] 'Management' is the way in which the state and other entities organize themselves to coordinate their own conduct (through administrative processes) and, in some cases, that of the administrators. 'Under the management of' would appear to mean that a contracting party has the capacity to exercise, directly or through a third party that is either dependent on it or under its supervision, acts of conservation and utilization of resources. Thus, the expression 'under the management of', says nothing about the legal status of the resources but, instead, refers to the handling of the material. A contracting party can certainly manage its own resources, and the resources of various third parties, in the sense described earlier. One may conclude from the first part of the definition of Article 11.2 that those resources that the contracting parties do not handle physically (or 'manage'), directly or by a third party under their instruction, do not form part of the multilateral system, regardless of the title that may be attributed to the 'managed' resources.

Apart from 'management', this rule requires that resources be under the 'control' of the contracting party. The ordinary meaning of 'control' in this context could be understood as 'dominance, command, preponderance'.[12] Although it would be easy to interpret that 'control' only reinforces the concept of 'management', meaning the possibility to handle (physically) the resources, the use of two different terms suggests to the interpreter that the inclusion of two different concepts was desired. It should be noted that at no point does Article 11 refer to the 'property' of the genetic resources. Paragraphs 2 and 3 in Article 11 refer to 'holders' and those 'who hold', respectively. In relation to the resources possessed by the IARCs, the term 'held' is used. However, the terms used do not completely eliminate the possibility that property rights may be claimed over the resources that a contracting party has in its possession.

'Possession' is a legal concept with Roman roots, and it is applicable to movable property as well as to real estate (Laquis, 1975, 139 and 246). Its use in Article 11 indicates the ability of a person to have something in his or her power. In the legal tradition, according to Savigny's doctrine, possession implies a *'corpus'* (power over something) and an *'animus'* (the intention of subjecting something to a property right) (ibid., 249). As a result, the legitimate possession of movable property creates a presumption of property over the object.

However, when the issue of property rights over plant genetic resources arises, the distinction must be established between rights over a physical entity as such (physical property) and over the genetic information contained in these resources (intangible property). It is the latter where the real value of the resources lies and also where the legal problems are especially complex. Concerning physical property, plant genetic resources can be the object of private or public property rights. Property can be linked to the land where the plants are located as a result of the application of the customary law principle, according to which everything that is affixed, or is destined to be affixed, to the land belongs to the owner of the land. After the plants (or their parts) are removed from the ground, they are subject to their own property regime, as movable property, even when transported outside the land of origin or to a different country (Correa, 1994). It should also be noted that plant genetic resources (in their physical appearance) are also susceptible to public property rights, as examined later in this chapter.

In any case, with the goal of interpreting the meaning of 'control' in Article 11.2, it seems sufficient to resort to the common meaning of the term, which does not require either 'possession' or 'property'. Instead, it calls for the capacity to exercise physical acts over the resources – a power generally associated with these concepts but that could also emerge from the mere tenancy of the resources without '*animus domini*'.[13] In other words, the use of the term 'control' suggests that the contracting parties are obligated to consider as part of the multilateral system those resources whether or not possession (in the legal meaning of the term) is exercised thereon. Holding the resources is sufficient. The following section examines whether this conclusion is nuanced or altered by the third element of the definition – 'public domain'.

Public domain

The ITPGRFA was negotiated and adopted 'in harmony' with the Convention on Biological Diversity (CBD) (Article 1.1) and specifically recognizes the sovereign rights of states regarding 'their own' plant genetic resources for food and agriculture (Article 10.1).[14] The existence of sovereign rights over national territory, including the natural resources therein, is a principle that is well established in international law. This principle means that a state has the power and jurisdiction to establish the manner in which these resources will be shared and used as well as whether they are the object of property rights (private or public) and the conditions under which this may occur.

Consequently, the recognition of the sovereign rights of contracting parties over their plant genetic resources for food and agriculture does not imply either the recognition of property rights over these resources or their legitimate possession. It simply expresses deference in favour of the decisions that the party may adopt, even relating to access to these resources, as provided for in Article 10.1 of the Treaty. In this context, there are two possible meanings for the concept of 'public domain'. The first meaning emerges from administrative law. In this framework, the concept of public domain (being equivalent to that of 'public property') describes a set of goods that belong to the general public and are dedicated to the public's use (for example, a navigable river bed in its entirety)[15] or a public service (Choisy, 2002, 2 and 24). Public property can be declared and exercised over quantified and individualized goods as well as over an indeterminate quantity of resources in a specific category. This is the case when public property is established, for example, over the water in rivers and, in some countries, over the hydrocarbons that exist within the territory of the state. All of the live material under the national maritime jurisdiction has also been declared public property, and its exploitation is the object of state concessions (Laquis, 1979, 476).

Public property over a good should be established by law in order to define its scope.[16] In some national legislation on access to genetic resources, specific reference is made to states' rights over genetic resources. The Andean Community's Decision no. 391, for example, states that genetic resources are assets belonging to the nation or state, but it permits the biological resources in which they are contained (that is, the 'materials') to be subjected to different property regimes (Articles 5 and 6).[17]

A second meaning of 'public domain' emerges from intellectual property rights. Here, the concept has been defined as information that is not subject to intellectual property rights. In other words, it is information that can be freely used without effectuating payment to third parties or obtaining authorization (Van Caenegem, 2002).[18] Public domain information is:

- information whose intellectual protection rights have expired;
- information for which protection would be appropriate, but which has been lost due to a failure to comply with certain formal requirements of intellectual property rights; and
- information that is outside the scope of legislation on intellectual property rights because it is not eligible for protection, according to the applicable law.

It has been observed that:

> information is not in the public domain because of its nature as a public good or even its governmental origin but as a result of a network of formal and informal social agreements, explicit or implicit but entrenched in the common law and in the culture of a society.
>
> (Forero-Pineda, 2004, 40)

Therefore, the breadth of the public domain spectrum can be greater or smaller, depending on the types and degree of appropriation determined by the states' law.

Some view the concept of 'public domain' within the confines of intellectual property, understood *stricto sensu*, to cover only that which was previously protected (in other words, which has 'fallen' into the public domain)[19] or that which could have had intellectual property protection (but for some reason was not acquired), with the exclusion of all material that was never eligible for protection (for example, purely fact-based information, unoriginal works and unpatentable techniques). They reserve the denomination of 'common fund' for this latter category (Choisy, 2002, 151).

Although this distinction is possibly worthy of attention in the specific confines of an author's rights, the interpretative method of the Vienna Convention requires one to find the ordinary meaning of the terms used in the ITPGRFA.[20] 'Public domain' is commonly used to allude to the entire pool of works and knowledge, including factual (McSherry, 2001, 191) and scientific (Reichman and Uhlir, 1999) information that is not subjected to intellectual property rights, including those that were not subjected in the past, nor could have been, because they were not eligible for protection. A generally accepted definition of 'public domain' is, in this sense, 'a collection of things available for all people to access and consume freely' (Kaul et al., 2003, 8).

There are some common elements and important differences between the two described meanings – that of administrative law and that of intellectual property law – for the concept of 'public domain'. Public use is an aspect common to both concepts, independent of the type of use. The main difference is that the state is not the owner of the works or information in the public domain in the framework of intellectual property rights (Choisy, 2002, 21, 32 and 41).[21] In the context of intellectual property, the concept of 'public domain' is similar (without being identical) to that of '*res communes*' or something that is not property and is available for common use.[22]

On the other hand, the public property or 'domain' in administrative law is susceptible to limits established by the state itself, such as in the case of the authorization given to a party to utilize assets under governmental control as part of the concession of public services. However, public domain, according to intellectual property law, is absolute and compulsory in principle (*ibid.*, 53), meaning that it cannot be the object of private appropriation unless a new law expands the limits of what may be appropriated (as EC Directive 96/9 on the Legal Protection of Databases did, for example, for the protection of unoriginal databases). It is also possible for information in the public domain to see its protection restored, as stipulated, for example, in Article 70.2 of the World Trade Organization's Agreement on the Trade-Related Aspects of Intellectual Property Rights.[23]

The key interpretative issue is which of the two meanings indicated is intended by the expression 'public domain' in Article 11.2. Several reasons seem to tip the balance towards the meaning linked to (the absence of) intellectual property protection. First, if the intention of those who wrote the ITPGRFA was to refer to 'public domain' as 'public property', they could have used the latter to avoid any doubt or ambiguity about the chosen concept. Second, in accordance with the interpretative principle of 'effectiveness', the interpretation that gives full meaning to all of the terms in the Treaty should be sought. If the intention of the negotiating parties had been to allude to the concept of 'public domain' derived from administrative law, it would seem superfluous to have made reference to 'management' and 'control' since these are normally powers that are derived from this domain. Third, as previously mentioned, the ITPGRFA recognizes 'the states' sovereign rights' over plant genetic resources for food and agriculture (Article 10.1). It would be a very risky conceptual jump to suppose that the same parties limited the multilateral system – a central component of the Treaty – to resources that are public property. Fourth, if this had been the intention, it is unlikely that the negotiating parties would have left the determination of which materials are, or are not, in the multilateral system basket to the total discretion of the parties. They would probably have established criteria for defining the scope of the actual public property. Strictly speaking, interpreting 'public domain' as the same as 'public property' could mean the emptying of the multilateral system, since every contracting party could, without being subjected to any international law, decide which resources to place in the basket and which to save for themselves. This has not been the spirit of the Treaty, nor does it emerge from the wording of Article 11.2 ('all plant genetic resources'). Finally, as evidenced by Article 12.3(d), the negotiating parties were wholly conscious of the implications of intellectual property for accessing resources in the multilateral system. It is reasonable to think that, when resorting to the concept of 'public domain', the negotiating parties attempted to make it clear that access to materials included in the system should not be blocked by such rights.

Conclusion

The concept of 'public domain' that is used in the ITPGRFA should be understood in the context of intellectual property rights. This includes (1) materials whose protection has ended (whatever the reason may be) and (2) materials that never were protected, nor will ever be protected, whether this is a result of not satisfying the respective formalities or of not fulfilling the substantive requirements to obtain protection. While there are aspects that still remain to be explored more carefully, the present analysis allows us to conclude that the obligation provided for in Article 11.2 of the Treaty includes all those materials that are the property of, held by or in the possession of the contracting parties, or that are under other forms of control or management of the contracting parties, with the sole exception of those resources under development or subject to intellectual property rights.[24] All of these resources are part of the multilateral system, without the need for any declaration or designation.

Notes

1 International Treaty on Plant Genetic Resources for Food and Agriculture, 29 June 2004, www.planttreaty.org/texts_en.htm (last accessed 30 October 2010) [ITPGRFA].
2 Ibid., Article 12.3(b) states: 'Access shall be accorded expeditiously, without the need to track individual accessions and free of charge, or, when a fee is charged, it shall not exceed the minimal cost involved.'
3 Ibid., Article 12.4. Standard Material Transfer Agreement, 16 June 2006, ftp://ftp.fao.org/ag/agp/planttreaty/agreements/smta/SMTAe.pdf (last accessed 30 October 2010) [SMTA].
4 Ibid., Article 12.3(d) states that 'recipients shall not claim any intellectual property or other rights that limit the facilitated access to the plant genetic resources for food and agriculture, or their genetic parts or components, in the form received from the Multilateral System.'
5 Utilizing the interpretive method codified in Articles 31 and 32 of the Vienna Convention on the Law of Treaties, 23 May 1969, 8 ILM 679 (1969) [Vienna Convention] – that is, a literal interpretation that takes into consideration the object and purpose of the ITPGRFA.
6 Given the use of the plural, it should be understood that the invitation should take place by consensus of the contracting parties (see ITPGRFA, supra note 1, Article 19.2, regarding the Governing Body's decision-making process).
7 Ibid., Article 30, is categorical: 'No reservations may be made to this Treaty.'
8 Unlike paragraph 5 in the same article, which refers exclusively to the ex situ resources preserved by the International Agricultural Research Centres (IARCs).
9 Following the procedures established in the national legislation, the said materials can be provided/supplied/transferred under the conditions of the SMTA, supra note 3, approved by the Governing Body.
10 See, in particular, clause 6.6 of the SMTA, which authorizes the imposition of a payment of monetary compensation.
11 According to the Concise Oxford Dictionary, 'management' means 'administration of business concerns or public undertakings; persons engaged in this', and 'administration' is the 'management (of business); management of public affairs, government'.
12 It is worth mentioning that the Vienna Convention, supra note 5, requires an interpretation based on the 'ordinary meaning' of the terms used in the ITPGRFA in their appropriate context.
13 'Tenancy' is deemed to exist when one physically holds the thing, but a third party's property is recognized.
14 Convention on Biological Diversity, 31 ILM 818 (1992).
15 Although specific volumes of water can be appropriated privately.
16 This assumes that a good leaving the public domain, when such an act is possible, should occur through a de-allocation act. There are goods that, by nature, should remain in the public domain (sometimes called 'natural public domain').

17 *Decision 391: Common Regime on Access to Genetic Resources*, Comunidad Andina, www.comunidadandina.org/ingles/normativa/d391e.htm (last accessed 30 March 2011).
18 Notably, as an exception, in some countries the reproduction of works of art in the public domain is subject to a payment to the state (*'domain public payant'*).
19 The Agreement on Trade-Related Aspects of Intellectual Property Rights, 15 April 1994, 33 ILM 15 (1994), only uses the expression 'public domain' on one occasion, in relation to protectable material that is no longer protected (Article 70.3).
20 'Public domain' means, according to the *Concise Oxford Dictionary*, 'belonging to the community, as a whole, esp. no longer copyright'.
21 Even in the case of 'public domain payant', which consists of a tax before exercising a property right.
22 The *res comunes* are by nature, completely unappropriable, such as air and water, but may be susceptible to partial appropriations.
23 Applying this article, for example, in the 1994 Uruguay Round Agreement Act, Pub. L. No. 103–465, 108 Stat. 4809, the United States restored authors' rights for foreign works, such as movies and music, which had not been protected in that country before.
24 As is the case with the varieties of plants protected by breeders' rights that are not in the 'public domain' but, rather, are accessible for the development of new varieties.

References

Choisy, S. (2002) *Le domain public en droit d'auteur*, Litec, Paris.
Correa, C. (1994) *Derechos soberanos y de propiedad intelectual sobre los recursos fitogenéticos*, Estudio Informativo no. 2, FAO, Rome.
Forero-Pineda, C. (2004) 'Scientific Research, Information Flows, and the Impact of Database Protection on Developing Countries', in J.M. Esanu and P.F. Uhlir (eds), *Open Access and the Public Domain in Digital Data and Information for Science: Proceedings of an International Symposium*, National Academies Press, Washington, DC.
Kaul, I., P. Conceicao, K. Le Goulven and R. Mendoza (2003) 'Why Do Public Goods Matter Today?' in I. Kaul, P. Conceicao, K. Le Goulven and R. Mendoza (eds), *Providing Global Public Goods: Managing Globalization*, Oxford University Press, New York.
Laquis, M. (1975) *Derechos Reales*, Vol. I, Ediciones Depalma, Buenos Aires.
——(1979) *Derechos Reales*, Vol. 2, Ediciones Depalma, Buenos Aires.
McSherry, C. (2001) *Who Owns Academic Work: Battling over the Control of Intellectual Property*, Harvard University Press, Cambridge, MA.
Reichman, J., and P. Uhlir (1999) *Database Protection at the Crossroad: Recent Developments and their impact on Science and Technology*, online: http://library.findlaw.com/1998/Dec/2/131341.html#Scene_1 (last accessed May 2009).
Van Caenegem, W. (2002) 'The Public Domain: Scientia Nullius', *European Intellectual Property Review* 24(6): 324–30.

[Handwritten note at top: Would have preferred an independent analysis of the Secretariat's functioning]

10

EFFORTS TO GET THE MULTILATERAL SYSTEM UP AND RUNNING

A review of activities coordinated by the Treaty's Secretariat

Selim Louafi and Shakeel Bhatti

Introduction

This chapter provides an overview of the Secretariat's work in supporting the implementation of the multilateral system of access and benefit sharing (multilateral system), five years after the entry into force of the International Treaty on Plant Genetic Resources for Food and Agriculture (ITPGRFA) and three years after the first meeting of the Governing Body.[1] The main focus of the Secretariat's work requested by the Governing Body has been 'to make the Multilateral System functional' (FAO, 2006, para. 54). It is indeed important to note that the multilateral system and the implementation of the Standard Material Transfer Agreement (SMTA)[2] are not self-executing in the sense that:

- the multilateral system is an *operational* system functioning at the global level, and the Secretariat has to manage the system and its operation to make it really functional for its users;
- the multilateral system is a *dynamic* framework, which comprises a number of elements, some of which are still being set in place.[3] In carrying out the decisions of the Governing Body and its subsidiary bodies, the Secretariat has consequently had to engage with the system to make sure that guidance is given by the Governing Body in order to ensure a coherent implementation process; and
- the multilateral system is a *global* and *innovative* framework, which regulates a wide range and category of users all around the world. As such, it requires a learning process and the strengthening of capacities of those who are implanting it and using its mechanisms. The Secretariat has had to provide minimum support to users in order to overcome various initial uncertainties and hesitancies.

Translating the text of the Treaty and the guidance given by the Governing Body into the day-to-day management of the Treaty's systems and interaction with stakeholder communities has consequently been the main challenge faced by the Secretariat during this start-up phase.

In practice, such goals have meant establishing a new global infrastructure for the multilateral system to work on a daily basis and creating a new network of institutions to ensure coherent global operations in multiple countries, jurisdictions, languages and pre-existing legal-technical operating systems – all of which must function in a coherent manner to operate the Treaty's multilateral system. It has been necessary to ensure that these activities have been undertaken in close cooperation with the contracting parties and the users of the multilateral system – that is, the providers and recipients of genetic resources.

The activities undertaken by the Treaty's Secretariat to make the operation and evolution of the multilateral system functional can be divided into three categories:

1. those servicing the day-to-day operations of the multilateral system and SMTA;
2. those generating policy support and further guidance on the development of the multilateral system by the contracting parties and the Governing Body;
3. those providing appropriate assistance to developing countries and promoting the exchange of experience among the users of the multilateral system and its SMTA.

Servicing day-to-day multilateral system and SMTA operations

The first role of the ITPGRFA's Secretariat with regard to the multilateral system and its SMTA is to manage its operation on a day-to-day basis. This management involves the following two dimensions: (1) increased coverage of the multilateral system at the national level by facilitating the inclusion of material and its documentation and (2) the standardized operations of the SMTA's use and management.

Coverage of the multilateral system

At its second session, the Governing Body 'requested the Secretary to continue gathering information on the assessment of progress in the inclusion of plant genetic resources in the Multilateral System' (FAO, 2007, para. 65). Having a clear and accurate picture of what is actually available in the multilateral system is of primary importance. This requires a number of things: first, that countries – particularly developing countries – take the legal and administrative steps to identify the materials in their countries that are part of the multilateral system, in accordance with the Treaty; and, second, that these steps be adequately documented so that they can be used by plant breeders, farmers, researchers and others.

The formal notification of plant genetic resources for food and agriculture (PGRFA) that are included in the multilateral system is one of the primary requirements of the system. For plant breeding and conservation, however – which are the purposes of facilitated exchange – full, easily consultable information on individual accessions is necessary or else such exchanges are not possible. In this sense, material can only be said to be effectively in the multilateral system if it is adequately and publicly documented.

For this reason, a page has been established on the Treaty's website, where notifications received from the contracting parties and natural and legal persons regarding material that they have included in the multilateral system are posted.[4] This webpage currently provides full-text copies of all of the notifications received as well as a downloadable letter of notification. This sample letter of notification (which is included in the Appendix at the end of this chapter) requests information for the website so that 'detailed data on the composition of the collection and user procedures to order samples are readily available', and thereby 'the website provides access to the collection's database'.[5]

Major international collections, and those of developed country contracting parties, usually have websites that provide such information, and even some of the larger developing countries currently provide such information for breeders.[6] However, it must be recognized that many smaller developing countries face substantial financial, technical and institutional limitations in providing public – preferably online – information on the resources that they are including in the multilateral system, and these countries will require assistance in order to improve the provision of such information in the future.

The multilateral system can be seen as a 'distributed gene bank' in the sense that it is not an institution with a headquarters and dedicated staff but, rather, relies on these entities to act for it. It comprises a very large number of plant genetic resources, which are held by a large number of governmental and private entities throughout the world. Article 11.1 of the Treaty provides facilitate access to plant genetic resources for food and agriculture, and to share, in a fair and equitable way, the benefits arising from the utilization of these resources, on a complementary and mutually reinforcing basis.[7]

For a plant breeder seeking useful materials, the multilateral system is only as good as the information systems that describe the materials that are available in it. Providing such information is a 'distributed' function – one that is not managed from the centre – and it is the task of gene bank and information system managers throughout the world. In implementing the multilateral system, the Governing Body draws on the support, creativity and goodwill of this community.

The multilateral system also relies heavily on persons who are not parties to the SMTA – in particular, the information system managers who 'publish' and make available information on PGRFA within the multilateral system for potential users. These information system managers are often part of the gene banks that are the potential providers of the material they document, but they are more increasingly provided within cross-institutional search tools and portals, which will be of especial importance in the implementation of the Treaty's Article 16 on international plant genetic resources networks and Article 17 on the global information system on plant genetic resources for food and agriculture. Through ever more efficient coordination and integration of existing information systems on agricultural plant genetic resources, the necessary documentation tools are being developed, in partnership with the Consultative Group on International Agricultural Research's (CGIAR) International Agricultural Research Centres (IARCs), the Global Crop Development Trust

(GCDT) and national and regional gene banks, to document and make visible the materials that are in the multilateral system, which is the *sine qua non* for addressing the challenges the world currently faces: climate change, population growth and persistent poverty.

In 2008, Bioversity International, the GCDT and the Secretariat of the ITPGRFA initiated a project to develop just such an integrated global information resource, bringing gene bank databases and existing crop and regional databases together in order to create a single information portal to facilitate the access to, and use of, accessions in *ex situ* gene banks. In May 2011, the portal known as GENESYS was officially launched. At the time, it contained 2.3 million accession records and some three million phenotypic records for 22 crops. GENESYS offers a new paradigm for access to, and use of, PGRFA data, not only because of the number of accessions it contains but also because any user can create custom queries across passport, phenotypic and environmental data categories to suit their specific requirements. A major challenge for this community will also be to assist plant genetic resources managers in developing countries to play a full role in developing the international system, adequately documenting the material they hold within the multilateral system and facilitating access.

In December 2008, a second Technical Consultation on Information Technology Support for the Implementation of the Multilateral System of Access and Benefit-Sharing was held, with support from the Centro di Ricerca per la Frutticoltura, on behalf of the Italian government (FAO, 2009b). Information specialists, including those from the contracting party gene banks and from the IARCs, discussed a number of current initiatives that aim at improving the coverage of information systems for PGRFA and at integrating wider sets of participating countries and institutions. They also discussed potential information technology tools to facilitate the use and management of the SMTA.

SMTA management

The SMTA is a cornerstone of the Treaty's multilateral system, and the Governing Body periodically reviews its implementation and operation. Each individual SMTA is a separate and whole contract and is established between the parties (the provider and the recipient) for that contract only. The day-to-day operations of the multilateral system generate a large number of actions, many of which have implications for the workload and efficiency of the Treaty's Secretariat. In the first nine months of the multilateral system's operation, more than 100,000 transfers were reported by the CGIAR system alone. A major functional responsibility of the Secretariat results from the receipt and storage of information that is to be provided to the Governing Body.

Through the regionally representative Ad Hoc Committee on the Third Party Beneficiary, the contracting parties have developed a decision containing a coherent draft set of procedures for the third party beneficiary, including recommendations for possible measures to minimize costs; provide information that is to be made available by the users; develop a reporting schedule; and establish a roster of experts. This decision was adopted by the third session of the Governing Body.

The ITPGRFA's Secretariat has also established information technology tools to assist the users of the SMTA in fulfilling their reporting obligations, which include the provisions of Article 5e (on information given by a provider to the Governing Body about material transfer agreements entered into); Article 6.4b (on notification by a provider that material has been transferred to a subsequent recipient); Article 6.5c (on the notification of the Governing Body, by a provider, about material transfer agreements entered into, in the specific case of PGRFA under development); and Article 6.11h (on notification by a recipient of having opted for payment in accordance with this article), as well as the provisions in Annex 2, paragraph 3 (on reports on payment due); Annex 3, paragraph 4 (on notification by the recipient of a decision to opt out of the alternative payment scheme provided for in Article 6.11); and Annex 4 (on the signed declaration of having opted for the alternative payment scheme provided for in Article 6.11).

Much of this information may be of prime importance to the third party beneficiary in initiating dispute settlement, should that become necessary, including the need to establish, when required, evidence of the chain of contractual rights and obligations. At its third session, the Governing Body gave guidance on the precise information that it wishes to receive.

Generating policy support and further guidance on the development of the multilateral system by the contracting parties and Governing Body

Guidance, monitoring, coordination and programming of a coherent implementation process by the Governing Body is particularly crucial for the coherent functioning of a global system such as the multilateral system. This is the case since the Treaty establishes concrete operational systems, which depend on day-to-day processes and procedures being carried out all over the world in a coherent manner, such as the application of the SMTA, the inclusion of materials in the multilateral system and the benefit-sharing operations under the four benefit-sharing mechanisms of Article 13.

The contracting parties and users of the multilateral system have asked the Secretariat many complex questions, including issues dealing with:

- the farmer who needs to be assured that the seeds that his community has developed over generations will benefit humanity and that he will, in his turn, have access to the seeds he needs in his farming system;
- the gene bank manager who needs to be convinced that his collections will also benefit from facilitated exchange;
- the user who wants to ship seeds, but who is told by her legal office that she first needs to understand the meaning of a particular clause in the SMTA;
- the researcher who worries about the intellectual property rights of his research results; and
- the breeding company that is perfectly willing to share benefits in accordance with the ITPGRFA but that wants legal certainty that it will not be accused of biopiracy.

These various concerns, expressed by different stakeholders, have often been translated into technical terms when interpreting the relevant provisions of the Treaty, and a number of the parties have asked for advice and assistance, which, as far as possible, the Secretariat has provided on an ad hoc basis. For example, a number of potential providers and recipients in the public and private sectors have brought to the attention of the Secretariat various legal and technical questions regarding the interpretation of their rights and obligations under the SMTA. This uncertainty has probably limited the number of people willing to make their materials available, or to receive materials, through the SMTA. It also possibly provides opportunities for operations that are not in line with the Treaty and the SMTA, which serves to increase legal and administrative confusion and leads to an overall weakening of the Treaty.

The ITPGRFA provides that the SMTA must be used for every transfer of PGRFA within the ambit of the multilateral system. The provider may be a government institution or a legal or natural person within the jurisdiction of a contracting party, in either the private or public sector, or an International Institution that has made an agreement with the Governing Body under Article 15 of the Treaty. The recipient may also be one of a very large range of private and public persons. In the sense that providers under the SMTA (the originators of the SMTAs) are scattered throughout the world and are not part of a single entity, the use of the SMTA is part of a 'distributed system'. Since this distributed system spans many languages, national and regional legal frameworks and institutional structures, the maintenance of operational coherence across the system represents a major challenge, and it requires active guidance by the Governing Body as well as the management of the operations of the SMTA and the multilateral system more generally.

During the 2008–9 biennium, and in preparation for the third session of the Governing Body, the Secretariat consulted with international technical experts on technical and legal matters that had been brought to his attention regarding the operation of the SMTA and the multilateral system generally. These experts were to assist him in responding to the contracting parties' requests as well as in developing information packages on key elements of the ITPGRFA as a means to enhance their implementation activities (FAO, 2007, 2009c).

At its third session, the Governing Body requested the Secretariat to give priority to assisting users of the SMTA to overcome any implementation problems. To this effect, the Governing Body requested the Secretariat to convene an Ad Hoc Advisory Technical Committee on the SMTA and the multilateral system, taking into account regional representation. The committee held two meetings in the biennium 2010–11, during which it advised the Secretariat on implementation questions raised by a variety of users of the SMTA and the multilateral system – that is, the contracting parties, the IARCs that signed agreements with the Governing Body and individual users. The questions covered a broad spectrum of components of the multilateral system, such as the reporting obligations of the SMTA parties, inclusion of material into the multilateral system by private holders and the transfer of PGRFA for non-food/feed uses.

In the reports produced by the committee, users can find responses to the various matters brought to the attention of the committee as well as opinions on specific questions. Where necessary, the committee brought questions regarding the SMTA and the multilateral system to the attention of the Governing Body for guidance. The third session of the Governing Body decided consequently that a special effort should be devoted in the forthcoming biennium to overcoming these problems, by rapidly addressing technical, legal and operational questions raised by the users of the SMTA and contracting parties; building capacity, raising awareness and promoting the exchange of experiences among those responsible for implementing the SMTA at the national level, particularly in developing countries and documenting best practices throughout the system. The Secretariat will also continue to collaborate with interested national providers and international institutions in developing information support tools for the electronic management of the SMTA and related reporting and information management.

Providing appropriate assistance to developing countries

As the multilateral system becomes operational, one of the main challenges for the Governing Body and its Secretariat is to advise and assist the providers and recipients of PGRFA to use the SMTA, and to efficiently implement the multilateral system, with the minimum transaction costs. In this context, there is evidence that the potential providers and recipients of plant genetic resources, especially in developing and least developed countries, are finding the operations of the SMTA overly complex, difficult and laborious and that this fact is slowing down the development of the multilateral system. Many such users do not regularly work with contracts (such as the SMTA), and they have underlined the usefulness of tools that could make it easier for them to apply the SMTA.

Such reactions have had a number of implications for the implementation and operation of the SMTA. The responsiveness of the ITPGRFA's system to the needs, requests and enquiries of its users will be critical for its credibility and will provide the basis for its consistent use. The differing levels of understanding of the various players must be allowed for, and a general process of outreach and awareness raising will be required for this purpose. The goodwill and diligence of this wide range of players will be crucial in the effective implementation of the SMTA.

Developing countries, in particular, will require assistance in factoring the SMTA into their administrative practice. With funds provided by the Swedish International Development Cooperation Agency, the ITPGRFA's Secretariat, the Food and Agriculture Organization and Bioversity International have implemented a joint capacity-building programme, which is aimed at government officials, politicians, farmers and other stakeholders and which extended into the 2010–11 biennium. The programme aimed at capitalizing on prior experience in order to prepare practical guidelines for the implementation of the multilateral system (FAO, 2009d).

During the 2008-9 biennium, the Secretariat also worked with the contracting parties and other users of the multilateral system to promote the exchange of experience and the documentation of best practices to help improve understanding of the multilateral system and the SMTA and to resolve problems at the implementation level that had been identified. During the 2010–11 biennium, this work continued as a priority, through a 'multi-stakeholder platform for users of the multilateral system', aimed in particular at strengthening national capacity to implement the multilateral system (FAO, 2009a).

Conclusion

The ITPGRFA's Secretariat has been engaged in a variety of endeavours in connection with the implementation of the multilateral system. Given the ongoing nature of the multilateral system and SMTA operations, the Governing Body of the Treaty has requested the Secretariat to assume certain responsibilities for the smooth performance of both mechanisms and its constituents. For the multilateral system, the Secretariat has a duty to manage it so that it functions on an ongoing basis at the international level as an operational system in order to bring growing benefits to its users. This is an evolving process, because the multilateral system is a dynamic mechanism that encompasses different but robust elements, posing a significant number of challenges for coherent implementation. One specific challenge is to make sure that the multilateral system serves as a global framework of innovation, entailing some regulatory requirements and growing demand for capacity building among the potential beneficiaries of the Treaty. In the start-up stage, the Secretariat has made a considerable effort to create an enabling environment that encourages coordination and interaction between the various stakeholders. It has accomplished this by identifying the outstanding issues and providing professional advice about the development of pertinent solutions. In some cases, such efforts have led to the creation of novel infrastructures across the world in order to ensure legal, administrative and technical conformity in the process of multilateral implementation.

Through its day-to-day operation, the Secretariat has been playing a pivotal role in facilitating the normal functioning of the multilateral system and the SMTA and in promoting the exchange of experiences and good practices among the users. It is also worth noting that the Secretariat has gathered a considerable amount of information on evaluating the prospects of including plant genetic resources in the multilateral system and has provided timely recommendations for the preparation of an appropriate context for the effective identification and adequate documentation of materials to be used by plant breeders, farmers and researchers. So far, the efforts of the Secretariat have produced impacts in raising the awareness of stakeholders on the importance of notification in the multilateral system. Building an information system is therefore an essential endeavour in consolidating a global network of resources for addressing the pressing issues that the world faces today, such as food crisis, climate change, population growth and persistent poverty.

The Secretariat believes that the SMTA is a cornerstone of the ITPGRFA's multilateral system and that the importance of the SMTA's management cannot be overemphasized. Given the fact that each individual SMTA is a separate contract in its own right, the management role of the Secretariat in this regard entails a significant workload, but it offers the potential to generate long-term and enduring benefits, including developing necessary information technology tools and providing technical support to developing countries. A widely accepted view is that guidance, monitoring, coordination and programming of a coherent implementation process by the Governing Body is crucial for the coherent functioning of the multilateral system. The Secretariat has been working to meet the various requests of the contracting parties, and it will continue to do so in order to better serve the interests of all agricultural stakeholders, including gene bank mangers, farmers, researchers and other users. Some of these services have taken the form of providing interpretation and technical analysis on the relevant provisions of the ITPGRFA, when requested. Since some countries and users are concerned about the complicated nature of the SMTA operation, the Secretariat has developed some information technology tools to help them apply the SMTA, as part of their outreach and awareness-raising work. Other endeavours include targeted programmes funded by various sources to build capacities in developing countries in order to better implement the multilateral system. It is expected that these efforts will continue in the future as a priority.

The multilateral system has so far experienced a positive start and a dynamic development with broad prospects for future growth. It is fair to predict that in the years to come, the multilateral system will play a bigger role in improving and sustaining the food security and agricultural development of its constituents. The Secretariat remains firmly committed to the goal of the multilateral system and is ready to cooperate more closely with all of the stakeholders for the effective and efficient realization of their shared visions and goals.

Appendix: Sample letter of notification of inclusion of material in the multilateral system

To the Secretary of the International Treaty on Plant Genetic Resources for Food and Agriculture

Mr. Shakeel Bhatti
Food and Agriculture Organization of the United Nations
Viale delle Terme di Caracalla 1,
00153 Rome, Italy

Subject: notification regarding the contribution of the [name of the CP/Natural or Legal person] to the Multilateral System

The International Treaty on Plant Genetic Resources for Food and Agriculture (the Treaty) has established a Multilateral System of Access and Benefit-sharing.

Regarding the coverage of the Multilateral System, Article 11 specifies that the Multilateral System shall include all plant genetic resources for food and agriculture listed in *Annex I* that are under the management and control of the Contracting Parties and in the public domain, and that Contacting Parties invite other holders of the plant genetic resources for food and agriculture listed in *Annex I* to include these in the Multilateral System.

Herewith, the [name of the CP/Natural or Legal person] wishes to notify to you that the following plant genetic resources for food and agriculture listed in *Annex I* and maintained in [name of the CP] have been included in the Multilateral System.

1. The collections held by [name of the collection centre], [name of the country], located in XX. Through the website [url address] detailed data on the composition of the collection and user procedures to order samples are readily available.
2. The [name of species] collection held by the [name of the collection centre] located in XX [and consisting of ...]. The website [url address] provides access to the collection's database.

Germplasm held in the collections listed above will be made available to users under the conditions of the Standard Material Transfer Agreement of the International Treaty on Plant Genetic Resources for Food and Agriculture.

Notes

1 International Treaty on Plant Genetic Resources for Food and Agriculture, 29 June 2004, www.planttreaty.org/texts_en.htm (last accessed 15 May 2011) [ITPGRFA].
2 Standard Material Transfer Agreement, 16 June 2006, ftp://ftp.fao.org/ag/agp/planttreaty/agreements/smta/SMTAe.pdf (last accessed 15 May 2011) [SMTA].
3 At its first session, for example, the Governing Body adopted the SMTA, and it has recently, during the last session held in June 2009, considered and adopted the third party beneficiary procedures.
4 ITPGRFA, *supra* note 1.
5 See the website at ftp://ftp.fao.org/ag/agp/planttreaty/agreements/models/inclu_e.doc (last accessed 15 May 2011).
6 See, for example, the International Rice Research Institute's database at www.iris.irri.org/ or the German National plant genetic resources system portal at www.genres.de/pgrdeu/ (last accessed 15 May 2011).
7 ITPGRFA, *supra* note 1.

References

Food and Agriculture Organization (FAO) (2006) *Report of the First Session of the Governing Body of the International Treaty on Plant Genetic Resources for Food and Agriculture*, Doc. IT/GB-1/06/Report, Madrid, Spain.
—— (2007) *Report of the Second Session of the Governing Body of the International Treaty on Plant Genetic Resources for Food and Agriculture*, Doc. IT/GB-2/07/Report, FAO, Rome.
—— (2009a) *Report of the Third Session of the Governing Body of the International Treaty on Plant Genetic Resources for Food and Agriculture*, Doc. IT/GB-3/09/Report, Tunis, Tunisia.
—— (2009b) *Report of the Expert Consultations on Information Technologies to Support the Implementation of the Multilateral System*, Doc. IT/GB-3/09/Inf.4.
—— (2009c) *Report by the Secretary*, Doc. IT/GB-3/09/4.
—— (2009d) *Report on the Status of Cooperation with Other International Organizations, Including Agreements between the Governing Body and the International Agricultural Research Centres of the Consultative Group on International Agricultural Research and Other Relevant International Institutions*, Doc. IT/GB-3/09/18.

Part III
Critical reflections

11

ASSESSMENT OF PROGRESS TO MAKE THE MULTILATERAL SYSTEM FUNCTIONAL

Incentives and challenges at the country level

Isabel López Noriega, Peterson Wambugu and Alejandro Mejías

Introduction

Almost 20 years after the adoption of the Convention on Biological Diversity (CBD), we are still learning how to translate countries' sovereign rights over genetic resources into workable laws and regulations that lead to the objectives of conservation and sustainable use of genetic resources.[1] The international recognition of countries' capacity to regulate the access to their genetic resources and the obligation of genetic resources users to share the benefits with provider countries represented a paradigm change to which we are still adapting. The vast literature on the development and implementation of international, regional and national access and benefit-sharing (ABS) regulations highlights the difficulties involved in getting such regulations developed in the first place and then known, understood and accepted by a broad range of genetic resources users who were used to collecting, exchanging and using genetic material without any relevant interference from public policies (Carrizosa et al., 2004; Glowka, 1998; Lewis-Lettington and Mwanyiki, 2006; Nnadozie et al., 2003; Pisupati, 2008).

The International Treaty on Plant Genetic Resources for Food and Agriculture (ITPGRFA) proposes an ABS system that, based on the CBD's general principles, has been designed taking into account the particular characteristics of plant genetic resources for food and agriculture (PGRFA) and the particular needs of PGRFA users.[2] This system is based on the multilateral facilitated access to a pool of genetic resources of certain crops (listed in Annex 1 of the Treaty) and the multilateral sharing of the monetary and non-monetary benefits arising from the use of such genetic resources in research and breeding for food and agriculture (see Chapter 7 by Daniele Manzella in this volume for a detailed description of the multilateral system on access and benefit sharing and its rationale).

The exchange of PGRFA within the Treaty's multilateral system is subject to the terms and conditions set out in a Standard Material Transfer Agreement (SMTA), whose text was negotiated and adopted by all of the Treaty members at the first session of the Governing Body of the ITPGRFA in June 2006.[3] Technically speaking, the movement of germplasm within the multilateral system could not have taken place before the SMTA was adopted. Therefore, the multilateral system has been in operation for four-and-a-half years at the time this chapter was written.

How far have countries gone in the implementation of the multilateral system? Have PGRFA users worldwide adopted it as the adequate ABS regime for PGRFA? What are the incentives that encourage PGRFA users to become full participants in the multilateral system? And what are the main disincentives or challenges they face to implement the system? In this chapter, we will try to answer these questions by describing the progress made so far and analysing the opportunities and challenges that the implementation of the multilateral system at the country level implies.

Methodology

This chapter relies on information that the authors have been gathering during the last two years. The main sources include the documents prepared by the Secretariat of the ITPGRFA with submissions from countries and international organizations for the last three sessions of the Governing Body of the Treaty; the national reports submitted to the Food and Agriculture Organization (FAO) for the preparation of the *Second Report on the State of the World's Plant Genetic Resources for Food and Agriculture* as well as the *Second Report* itself (FAO, 2010); four country case studies (Kenya, Morocco, Peru and the Philippines) developed as part of the project entitled Global Public Goods, Phase 2 (SGRP, 2010) of the Consultative Group on International Agricultural Research (CGIAR), and the questionnaires and internal reports prepared in the context of the FAO/ITPGRFA Secretariat/Bioversity Joint Capacity Building Programme for Treaty Implementation (which took place from 2007 to 2010). Published records have been completed using informal interviews with a number of the ITPGRFA's national focal points and continued communications with stakeholders in various countries, individuals in the Treaty Secretariat and experts in the International Agricultural Research Centres (IARCs) of the CGIAR as well as other international agricultural research centres. In addition, in order to put into context our analysis of the advances, or lack thereof, in the implementation of the Treaty, we have consulted relevant literature on the conservation and use of PGRFA and, in particular, the policy and legal issues involved.

An important limitation that we have faced is the scarcity of official published documentation about countries' activities related to the implementation of the multilateral system. Bearing in mind that issues around the conservation and use of PGRFA are highly politicized these days, the authors did not always feel comfortable using information obtained from unofficial sources and/or presenting individuals' impressions and opinions that could not be backed up by publicly available records.

The preparation of this chapter involved a second important difficulty, which derives from the dynamic nature of the subject matter. The implementation of the multilateral system of the ITPGRFA is a process where the elements change almost every day (as well as the available information about the changes). Some of the facts we present in the following sections may be different when this volume is eventually published or shortly thereafter. Some of the difficulties and opportunities for the implementation of the multilateral system might not be so relevant in some countries in a few years' time. The purpose of this chapter is to describe the advances made so far and to document the challenges at this point in time.

Overview of the national measures for the implementation of the multilateral system

Treaty members have made progress in putting the system into action, yet the efforts and practical results differ greatly from one country to another.

Designation of the focal point and the national authority in charge of the implementation of the Treaty and the multilateral system

According to the information available on the ITPGRFA's website, 92 out of 127 countries have designated the national focal point for the Treaty.[4] Often, the focal points are in ministerial bodies or research institutes under the Ministry of Agriculture (or the equivalent). Heads of gene banks, directors of genetic resources institutes and specialized personnel from plant variety protection and seed quality offices are abundant in the list of focal points. Brazil is the only party in which the focal point belongs to the Ministry of External Relationships.

The leading and coordinating role of a given national authority in the implementation of the Treaty is either explicitly stated or implicitly recognized by the national legislation, based on the body's traditional responsibilities with regard to PGRFA. In some countries, however, there is still confusion around who should lead and facilitate the implementation of the ITPGRFA at the national level. In Peru, for example, the National Institute for Agricultural Research (INIA), which is the natural leading authority, is still waiting for the Ministry of Agriculture to officially sanction its mandate. This circumstance has not prevented the INIA from taking action towards implementing the Treaty, but the uncertainty has hindered the institution from moving ahead confidently (personal communications with Manuel Sigüeñas in 2009 and 2010). Changes in the structure and duty of various ministries have also caused important delays in some countries, such as Morocco, where the long process of restructuring the Ministry of Agriculture in 2008 and 2009 led to a situation where the originally designated focal point and leading authority for the implementation of the Treaty was not sure about its actual role for more than one year (personal communications with Amar Tahiri, 2009). The lack of clarity around the actual mandate of national focal points with regard to Treaty implementation is one of the most limiting factors in the operation of the multilateral system at the country level, and it can be easily overlooked by representatives of countries where the mandate, capacity and resources of the national focal point are already well ironed out. In the process of drafting the necessary legislation for the implementation of the Treaty in the Philippines, one of the biggest concerns of the authorities in charge of such process was to reinforce its own mandate and capacity to coordinate the implementation activities. Getting the draft law considered and approved by governmental organizations and research institutes involved in PGRFA conservation and use has been a necessary step for the Philippines' Treaty focal point to move ahead with the practical implementation of the multilateral system (Bioversity International, 2011).

Awareness raising and training activities

Many countries, including Benin, Brazil, Canada, Ethiopia, Germany, Ghana, Guatemala, Kenya, Madagascar, the Netherlands, Peru, the Philippines, Spain, Switzerland, Sudan and Zambia, have arranged workshops and meetings to raise awareness about the ITPGRFA among a diversity of stakeholders and to discuss actions to advance the implementation of the multilateral system. In some developing countries, international agencies such as the FAO and Bioversity International have assisted national authorities by channelling funds provided by donor countries and international organizations, inviting international experts to the workshops and preparing guidelines and learning modules for training activities.[5] These workshops have proved to be helpful in setting the foundations for further implementation activities. However, some of these initiatives have ended up being isolated events with no continuation, and they have not had an impact in the long term.

Establishment of consultative intersectoral bodies

In a number of countries, the leading authorities have established consultative interdepartmental bodies to discuss and coordinate actions leading towards the implementation of the ITPGRFA. In Kenya, for example, the Ministry of Agriculture has created the National Plant Genetic Resources Committee, a multidisciplinary and multisectoral body where a range of public institutions, as well as some private ones, are represented. The committee is in charge of domesticating the Treaty by engaging relevant national programmes and stakeholders (Wambugu and Muthamia, forthcoming). The Department of Agriculture in the Philippines has also established a Technical Support Working Group for the Treaty's implementation, which includes representatives from national gene banks and academia (personal communications with Amparo Ampil, 2010). Similar bodies have been established in Madagascar, Guatemala, Egypt, Jordan and South Arabia (surveys conducted by the Joint Capacity Building Programme, 2008; personal communications with focal points, 2009). The experience of these consultative bodies have evidenced the need of involving officials from the ministries of environment in the implementation of the multilateral system, particularly in those countries where, since the negotiation and adoption of the CBD, the development of laws and regulations on access to genetic resources has traditionally been considered a duty of the governmental branches that deal with the conservation and use of biodiversity and (wild) genetic resources. Getting the ministries of environment (or analogous national bodies) to recognize the particularities of PGRFA and the need to respect and support a specialized ABS regime for such resources is still a pending task in many countries (Lewis-Lettington and Mwanyiki, 2006, 1).

Adoption of legislative and administrative measures

Determining whether there is legal space within the national policy and legal landscape for implementing the ITPGRFA is one of the first important issues that governments need to consider. Particular attention needs to be paid to the existing national ABS legal framework, as it may contradict, or not sufficiently recognize, access and benefit-sharing rules under the multilateral system. A study recently published by the FAO Commission on Genetic Resources for Food and Agriculture shows that most ABS laws do not apply specialized rules to different types of resources (plant, animal, microbial and so on) and do not differentiate between the sectors that make use of such resources (agriculture and pharmacy/cosmetics, for example) and the final purpose of the use (fundamental research, applied research, development of commercial product and so on) (Nijar et al., 2009). However, the number of countries that provide for a particular treatment for PGRFA in their ABS frameworks is increasing. Shortly before or after their ratification of the Treaty, several countries, such as Australia, Brazil, Ethiopia, Peru, the Philippines and Spain passed ABS laws and regulations that exempt, implicitly or explicitly, PGRFA from the general rules of ABS.[6] The scope of such an exemption differs from country to country. For some, it applies to PGRFA of crops listed in Annex 1 of the Treaty, while, for others, it applies to PGRFA conserved in *ex situ* conditions. In some cases, the exclusion refers, in general, to genetic resources whose access and use is explicitly regulated by international instruments to which the country is a party. So far, there are no instances of national legislation that set out detailed procedures for dealing with ABS under the multilateral system (FAO, 2010, 173), but some countries such as Peru, the Philippines, Sudan and Syria are in the process of drafting legal provisions in this regard, either as part of their general laws on (plant) genetic resources or in the form of regulations that include specific processes and responsibilities.

Some governments consider that the implementation of the multilateral system within their countries does not require adopting national legislation or amending the existing one, and they have dealt with the different issues involved in the implementation through administrative measures coordinated by the implementing authority. This is the approach adopted, for example, by Germany, the Netherlands, Spain and Switzerland (FAO, 2010, 172; personal communications with Luis Ayerbe and François Pythoud, 2009).

Identification of PGRFA included in the multilateral system of the ITPGRFA

According to Article 11.2 of the Treaty, PGRFA that are automatically included in the multilateral system are those of the crops listed in Annex 1 that are (1) under the management and control of contracting parties and (2) in the public domain. It is not very clear whether all of the country members have gone through the exercise of identifying which national PGRFA collections are automatically included in the multilateral system according to Article 11.2, but, for those who have, the interpretation of these two criteria does not seem to have been particularly problematic.[7] In the opinion of the ad hoc technical advisory committee on the SMTA and the multilateral system, 'under the management and control of the contracting party' refers to materials for which the central national administration has 'the power to undertake acts of conservation and utilization' (ITPGRFA, 2010a, 10). The notifications sent to the Secretariat of the Treaty on the inclusion of PGRFA in the multilateral system (ITPGRFA website) as well as the studies conducted during the Joint Capacity Building Programme for the Treaty's implementation (Nijar, 2012; Sigüeñas Saavedra et al., 2011) suggest that there is a general understanding that germplasm collections held by national agricultural research institutes and by other research organizations that are subject to a ministry's policy guidelines and/or budget are 'under the management and control of the contracting party' and therefore automatically included in the multilateral system. Regarding the second criteria, not all countries have adopted the same meaning of 'in the public domain'. A number of parties have interpreted the term in the context of intellectual property rights and, therefore, have concluded that it applies to PGRFA that are not subject to such rights. This view is in accordance with the technical advisory committee's opinion (ITPGRFA, 2010a, 11). Other countries have translated 'in the public domain' to mean public property. The United Kingdom, for example, has qualified as 'public domain' the germplasm conserved in collections that are entirely or partially supported by public funds (ITPGRFA, 2010b).

For PGRFA of the crops and forages listed in Annex I that are not 'under the management and control' of the government and 'in the public domain', the ITPGRFA states that contracting parties agree to take appropriate measures to encourage natural and legal persons responsible for such PGRFA to put them in the multilateral system (Article 11.3). In countries with a federal system of government, such as Australia, Brazil, Germany and Spain, collections that are not under the administration of the central government fall under Article 11.3 of the Treaty – that is, they are not automatically included in the multilateral system. Countries' notifications and reports indicate that, in general, germplasm collections held by public conservation and research institutions such as universities and botanical gardens have not yet been identified as part of the multilateral system. Unless this situation is revised, it will leave outside the multilateral system a number of PGRFA collections that may be crucial because of their size, diversity or uniqueness. Representatives from many countries have reported successful experiences in getting federal and autonomous public institutions' germplasm collections included in the multilateral system. For example, in Switzerland, as part of the national action plan for the conservation of PGRFA, all of the organizations that are involved in the plan and that hold Annex 1 PGRFA must put their material in the multilateral system. These organizations include a number of institutions that are not under the management and control of the Swiss government (European Regional Group, 2011).[8]

Assessment of progress to make the multilateral system functional **205**

Getting accurate information about the material currently available in the multilateral system is not easy because many countries have not formally confirmed which PGRFA collections are included in the system according to Article 11.2 of the ITPGRFA. Public records, including countries' notifications to the Treaty's Secretariat, indicate that around 460,000 accessions of germplasm held by national entities are available through the multilateral system (Argumedo, 2009; ITPGRFA website; European Regional Group, 2011). However, much more material may be available, practically speaking. Out of the 460,000 accessions, countries in the European regional group have contributed 319,000, or around 70 per cent, of the total.[9] Most of these accessions are documented in EURISCO, a web-based catalogue that provides information about *ex situ* plant collections maintained throughout Europe. The catalogue has been recently amended to include information about the accessions' status with regard to the multilateral system.[10] Germany, with 108,675 accessions, the United Kingdom, with 42,722, Switzerland, with 33,926, and the Czech Republic, with 32,616, are the largest European contributors. With more than 62,000 accessions each, the collections of wheat and barley are larger than any other crops included by countries in the European region (European Regional Group, 2011). Canada

International Gene Banks and Research Centres		Approximate number of accessions
CGIAR Centres	Africa Rice Center	693,000
	Bioversity International	
	International Centre for Tropical Agriculture (CIAT)	
	International Maize and Wheat Improvement Center (CIMMYT)	
	International Centre for Agricultural Research in the Dry Areas (ICARDA)	
	International Crops Research Institute for the Semi-Arid Tropics (ICRISAT)	
	International Institute for Tropical Agriculture (IITA)	
	International Livestock Research Institute (ILRI)	
	International Potato Center (CIP)	
	International Rice Research Institute (IRRI)	
	World Agroforestry Centre	
Centre for Pacific Crops and Trees (CePaCT)-SPC Community		15,000
Tropical Agricultural Research and Higher Education Center (CATIE)		11,400
Mutant Germplasm Repository of the FAO/IAEA Joint Division		2,500
International Cocoa Gene Bank		2,000
International Coconut Gene Banks		158

FIGURE 11.1 Number of accessions held by international gene banks and research centres.

has notified that the collections maintained by the Department of Agriculture and Agri-Food Canada are included in the multilateral system. These represent around 110,000 accessions or almost 25 per cent of the total number of national accessions identified so far.

To date, the inclusion of collections held by private institutions has been modest. In addition to the Swiss associations and foundations mentioned earlier, two French associations have put a total of 2,300 accessions of maize and bread wheat into the multilateral system,[11] and the ANDES association, on behalf of the Potato Park in Peru, has declared that the park's germplasm collection (representing around 1,300 varieties of potato, according to ANDES) has also been included (Argumedo, 2009). The private seed industry has made no contribution to the multilateral system so far.

In addition to public and private national collections, the multilateral system includes PGRFA held by international organizations. Various international institutes, including 11 IARCs under the CGIAR, have signed agreements with the Governing Body of the Treaty to put their germplasm collections into the multilateral system (see Figure 11.1 above).

For some countries' representatives, PGRFA found *in situ* are not automatically included in the multilateral system, but this interpretation is in contradiction with the provisions of the ITPGRFA, which do not exclude *in situ* PGRFA from the multilateral system as long as they meet the criteria defined by Article 11.2 (Correa, 2008; ITPGRFA, 2010a, 10). In some countries, such as Malaysia, legal experts are currently conducting studies on how the criteria 'under the management and control of the contracting parties' apply to PGRFA found in national parks and natural reserves, with the aim of identifying possible multilateral system germplasm held *in situ*, but, as of today, there is no evidence that any government or public or private entity has identified *in situ* germplasm as being included in the multilateral system of the Treaty, except for the Potato Park mentioned earlier. The actual accessibility of the park's PGRFA is still not clear though.

The material currently included in the system according to publicly available records is close to 23 per cent of the almost 5 million accessions of Annex 1 crops

Location of PGRFA identified as included in the multilateral system (2011)

- International genebanks and research centres
- European Region
- Other countries (including Canada)

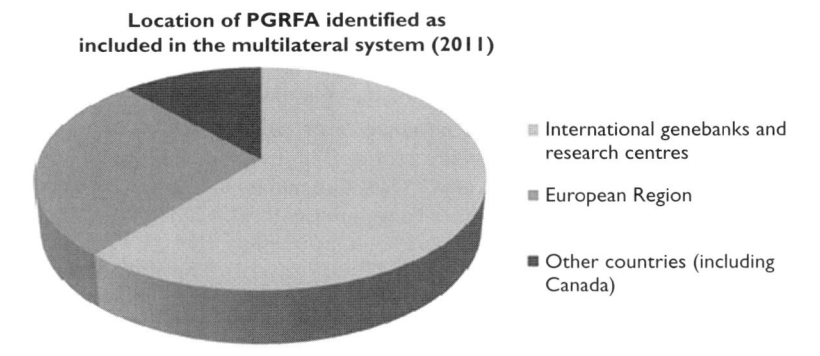

FIGURE 11.2 Location of PGRFA identified as included in the multilateral system, 2011.

Assessment of progress to make the multilateral system functional **207**

■ Portion of Annex 1 samples that have been identified as included in the multilateral system (2011)

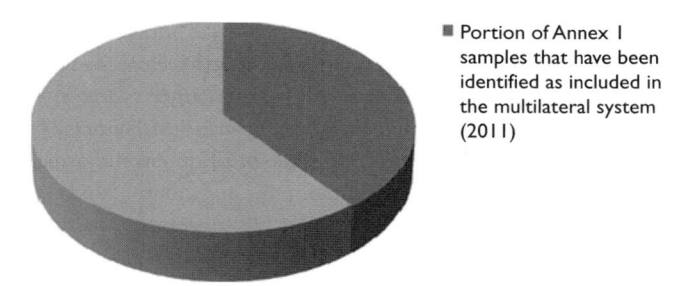

FIGURE 11.3 Portion of Annex 1 accessions that have been identified as included in the multilateral system, out of the total of accessions held worldwide.

that have been calculated to be conserved in more than 1,240 gene banks worldwide, and around 35 per cent of the approximately 3 million conserved in national gene banks of the current ITPGRFA contracting parties and the IARCs (FAO, 2010, 55). These figures give us an idea of the total number of accessions potentially available in the multilateral system.

Countries and organizations are not obliged to formally report on the PGRFA collections included in the multilateral system, but the actual functioning of the system depends, to a great extent, on the information available to PGRFA users about which materials are actually included in the system. The Governing Body has stressed this point several times (ITPGRFA, 2009c, Appendix A, 20, 2011c, para. 2). With the purpose of facilitating information flow between PGRFA providers and users in the multilateral system, the Treaty's Secretariat has formulated and made available a sample notification letter that institutions can use to inform about the germplasm collections they have put into system. Whenever the Secretariat receives notifications from governmental agencies and research institutions, it posts them on the Treaty's website. The information available at the moment refers to collections maintained by national institutions in 20 countries.[12] The amount and quality of the information that potential PGRFA users can get through this means varies a lot from country to country. While for the Nordic countries and the Czech Republic, for example, the institutions' websites provide, at least, complete passport data for most of the germplasm accessions they hold, for other countries, such as Brazil and Zambia, there is no information available online. In practice, this limitation greatly restricts the use of such collections by scientists, breeders and farmers abroad.

Use of the benefit-sharing fund

Those who commercialize a PGRFA that incorporates genetic material from the multilateral system are required to transfer 1.1 per cent (minus 30 per cent) of the sales of such PGRFA to a benefit-sharing fund managed by the governing body of the Treaty, whenever the commercialized PGRFA is not available for research and breeding. Alternatively, the recipient can pay 0.5 per cent of sales of any PGRFA that are of the same crop as the germplasm received from the multilateral system. So far, the benefit-sharing fund has not received any contribution through this channel, and it would be unrealistic to expect that this will happen for the next few years as the development of a marketable plant variety takes from 5 to 15 years (Halewood and Nnadozie, 2008). The money currently available in the benefit-sharing fund comes from voluntary contributions from countries (Australia, Italy, Norway, Spain and Switzerland) as well as from the United Nations Development Programme, totalling approximately US $13.4 million. The aim of the fund is to support conservation and crop improvement efforts, especially in developing countries and economies in transition. The ITPGRFA's governing body establishes priorities, criteria and conditions for the use of the benefit-sharing fund and selects projects for funding. So far, there has been one round of projects funded, while the second round of projects is expected to be approved in the near future. In the first round, which started in December 2008, the fund provided a total of US $580,000 to 11 projects, which were selected from several hundred proposals, whose total cost would have been approximately US $20 million.[13] The second round of projects was opened in December 2010, with much more money in the fund and with a target of US $10 million invested in the projects. The participation of the various countries in both rounds has been impressive. The enormous number of proposals submitted demonstrates the need for such a fund, its success as a multilateral financial mechanism and its effectiveness in making the multilateral system known among national actors. However, the functioning of the fund is hindered by some serious weaknesses. In the first place, the fund is entirely dependent on countries' and international organizations' voluntary contributions. So far, member countries' donations to the fund have been quite limited. Second, it does not have a clear mechanism to evaluate how the funded projects are actually contributing to PGRFA conservation and food security. The fund has been the target of critics from civil society and farmers' organizations, who have expressed their discontent about how the Governing Body selects the projects to be funded (Kastler et al., 2009) (we will address this issue later in this chapter). All of these elements confirm that the benefit-sharing tool of the Treaty can certainly be improved.

Analysis of the use of the SMTA

The number of SMTAs issued to exchange germplasm and the amount of different SMTA users are critical indicators of the level of implementation of the multilateral system. In order to assess the level of utilization of the SMTA, we have analysed the information provided by the member countries and international research institutions in reports submitted to the second, third and fourth sessions of the Governing Body of the Treaty regarding the implementation of the multilateral system and the use of the SMTA (ITPGRFA, 2007a, 2009a, 2009b, 2011a, 2011b; European Regional Group, 2011). It is important to note that, so far, only the European Regional Group has submitted a report detailing the use of the SMTA by organizations in the European member states. In addition, we have gathered publicly available information from other national and international centres and gene banks, including NordGen, the international cocoa gene bank and one of the international coconut gene banks.[14]

When analysing the available data, we have to bear in mind that the sources we have used do not provide a complete picture of the situation. A number of PGRFA transfers within the multilateral system remain outside these records because countries do not publish or share the information about how national institutions are using (or not using) the SMTAs. In informal communications with members of the delegations of Iran, Jordan and Zambia, we have been told that some national institutions have started to transfer PGRFA using the SMTA but that detailed information has not yet been made publicly available. Similarly, in their submissions to the Governing Body, Canada has indicated that the SMTA started to be used in the country in July 2008, but it has not provided details about how much it has been used (ITPGRFA, 2009d). In addition, a number of sources from national research institutes (who we prefer to treat anonymously) have confirmed that the SMTA has not been used by organizations or individuals in their countries, except on those occasions in which they have been invited to do so by the provider or the requester of the PGRFA under transaction, commonly a centre of the CGIAR.

The available information is therefore insufficient to conduct a systematic analysis of the use of the SMTA as well as the typology of providers and recipients of PGRFA within the multilateral system. That said, we can extract some interesting figures and present some preliminary conclusions. According to the sources mentioned earlier, around 1,222,000 samples of accessions of PGRFA have been transferred using the SMTA between January 2007 and December 2010.[15] Approximately 95 per cent of the samples were transferred by the IARCs, and the majority of these were sent to developing countries (around 85 per cent) (ITPGRFA, 2007a; 2009a; 2011a). About 60,000 samples were distributed by national institutions, mainly European ones. This amount represents less than 5 per cent of the total germplasm distributed in the multilateral system and a very small portion of the samples that are regularly distributed by countries outside the multilateral system, according to the information about germplasm distribution available in the *Second Report on the State of the World's Plant Genetic Resources for Food and Agriculture* (FAO, 2010). However, the most recent data show that countries' distributions under the SMTA tend to increase.

The figures on germplasm acquisition presented by the CGIAR IARCs' reports to the Governing Body show that the use of the SMTA by national institutions when distributing their PGRFA is still modest. According to such figures, roughly 24,000 samples have been acquired without the SMTA during 2007–9. This amount triples the quantity of PGRFA that the centres have acquired under the SMTA from countries for the same period. These figures have to be interpreted with caution, taking into consideration that much of the PGRFA that the IARCs receive come from countries that are not parties to the Treaty. It is also important to note that some of the materials transferred to the IARCs' gene banks are done without an SMTA but, rather, with the understanding that the centre may distribute it under the SMTA (ITPGRFA, 2011b). At least one national institution has had a similar experience. In a collecting mission on wild spinach conducted by the Dutch gene bank CGN in Uzbekistan and Tajikistan (which are not parties to the ITPGRFA), the countries' representatives agreed with CGN that the collected material would be shared between the countries and the CGN and that the Dutch gene bank could transfer such materials under the conditions of the SMTA (Van den Hurk, 2011). In this context, it is interesting to mention the experience of the International Network for the Genetic Evaluation of Rice (INGER). In the mid-1990s, shortly after the IARCs (including the International Rice Research Institute (IRRI)) signed agreements with the FAO to put their PGRFA collections in trust for the world community, the national research institutes that were part of INGER decided that they would only exchange rice germplasm through the network on the condition that it was not included in the gene bank of the IRRI, where it would be made available for international distribution under the IARC's material transfer agreement pursuant to the FAO-CGIAR agreements. After the Governing Body of the ITPGRFA adopted the SMTA in 2006, and after the IRRI signed its agreement with the Governing Body, putting the IRRI's rice collection under the Treaty's multilateral system, INGER members decided to reverse the decision they made in the 1990s, welcoming the benefit-sharing provisions of the SMTA. Since 2006, the INGER countries have made contributions of national germplasm to the network again, allowing the IRRI to deposit such material in the gene bank and distribute it to other countries using the SMTA (see Chapter 5 by Michael Halewood et al. in this volume). All of these examples show that there is support for using the SMTA for PGRFA exchange, even among stakeholders in countries that are not parties to the Treaty.

Following the decision adopted by the Treaty's Governing Body on the material transfer agreement to be used for the transfer of non-Annex 1 PGRFA by international organizations that have signed agreements with the Governing Body, these international gene banks and research centres began to use the SMTA to transfer the germplasm of crops and forages not included in Annex 1, extending, in this way, the multilateral system's terms and conditions to genetic material beyond the initial scope of the system. The IARCs distributed approximately 15,000 accessions of non-Annex-1 materials between August 2007 and December 2009 (ITPGRFA, 2009a, 2011a). The Tropical Agricultural Research and Higher Education Centre (CATIE), the Centre for Pacific Crops and Trees and the International Cocoa Genebank have used the SMTA to transfer samples of cacao, coffee, taro, capsicum, cucurbits and other crops that are not included in Annex 1 (ITPGRFA, 2011b; personal communication with Pathmanathan Umaharan, December 2010). At the country level, the government of the Netherlands has officially stated that non-Annex 1 PGRFA conserved in one of the national collections will be also made available through the SMTA (Netherlands Ministry of Agriculture, Nature and Food Quality, 2008). At the regional level, the European institutions involved in the European Genebank Integrated System (AEGIS) have committed to make both Annex 1 and non-Annex 1 PGRFA available under the SMTA.[16]

Germplasm in the multilateral system is being sent to non-parties of the ITPGRFA. For example, according to the information provided by the IRRI on the

Treaty's website, in the first five months of 2009 around 19,500 samples of rice were sent using the SMTA to countries that were not parties to the Treaty, including the United States (10,185 samples), China (4,300 samples), Thailand (2,095 samples), South Korea (1,170 samples), Japan (900 samples) and Sri Lanka (870 samples) (all amounts are approximate only). Before the Treaty was adopted, access to germplasm conserved in trust by the IARCs was already subject to facilitated access, through the material transfer agreement whose text had been agreed upon by the FAO's Commission on Genetic Resources. The main difference is that now all of the recipients, both those in contracting parties and those in non-contracting parties, are subject to the benefit-sharing conditions of the multilateral system (see Chapter 6 by José Esquinas-Alcazar, Angela Hilmi and Isabel López Noriega). By using the SMTA to transfer materials to non-parties to the Treaty, the IARCs are extending the multilateral system's terms and conditions to PGRFA users in such countries, increasing in this way the possibilities of getting monetary benefits flowing back to the benefit-sharing fund. On the other hand, this non-discriminatory treatment may have the effect of discouraging countries from becoming parties to the Treaty or from implementing it effectively, as we will explain later in this chapter.

Around 75 per cent of the PGRFA distributed by the IARCs are materials that the centres had been involved in improving (ITPGRFA, 2007a, 2009a, 2011a). This figure confirms the observed general trend that improved materials and breeding lines are exchanged more frequently than materials held in gene banks (FAO, 2010, 96).

The ITPGRFA's Governing Body has stressed the importance of documenting the use of the SMTA (ITPGRFA, 2011c, para. 16). During the third session of the Governing Body, member countries agreed on the reporting schedule and the format for PGRFA providers using the SMTA to notify the third party beneficiary, pursuant to Article 5.e of the SMTA (see Chapter 8 by Gerald Moore in this volume). Although this information would be extremely useful to evaluate the functioning of the multilateral system across countries and PGRFA users worldwide and, in general, to analyse how the system actually contributes to food security in different countries, the Governing Body has decided to treat such information as confidential, which means that the information gap will continue to exist at least until the moment countries decide to publish or make available their records on the use of the SMTA (ITPGRFA, 2009c, Appendix A, 25–32).

Incentives and challenges in the implementation of the multilateral system

Incentives

All countries rely on PGRFA coming from abroad to adapt crops to the different variables that intervene in the food production and supply chain – from the environmental conditions in the production fields to the crop transportation and storage capacities and the consumers' preferences (Flores Palacios, 1998; Gepts, 2004; Kloppenburg and Kleinman, 1987). The multilateral system of the ITPGRFA provides for facilitated access to a diverse pool of PGRFA that scientists and farmers can use to develop plant varieties suited to the existing conditions. Such facilitated access is the greatest and most obvious incentive for countries to join the Treaty and implement the multilateral system.

Countries' consciousness that their agricultural production depends on genetic resources conserved in other countries tends to increase when they face episodes of devastating pests and climate conditions threatening the whole national production of staple crops (Joshi et al., 2010; Taylor, 2011). Enhancements to national research and breeding capacities have contributed to raising countries' awareness of the importance of having access to other countries' reserves of plant diversity. The steady rise in research on the most important crops in developed and developing countries during the last decades (rice, wheat and maize) and the rapid distribution of new varieties of these crops have evidenced the extent to which national agriculture production relies on genetic resources coming from other countries (Evenson and Gollin, 2003, 7–38). Representatives from China and Brazil, two countries that have experienced a rapid economic growth and that have made impressive investments in agricultural research in the last 10–15 years, have repeatedly declared that their crop improvement programmes rely heavily on germplasm coming from abroad (Brazil, 2009, 8; Nature Editorial, 2010; People's Republic of China, 2008, 55). Another factor that helps national actors recognize the need for having access to other countries' genetic resources is their involvement in regional and international networks and initiatives for plant genetic resources conservation and use. Networks that facilitate the exchange of germplasm between their members show that only by sharing their own germplasm can countries have access to other's genetic resources. This is the case in regional networks such as the European Cooperative Programme for Plant Genetic Resources and the Pacific Agricultural Genetic Resources Network (Taylor, 2011) as well as crop networks such as the INGER.

For many national experts, the multilateral system represents an opportunity for countries to improve their national genetic resources systems and to become more visible among decision makers. The actions needed to effectively implement the ITPGRFA, and the multilateral system in particular, are closely related to the enhancement of different components of the national systems on genetic resources, such as *ex situ* collections, breeding programmes, PGRFA documentation systems and coordination tools among different entities working on genetic resources. In the course of developing mechanisms to implement the Treaty, stakeholders can contribute to 'putting things in order' in their own genetic resources systems and strengthening national capacities to conserve and use genetic resources. In Peru, for example, one of the results that has developed from identifying PGRFA conserved by national institutions that could be included in the multilateral system has been that the INIA has recognized and classified almost 8,000 accessions that were held by public research entities (including the INIA's decentralized stations) and that had not been clearly included in the INIA's records until now (Sigüeñas Saavedra et al., 2011).

Another important motivation for countries to become active members in the multilateral system is the possibility of getting support from different financing mechanisms, such as the Global Crop Diversity Trust (GCDT) and the benefit-sharing fund of the ITPGRFA. The former provides funds for countries to upgrade valuable PGRFA collections through regeneration and safety backups of germplasm under the condition that such germplasm will be made available in the multilateral system of the Treaty (GCDT, 2010, 11). So far, over 100 institutions in more than 80 countries have received support from the GCDT to improve the conservation and management of PGRFA collections (*ibid.*, 9). It is predicted that around 90,000 accessions of at-risk, unique germplasm will be made publicly available under the SMTA as a result of the GCDT-supported regeneration work (Fowler and Hawtin, 2011). The fact that money is actually available in the Treaty's benefit-sharing fund and that it is being distributed among various national projects has increased members' trust in the system's ability to balance access to genetic resources and access to the financial support necessary to conserve and use the genetic resources (see Chapter 12 by Godfrey Mwila in this volume). In addition, PGRFA listed in Annex 1 of the Treaty resulting from projects funded by the benefit-sharing fund shall be made available according to the terms and conditions of the multilateral system, and information generated by such projects shall be made publicly available within one year of the completion of the project (ITPGRFA, 2009c, Appendix A, 4).

In general, countries appreciate the clarity and legal certainty that the ITPGRFA provides for the exchange of PGRFA within the multilateral system. Many governments have struggled to develop and approve access and benefit-sharing regulations. In a number of countries, the legal status of genetic resources and how they can be accessed and used is not clear and both potential users and providers get lost in the complexity of laws, regulations and authorities involved in the process of granting access permits. More than 35 national reports submitted to the FAO for the preparation of the *Second Report on the State of the World's Plant Genetic Resources for Food and Agriculture* state that governments lack the necessary multidisciplinary scientific, institutional and legal capacity to develop a satisfactory system of ABS, given the inter-related dimensions of access, benefit-sharing possibilities, protection of local community rights and traditional knowledge, and connected issues of intellectual property. In some countries, the national authorities are reluctant to negotiate and grant access permits because they are afraid of doing things incorrectly (Lapeña et al., 2010; Wambugu and Muthamia, forthcoming). In this scenario, the availability of a standard ABS contract (the SMTA) that is agreed upon internationally, that can be signed without the intervention of governmental agencies and that saves providers and recipients the risk of writing inappropriate clauses has relieved many actors.

Finally, national stakeholders – in particular, those in developing countries – have big hopes that the multilateral system will give them access to knowledge and technologies, including those protected by intellectual property rights. Active participation in the multilateral system has been identified as an avenue for opening up greater regional and international collaboration and partnerships in various areas of germplasm conservation and use. These partnerships are important for germplasm exchange and access to scientific information and technology.

Challenges

Probably the biggest challenge in the implementation of the ITPGRFA in many countries is the lack of awareness and knowledge of the Treaty. National policy makers lack an appreciation for the national dependence on foreign germplasm and are not aware of the benefits that the country can gain by becoming an active member in the multilateral system. Without the support of their governments, Treaty focal points find it difficult to move the implementation agenda ahead (Altoveros et al., forthcoming; Lapeña et al., 2010; Sadiki et al., forthcoming; Wambugu and Muthamia, forthcoming).

In various countries, such a lack of awareness is a problem not only among high level decision makers but also among the PGRFA users themselves. The multilateral system has started to operate in an environment where genetic resource issues are highly politicized. In the last few decades, public attention has been more and more fixated on the activity of private companies that increasingly concentrate seed commercialization worldwide as well as by how these companies commercialize new varieties under very strict schemes of intellectual property rights. Civil society organizations and governmental entities have evidenced cases in which genetic resources found in developing countries have been misappropriated by industry in the developed world. There have been an increasing number of media reports about alleged cases of bio-piracy especially levelled against Western countries. While, at times, it is clear that these reports are founded on shaky factual foundations, such assertions resonate with the public and create a misleading and incorrect impression about the prevalence of this activity (Finston, 2005). In addition, they have contributed to shaping the perceptions about the potential value of germplasm, making scientists, policy makers and civil society organizations feel that all germplasm has a high potential economic value and that they need to guard against losing control over their germplasm and, thus, should protect their capacity to negotiate the conditions of access to every single sample of germplasm (Tumushabe and Mugoya, 2004).[17] Policies on access to genetic resources have been the target of politics, misconceptions and negative publicity (Carrizosa et al., 2004), and bilateral access and benefit-sharing agreements that were initially formulated in a collaborative and transparent manner have ended up being controversial due to the political context in which they have taken place (Rosenthal and Katz, 2004). With this background of mistrust and misunderstanding, it is not surprising that some sectors see the multilateral system of the ITPGRFA as a tool for industrialized countries to get easy access to poor countries' germplasm (Egziabher et al., 2011). The implementation of the multilateral system will be hard until this and other similar opinions are overcome, and it can only take place if, in turn, the multilateral system is put in place effectively. These difficulties are aggravated in those countries where there was very little consultation before the decision to ratify the Treaty was made. Now, at the implementation stage, focal points realize that they have to begin to raise awareness from scratch.

Many actors have repeatedly stated that the implementation of the multilateral system should take place in a balanced way, meaning, first, that the different elements of the ITPGRFA (the multilateral system, farmers' rights, the conservation of PGRFA, the sustainable use of PGRFA and so on) should be paid equal attention at both the Governing Body and individual country levels (Chapter 12 by Godfrey Mwila in this volume; Egziabher et al., 2011). Second, it means that, together with PGRFA, parties to the Treaty should put financial resources into the multilateral system to support developing countries' efforts to implement the Treaty (ITPGRFA, 2007b, para. 49, 2009c, Appendix A, 21–22). Third, it means that the sharing of non-monetary benefits such as information, knowledge, capacities and technology should be considered as important as the monetary benefit sharing and should be properly addressed by both the Governing Bodies' and the member countries' work plans for the implementation of the multilateral system (ITPGRFA, 2011c). Some governments' lack of will to make their PGRFA available in the multilateral system is partially because they feel that developed countries have neither submitted enough funds to the Treaty's funding strategy nor adopted any clear measure to share information, technologies and capacities for PGRFA conservation and use with developing countries (Chapter 12 by Godfrey Mwila in this volume).

Another element that limits some countries' motivation to implement the multilateral system is that it is difficult to comprehend its potential benefit for poor farmers. Many perceive the multilateral system to be focused mostly on *ex situ* conservation and meeting the needs of gene bank managers and plant breeders. They feel that it pays little attention to *in situ* conservation and the needs of other actors who play an important role as conservers and generators of crop diversity – that is, the farmers and, in particular, the low income farmers in developing countries. These perceptions are based on different facts. First, the majority of national delegations that were involved in the negotiations of the Treaty and that are currently involved in the Treaty's Governing Body come from agricultural research institutes and are naturally biased by their own institutes' reality and problems, which are not necessarily the same as farmers'. Some non-governmental organizations representing farmers' interests were very active during the Treaty negotiations and some national delegations were strongly committed to supporting farmers' interests, but their efforts focused on getting farmers' rights included in the Treaty, rather than bringing the multilateral system closer to farmers' priorities and *modus operandi* (Chapter 6 by José Esquinas-Alcazar, Angela Hilmi and Isabel López Noriega in this volume; Egziabher et al., 2011). For these reasons, the negotiations have resulted in a multilateral system on ABS that seeks to serve poor farmers' well-being in the long run but whose operative elements (*ex situ* collections, 22 page-long standard material transfer agreement, funding strategy, international benefit-sharing fund, third party beneficiary and international arbitration) are considerably removed from the reality of most farmers in developing countries.

Second, many if not most of the rural communities in developing countries continue to use traditional or informal sources to meet their seed needs. Common figures suggest that the formal seed production and distribution system provides for around 15 per cent of the total seed used by farmers in developing countries (Cooper, 1993; FAO, 1998, 2010; Hodgkin et al., 2007), although the situation varies by crop and region. Recent studies show that in Morocco, for example, the rate of certified seed used by farmers in grain and legume production ranges from 1–13 per cent (Sadiki et al., forthcoming). The rest may have been selected from the farmers' own crop in the preceding season, exchanged or purchased from other farmers or institutions or be a mixture of seeds from a combination of sources (Bellon and Risopoulos, 2001; Jarvis and Ngung'u-Skilton, 2000; Sperling and Mcguire, 2010). Only a certain number of improved varieties produced by breeders enter into these informal circuits of seed production and distribution, for different reasons that include weak or non-existent formal seed production and distribution mechanisms (such as extension services, seed industry and seed market) and farmers' lack of trust in seed sources that are different from traditional ones (Jarvis et al., 2011; Lapeña et al., 2010).

In addition, it has been recognized that modern improved varieties released by breeders do not often meet environmental conditions and production techniques of farms in remote and marginal areas (Cooper, 1993; Jarvis et al., 2011). Taking into account that, at the present time, the main users of PGRFA in the multilateral system are public sector breeders aimed at producing new varieties that farmers can acquire largely through formal seed production and distribution systems, we can understand why some stakeholders feel that the multilateral system does not properly meet the needs of farmers. The multilateral system's benefits for such farmers may continue to be uncertain until the linkages between formal and informal research, development and seed distribution systems are enhanced in ways that allow farmers to benefit from modern varieties as much as they benefit from the traditional ones. The tasks of establishing and maintaining appropriate seeds systems are beyond the scope of the multilateral system, but we cannot ignore the fact that the ability of the Treaty to achieve its final objectives of sustainable agriculture and food security depend, to a great extent, on the accomplishment of such tasks.

In regard to monetary benefit sharing, international farmers' organizations have expressed their disappointment about the way the money in the benefit-sharing fund is being spent. After the first round of projects was approved, they complained about the fact that the bulk of the fund went to governmental and non-governmental organizations, instead of farmers, and that most of the selected projects were not for on-farm conservation (Kastler et al., 2009). One of the challenges being posed to the parties to the ITPGRFA and international organizations from civil society organizations is to find ways to facilitate farmers' access to germplasm in the multilateral system and to monetary and non-monetary benefits arising from its use. The apparent distance between the multilateral system and farmers is seen as a disincentive for the Treaty's implementation, but it can also be taken as an opportunity for countries and international organizations to revisit the role and recognition of farmers in plant conservation, research and development schemes to elevate their profile.

Another stakeholder group that has expressed dissatisfaction with the multilateral system is the private sector. Although the seed industry strongly supports the multilateral system's principles and the adoption of standard procedures for facilitated access to germplasm, they complain about various terms and conditions in the SMTA. These include the fact that the multilateral system does not cover all crops that are important for food and agriculture; the absence of a threshold for the level of incorporation of accessed material in the final product; and the ambiguity in regard to the duration of mandatory benefit sharing in the case of restrictions for further research and breeding (International Seed Federation, 2007). In addition, some companies have expressed concern with regard to the non-harmonized and slow adoption of the multilateral system by countries (Lambalk, 2009). For all of these reasons, it is possible that the most likely candidates for making mandatory payments according to the benefit-sharing provisions of the Treaty and the SMTA may seek PGRFA from sources outside the multilateral system, at least for the time being (Halewood and Nnadozie, 2008).

Figures show that there has been an increase in the number of breeding programmes, particularly in the private sector. However, the allocation of resources to plant breeding has not increased substantially in the last decade, and stakeholders from all of the regions, except Europe, report constraints in human resources, funds and facilities (FAO, 2010, 98–101; GIPB, 2011). There is a common understanding that improved research and breeding capacities in developing countries would lead to an increased demand for genetic materials and, subsequently, would develop a greater interest in the multilateral system among PGRFA users in those countries (Altoveros et al., forthcoming; Lapeña et al., 2010; Sadiki et al., forthcoming; Wambugu and Muthamia, forthcoming). On the other hand, breeders' uses of PGRFA conserved in national gene banks is limited, largely because of the difficulty of transferring traits from non-adapted environments and the fact that germplasm collections often lack useful characterization or evaluation data. They tend to get their parental materials from their own or others' working collections and from nurseries supplied by international institutions such as the IARCs (FAO, 2010, 96). Since such international institutions provide materials to PGRFA users in any country – whether or not they are parties to the ITPGRFA and whether or not they have made materials available under the multilateral system – there is not much incentive for countries to ratify the Treaty or become providers of PGRFA under the multilateral system.

Through informal consultations with countries' representatives (we prefer not to cite the sources), we have learned of some people's reluctance to share PGRFA through the multilateral system because the system does not allow providers of high quality PGRFA to negotiate some type of bilateral benefit sharing. In some countries, including developed countries in south and eastern Europe, resources are scarce, and public institutions have to make an enormous effort to be able to maintain international gene bank standards and to provide PGRFA with high rates of viability that are accompanied by useful information, such as the data resulting from initial characterization and evaluation activities. Representatives of those institutions find it discouraging and frustrating to see their materials transferred to countries with much greater conservation and research capacities without the possibility of obtaining some kind of financial compensation for the efforts they have made under difficult circumstances.

The use of the SMTA represents a big change in many PGRFA users' usual *modus operandi*, and it is understandable that the general reaction to the long and complex text of the SMTA has been negative. This tendency is exacerbated in a number of institutions where germplasm has always been transferred without any formal material transfer agreement and on the basis of individual's personal relationships. Uncertainties around some SMTA provisions and the practical enforcement of its conditions make some providers feel uncomfortable when using it. There have been cases in which the access to PGRFA in the multilateral system has been denied on the basis that the SMTA contains a number of loopholes. It is reasonable to expect that ambiguities in the text of the SMTA and doubts about the efficiency of the third party beneficiary as the main tool for monitoring whether PGRFA users observe the multilateral system's rules may be clarified as the multilateral system continues to operate. Outstanding questions such as that of Article 6.2 (on the obligation to not get intellectual property rights over the PGRFA in the form received from the multilateral system) could be addressed by arbitration courts in the context of particular cases of intellectual property claims over PGRFA acquired from the multilateral system (Gerstetter et al., 2007; Halewood and Nnadozie, 2008).[18] Monitoring and enforcement issues in general, and the role and actual capacities of the third party beneficiary in particular, will hopefully become more clear when PGRFA users gain more experience in exchanging germplasm in the multilateral system. It may be possible that some of the principles and procedures that were initially established need to be revised in order to increase transparency.

Responsibilities with regard to the signing and archiving of SMTAs are not always clear. This is particularly true in those countries where the legislation on access and benefit sharing has deposited all of the responsibilities to grant access in a unique public authority or where such legislation is so confusing that the public authorities and their capacities are not clear any more. In such a scenario, institutions with the mandate to conserve and distribute PGRFA feel uncertain about their own capacities to sign SMTAs.

Finally, many countries' representatives have reported a general lack of human, financial and technical capacities needed to effectively implement the multilateral system. Some focal points that have a very broad mandate and count on very limited resources feel discouraged by the long list of complex tasks they have in front of them, including appointing the authorities that will be in charge of the implementation and those that will be responsible for signing the SMTA; clearing sufficient legal space for the operation of the Treaty by clarifying the relationships between different ministries' mandates and capacities; identifying the collections and materials in the multilateral system; developing the capacity of gene bank curators and scientists working with PGRFA; amending or developing the necessary laws and regulations; and enhancing PGRFA documentation systems and making information on PGRFA available. This last aspect is crucial for making the multilateral system work. One of the most significant obstacles for the greater use of PGRFA is the lack of adequate characterization and evaluation data and the capacity to generate and manage such data (FAO, 2010, 96).

Box 11.1 Capacity development needs

In 2007, the Secretariat of the Governing Body of the ITPGRFA conducted a survey on capacity-building needs among 64 different countries. The survey was distributed to key national stakeholders The vast majority of the responses were provided by representatives from the government, national research institutes and national gene banks (82 per cent), followed by universities (6 per cent) and the private sector (2 per cent). This synthesis of results reflects the major common needs on these fields in ranked order:

- Assessment of existing laws, administrative practices and their conformity with the ITPGRFA.
- Human resources development, research and studies. Technical assistance is essential to strengthen this need.
- Holding comprehensive consultations among key stakeholders.
- Establishment of a comprehensive legal framework.
- Establishment of functional and administrative measures.

Source: Survey on capacity-building needs and priorities for the implementation of the ITPGRFA, available at ftp://ftp.fao.org/ag/agp/planttreaty/gb3/cbcm1/CBCM1_3e.pdf (last accessed January 2011).

Conclusions

Since the text of the SMTA was adopted in Madrid in 2006, there has been relevant progress in the implementation of the multilateral system, but the commitment, efforts and resources to make the multilateral system work varies considerably from country to country. Publicly available data show that both the amount of PGRFA in the system and the amount of PGRFA exchanged under the SMTA are increasing. However, much more effort has to be made to engage the variety of national actors involved in germplasm conservation and use in the day-to-day implementation of the multilateral system. On the benefit-sharing side, advances have been made to feed the benefit-sharing fund and to make it work effectively, but the exclusive focus on monetary benefit sharing has minimized the attention given to other types of benefit sharing such as information exchange, technology transfer and capacity building, despite the fact that these benefits are of equal or more importance for the effective conservation and use of PGRFA for sustainable agriculture and food security.

Making the ITPGRFA's multilateral system fully operational requires fundamental changes in the mindset of those stakeholders who continue to be anchored by the relaxed rules that governed the access and use of genetic resources several decades ago and also of those who feel that a fierce control over every sample of PRGFA is the only way to protect countries' sovereign rights over their genetic resources and the farmers' rights to have access to the benefits arising from the use of the resources.

Despite the difficulties involved in the implementation of the system where financial and human resources are scarce and where dealing with genetic resource issues is a risky business, a number of developing countries have made impressive advances and have shown a high level of faith in, and commitment to, the multilateral system. The presence of individuals that are knowledgeable about, and devoted to, the Treaty within the national institution in charge of the multilateral system implementation is key to moving things ahead.

The success of the multilateral system relies on the combination of supportive and determined actions at the policy level, effective awareness-raising and capacity-building activities and the adoption of appropriate supporting technology. Ultimately, individual countries' advances in the implementation of the multilateral system at the national level, together with progress made by countries collectively at the international level through the Governing Body of the Treaty, should build trust in the multilateral system among national PGRFA conservers and users and increase the number of PGRFA users who are confident enough to exchange genetic material within the multilateral system. Only at such a stage can the multilateral system start to function as a central component of a global system of PGRFA conservation and use where different stakeholders in different countries can collaborate with each other in sharing genetic resources, knowledge, technical capabilities and funds.

Notes

The authors would like to thank Amar Tahiri, Amparo Ampil, Brad Fraleigh, Eng Siang Lim, Daniele Manzella, Fernando Latorre, Francois Pythoud, Gerald Moore, Javad Mozafari Hashjin, Jean Luis Konan, Luis Ayerbe, Mary Taylor, Michael Halewood, Nelly Vasquez, Roberto Cobaquil, Selim Louafi, Zachary Muthamia and Zofia Bulinska.

1 Convention on Biological Diversity, 31 ILM 818 (1992).
2 International Treaty on Plant Genetic Resources for Food and Agriculture, 29 June 2004, www.planttreaty.org/texts_en.htm (last accessed 15 May 2011) [ITPGRFA].
3 Standard Material Transfer Agreement, 16 June 2006, ftp://ftp.fao.org/ag/agp/planttreaty/agreements/smta/SMTAe.pdf (last accessed 15 May 2011).
4 ITPGRFA, *supra* note 2, www.planttreaty.org/members_en.htm (last accessed January 2011).
5 See, for example, the recently launched Learning Module on the Treaty on Plant Genetic Resources for Food and Agriculture, Bioversity International, www.bioversityinternational.org/training/training_materials/international_treaty/treaty_module.html (last accessed 15 May 2011).
6 Australia, Environment Protection and Biodiversity Conservation Amendment Regulations 2005 (No. 2), 10 November 2005 (Provision 8A.05); Brazil, Medida Provisoria no. 2 186–16, 23 August 2001 (Article 19, para. 2); Ethiopia, Proclamation no. 482/2006 on Access to Genetic Resources and Community Knowledge, and Community Rights, 27 February 2006 (Article 15); Peru, Reglamento de Acceso a Recursos Genéticos, aprobado por Resolución Ministerial No. 087–2008 y ratificado por Decreto Supremo 003–2009 MINAM, 31 December 2008 (Article 5.c); Philippines, Joint DENR-DA-PCSD-NCIP Administrative Order No. 1 Series of 2005, Guidelines for Bioprospecting Activities in the Philippines, 14 January 2005 (Section 3.1.g); Spain, Ley 30/2006, de semillas y plantas de vivero y de recursos fitogenéticos, 26 July 2006 (Article 45.3).

7 Authors' communications with focal points and members of the delegations of Canada, Kenya, the Philippines, Poland, Spain and Switzerland in 2009 and 2010 and the reports of the Joint Capacity Building Programme from 2008 to 2011 have confirmed this impression.
8 Such as ProSpecieRara, FRUCTUS Association Suisse pour la Sauvegarde du Patrimoine Fruitier and the Association de l'Arboretum du Vallon de l'Aubonne, among others.
9 Composition of the European regional group: Albania, *Armenia, Austria*, Belgium, Bulgaria, Croatia, Cyprus, *Czech Republic*, Denmark, *Estonia, Finland, France, Germany*, Greece, Hungary, *Iceland, Ireland*, Italy, Latvia, Lithuania, Luxembourg, Montenegro, *Netherlands, Norway, Poland, Portugal, Romania*, Slovenia, Slovakia, *Sweden, Spain, Switzerland*, Turkey, *United Kingdom* (countries in italics are those that have reported that they have put PGRFA in the multilateral system).
10 EURISCO, http://eurisco.ecpgr.org/static/index.html (last accessed January 2011).
11 Association pour l'Etude et l'Amélioration du Maïs (Pro-Maïs) and Association Française des Semences de Céréales à paille et autres espèces Autogames (May 2009).
12 ITPGRFA, *supra* note 2, www.planttreaty.org/inclus_en.htm (last accessed January 2011).
13 Of these proposals, 31.7 per cent came from Africa, 28.3 per cent from Asia, 23.6 per cent from Latin America and the Caribbean, 7.5 per cent from the Near East, 6.2 per cent from Europe, 1.5 per cent from the southwest Pacific and 1.1 per cent from North America.
14 NordGen (Tilander, 2009), International cocoa genebank (personal communication with Path Umaharan, December 2010) and the international coconut gene bank for Africa and the Indian Ocean (personal communication with Jean Luis Konan, February 2011).
15 The reports of the IARCs and CATIE do not cover transactions in the year 2010 and do not include all distributions and acquisitions by all of the centres. The report of the European Regional Group provides information about the number of accessions for Germany only, while for the other countries it presents the number of SMTAs signed only, not the number of samples of accessions transferred. The reader must also bear in mind that, on the other hand, this figure might include some duplication. For example, a transaction of germplasm between CATIE and a CGIAR centre could have been reported twice – as a transfer by CATIE and as an acquisition by the CGIAR centre. This situation would be rare though.
16 To date, the following are member countries of the European Gene Bank Integrated System (AEGIS): Albania, Azerbaijan, Bosnia and Herzegovina, Bulgaria, Croatia, Cyprus, Czech Republic, Denmark, Estonia, Finland, Georgia, Germany, Iceland, Ireland, Lithuania, Montenegro, Netherlands, Norway, Poland, Portugal, Romania, Slovakia, Slovenia, Switzerland, United Kingdom and Ukraine. See AEGIS, www.aegis.cgiar.org (last accessed 5 June 2011).
17 This principle was in many country representatives' minds during the Treaty negotiations on the Annex 1 list. See Chapter 14 by Bert Visser in this volume.
18 As Article 6.2 states, '[t]he Recipient shall not claim any intellectual property or other rights that limit the facilitated access to the Material provided under this Agreement, or their genetic parts or components, in the form received from the Multilateral System.'

References

Altoveros, N.C., T.H. Borromeo, N.A. Catibog, H.R. de Chavez, M.H.F. Dayo and M.L.H. Villavicencio (forthcoming) *Challenges and Opportunities for the Philippines to Implement the Multilateral System of Access and Benefit Sharing of the International Treaty on Plant Genetic Resources for Food and Agriculture*, Bioversity International, Rome.

Argumedo, A. (2009) 'Customary Laws for Traditional Knowledge Protection and ABS', in International Institute for Environment and Development, *Protecting Community Rights over Traditional Knowledge: Implications of Customary Laws and Practices,* online: www.iied.org (last accessed 5 June 2011).

Bellon, M.R., and J. Risopoulos (2001) 'Small-Scale Farmers Expand the Benefits of Improved Maize Germplasm: A Case Study from Chiapas, Mexico', *World Development* 29(5): 799–811.

Bioversity International (2011) *Implementation of the International Treaty on Plant Genetic Resources for Food and Agriculture*, Joint Capacity Building Programme, Final Report, Bioversity International and FAO, Rome.

Brazil, Ministry of Agriculture, Livestock and Food Supply (2009) *Country Report on the State of Plant Genetic Resources for Food and Agriculture*, FAO, Rome.

Carrizosa, S., S.B. Brush, B.D. Wright and P.E. McGuire (2004) *Accessing Bioversity and Sharing the Benefits: Lessons from Implementation of the Convention on Biological Diversity*, International Union for the Conservation of Nature, Gland, Switzerland, and Cambridge, United Kingdom.

Cooper, D. (1993) 'Plant Genetic Diversity and Small Farmers: Issues and Options for IFAD', Staff Working Paper no. 13, International Fund for Agricultural Development, Rome.

Correa, C. (2008) 'Recursos fitogenéticos bajo la administración y control de las partes contratantes y en el dominio público: Cuán rica es la canasta del sistema multilateral del Tratado Internacional', *Recursos Naturales y Ambiente. Informe Especial Recursos Fitogenéticos* 53: 118–25.

Egziabher, T.B.G, E. Matos and G. Mwila (2011) 'The Africa Regional Group: Creating Fair Play between North and South', in C. Frison, F. Lopez and J. Esquinas (eds), *Plant Genetic Resources and Food Security. Stakeholder Perspectives on the International Treaty on Plant Genetic Resources for Food and Agriculture*, Earthscan, London.

European Regional Group (2011) *Report of the European Regional Group on the Implementation of the Multilateral System on Access and Benefit-Sharing*, Document submitted to the Treaty Secretariat in preparation for the fourth session of the Governing Body of the Treaty, Bali, 14–18 March 2011.

Evenson, R.E., and R. Gollin (2003) 'Crop Genetic Improvement in Developing Countries: Overview and Summary', in R.E. Evenson and R. Gollin, *Crop Variety Improvement and Its Effect in Productivity: The Impact of International Agricultural Research*. CABI publishing, Oxford.

Food and Agriculture Organization (FAO) (1998) *State of the World's Plant Genetic Resources for Food and Agriculture*, FAO, Rome.

——(2010) *Second Report on the State of the World's Plant Genetic Resources for Food and Agriculture*, FAO, Rome.

Finston, S. (2005) 'The Relevance of Genetic Resources to the Pharmaceutical Industry: The Industry Viewpoint', *Journal of World Intellectual Property* 8(2): 141–55.

Flores Palacios, X. (1998) *Contribution to the Estimation of Countries' Interdependence in the Area of Plant Genetic Resources*, Report no. 7, Rev. 1, Committee on Genetic Resources for Food and Agriculture (CGRFA), FAO, Rome.

Fowler, C., and G. Hawtin (2011) 'The Global Crop Diversity Trust: An Essential Element of the Treaty's Funding Strategy', in C. Frison, J. Esquinas and F. Lopez (eds), *Plant Genetic Resources and Food Security. Stakeholders' Perspectives on the International Treaty on Plant Genetic Resources for Food and Agriculture*, Earthscan, London.

Gepts, P. (2004) 'Who Owns Biodiversity and How Should the Owners Be Compensated?' *Plant Physiology* 134: 1295–1307.

Gerstetter, C., B. Görlach, K. Neumann and D. Schaffrin (2007) 'The International Treaty on Plant Genetic Resources for Food and Agriculture within the Current Legal Regime Complex on Plant Genetic Resources', *Journal of World Intellectual Property* 10(3–4): 259–83.

Global Crop Diversity Trust (GCDT) (2010) *Annual Report 2009*, GCDT, Rome.

Global Partnership Initiative for Plant Breeding Capacity Building (GIPB) (2011) *Plant Breeding and Related Biotechnology Capacity Assessments*, GIPB, online: http://gipb.fao.org/Web-FAO-PBBC/index.cfm?where=01 (last accessed 15 January 2011).

Glowka, L. (1998) *A Guide to Designing Legal Frameworks to Determine Access to Genetic Resources*, International Union for the Protection of Plant Varieties, Gland, Switzerland.

Halewood, M., and K. Nnadozie (2008) 'Giving Priority to the Commons: The International Treaty on Plant Genetic Resources for Food and Agriculture', in G. Tansey and T. Rajotte (eds), *The Future Control of Food: A Guide to International Negotiations and Rules on Intellectual Property, Biodiversity and Food Security*, Earthscan, London.

Hodgkin, T., R. Rana, J. Tuxill, B. Didier, A. Subedi, I. Mar, D. Karamura, R. Valdivia, L. Collado, L. Latournerie, M. Sadiki, M. Sawadogo, A.H.D. Brown and D. Jarvis (2007) 'Seed Systems and Crop Genetic Diversity in Agroecosystems', in D.I. Jarvis, C. Padoch, and D. Cooper (eds), *Managing Biodiversity in Agricultural Ecosystems*, Columbia University Press, New York.

International Seed Federation (ISF) (2007) *Position paper on Plant Genetic Resources for Food and Agriculture*, ISF, Christchurch.

International Treaty on Plant Genetic Resources for Food and Agriculture (ITPGRFA) (2007a) *Experience of the Centres of the Consultative Group on International Agricultural Research with the Implementation of the Agreements with the Governing Body, with Particular Reference to the Standard Material Transfer Agreement*, Information document prepared for the Second Session of the Governing Body, Rome, 29 October–2 November 2007, Doc. IT/GB-2/07/Inf.11, FAO, Rome.

——(2007b) *Report of the Second Session of the Governing Body of the Treaty on Plant Genetic Resources for Food and Agriculture*, Rome, 29 October–2 November 2007, Doc. IT/GB-2/07/Report, FAO, Rome.

——(2009a) *Experience of the International Agricultural Research Centres of the Consultative Group on International Agricultural Research with the Implementation of the Agreements with the Governing Body, with Particular Reference to the Use of the Standard Material Transfer Agreement for Annex 1 and Non-Annex 1 Crops*, Information document prepared for the Third Session of the Governing Body, Tunis, 1–5 June 2009, Doc. IT/GB-3/09/Inf, FAO, Rome.

——(2009b) *Review of the Implementation and Operation of the SMTA*, Document prepared for the Third Session of the Governing Body of the International Treaty on Plant Genetic Resources for Food and Agriculture, Tunis, Tunisia, 1–5 June 2009, Doc. IT/GB-3/09/14, FAO, Rome.

——(2009c) *Report of the Third Session of the Governing Body of the Treaty on Plant Genetic Resources for Food and Agriculture*, Tunis, 1–5 June 2009, Doc. IT/GB-3/09/Report, FAO, Rome, online: ftp://ftp.fao.org/ag/agp/planttreaty/gb3/gb3repe.pdf (last accessed 5 June 2011).

——(2009d) *Compilation of Submissions Made by Contracting Parties on the Implementation of the Funding Strategy*, Document prepared for the Third Session of the Governing Body of the International Treaty on Plant Genetic Resources for Food and Agriculture, Tunis, Tunisia, 1–5 June 2009, Doc. IT/GB-3/09/Inf.9, FAO, Rome.

——(2010a) *Report of the First Meeting of the Ad Hoc Advisory Technical Committee on the Standard Material Transfer Agreement and the Multilateral System of the Treaty*, Rome, 18–19 January 2010, Appendix 3, Doc. IT/AC-SMTA-MLS 1/10/Report, FAO, Rome.

——(2010b) *Legal and Administrative Measures to Encourage Natural and Legal Persons to Voluntarily Place Material in the Multilateral System*: Background paper prepared for the First meeting of the Ad Hoc Advisory Technical Committee on the Standard Material Transfer Agreement and the Multilateral System of the Treaty, Rome, 18–19 January 2010, Doc. IT/AC-SMTA-MLS 1/10/5, FAO, Rome.

——(2011a) *Experience of the IARC of the CGIAR with the Implementation of the Agreements with the Governing Body, with Particular Reference to the Use of the Standard Material Transfer Agreement for Annex 1 and Non-Annex 1 Crops*, Information document prepared for the Fourth Session of the Governing Body, Bali, 14–18 March 2011, Doc. IT/GB-4/11/Inf. 05, FAO, Rome.

——(2011b) *Experiences of International Institutions with the Implementation of the Agreements with the Governing Body under Article 15 of the Treaty, with Particular Reference to the Use of the Standard Material Transfer Agreement for Annex I and non-Annex I Crops*, Information document prepared for the Fourth Session of the Governing Body, Bali, 14–18 March 2011, Doc. IT/GB-4/11/Inf. 10, FAO, Rome.

——(2011c) *Resolution 4/2011 on the Implementation of the Multilateral System*, Fourth Session of the Governing Body of the International Treaty on Plant Genetic Resources for Food and Agriculture, Bali, Indonesia, 14–18 March 2011, FAO, Rome, online: www.itpgrfa.net/International/sites/default/files/R4_2011_en.pdf (last accessed 5 June 2011).

Jarvis, D., and J. Ngung'u-Skilton (2000) 'IPGRI in Situ Project: Research and Institutions Supporting Local Management of Agrobiodiveristy', in C. Almekinders and W. de Boef (eds), *Encouraging Diversity: The Conservation and Development of Plant Genetic Resources*, Intermediate Technology Publications, London.

Jarvis, D., T. Hodgkin, B. Sthapit, C. Fadda and I. López Noriega (2011) 'A Heuristic Framework for Identifying Multiple Ways of Supporting the Conservation and Use of Traditional Crop Varieties within the Agricultural Production System', *Critical Reviews in Plant Science* 30(1, 2): 125–76.

Joshi, A.K., M. Azab, M. Mosaad, M. Moselhy, M. Osmanzai, S. Gelalcha, G. Bedada, M.R. Bhatta, A. Hakim, P.K. Malaker, M.E. Haque, T. P. Tiwari, A. Majid, M.R.J. Kamali, Z. Bishaw, R.P. Singh, T. Payne, H.J. Braun (2010) 'Delivering Rust Resistant Wheat to Farmers: A Step towards Increased Food Security', *Euphytica* 179: 187–96.

Kastler, G., V. Campesina and P. Mooney (2009) *Small Coins at Carghage: FAO Plant Treaty off to an Embarrassing Start*, Civil Society Press Release, 3 June 2009, online: www.viacampesina.org/en/index.php?option=com_content&view=article& id=735:small-coins-at-carghage-fao-pland-treaty-off-to-an-embarassing-start&catid=22:biodiversity-and-genetic-resources&Itemid=37 (last accessed 1 July 2011).

Kloppenburg, J., and D.L. Kleinman (1987) 'Analyzing Empirically the Distribution of the World's Plant Genetic Resources: The Plant Germplasm Controversy', *Bioscience* 37(3): 190–98.

Lambalk, J. (2009) 'Working with the Multilateral System: Perspective of a Seed Company (Enza Zaden)', in International Union for the Protection of New Plant Varieties (UPOV), *Responding to the Challenges of a Changing World: The Role of New Plant Varieties and High Quality Seed in Agriculture*, UPOV, Geneva, Switzerland.

Lapeña, I., M. Sigüeñas, I. López Noriega and M. Ramírez (2010) *Incentivos y desincentivos para la participación del Perú en el sistema multilateral del Tratado Internacional sobre Recursos Fitogenéticos para la Agricultura y la Alimentación*, Bioversity International, Rome.

Lewis-Lettington, R.J., and S. Mwanyiki (eds) (2006) *Case Studies on Access and Benefit-Sharing*, International Plant Genetic Resources Institute, Rome.

Nature Editorial (2010) 'How to Feed a Hungry World', *Nature* 466(7306): 531–32.

Netherlands Ministry of Agriculture, Nature and Food Quality (2008) *The Netherlands, Second National Report on Plant Genetic Resources for Food and Agriculture*, Country report for the FAO Second State of the World's Plant Genetic Resources for Food and Agriculture, The Hague, Netherlands.

Nijar, G.S., G.P. Fern, L.Y. Harn and C.H. Yun (2009) *Framework Study on Food Security and Access and Benefit-Sharing for Genetic Resources for Food and Agriculture*, Background study paper prepared for the twelfth regular session of the Commission on Genetic Resources for Food and Agriculture, Rome, 19–23 October 2009, FAO, Rome.

Nijar, G.S. (2012) *Malaysia's Implementation of the Multilateral System of Access and Benefit Sharing*, Bioversity International and the Malaysian Agricultural Research and Development Institutue, Rome and Kuala Lumpur.

Nnadozie, K., R. Lettington, C. Bruch, S. Bass and S. King (2003) *African Perspectives on Genetic Resources: A Handbook on Laws, Policies and Institutions*, Environmental Law Institute, Washington, DC.

People's Republic of China, Ministry of Agriculture (2008) *Country Report on the State of Plant Genetic Resources for Food and Agriculture*, FAO, Rome.

Pisupati, B. (ed.) (2008) 'Special Issue on Access and Benefit-Sharing', *Asian Biotechnology and Development Review* 10(3).

Rosenthal, J.P., and F.N. Katz (2004) 'Natural Products Research Partnerships with Multiple Objectives in Global Biodiversity Hotspots: Nine Years of the International Cooperative

Biodiversity Groups Program', in A. Bull (ed.), *Microbial Diversity and Bioprospecting*, ASM Press, Washington, DC.

Sadiki, M., A. Tahiri and I. López Noriega (forthcoming) *Incentives and Disincentives for Morocco's Participation in the Global System of Conservation and Use of Plant Genetic Resources for Food and Agriculture and Proposals to Increase Participation in the Global System and, in particular, in the Multilateral System of Access and Benefit Sharing under the International Treaty on Plant Genetic Resources for Food and Agriculture*, Bioversity International, Rome.

Sigüeñas Saavedra, M., R. Becerra Gallardo and J. Ticona Núñez (2011) *Implementación del Tratado Internacional sobre los Recursos Fitogenéticos para la Agricultura y la Alimentación*, Instituto Nacional de Innovación Agraria (INIA), Lima, Peru.

Sperling, L., and S. Mcguire (2010) 'Understanding and Strengthening Informal Seed Markets', *Experimental Agriculture* 46(2): 119–36.

System-wide Genetic Resources Programme (SGRP) (2010) *Global Public Goods Project Phase 2 – Final Report*, SGRP, Bioversity International, Rome.

Taylor, M. (2011) 'South West Pacific Region. A view from the Pacific Island Countries and Territories', in C. Frison, F. Lopez, and J. Esquinas (eds), *Plant Genetic Resources and Food Security: Stakeholder Perspectives on the International Treaty on Plant Genetic Resources for Food and Agriculture*, Earthscan, London.

Tumushabe, G., and C. Mugoya (2004) *Status and Trends in Access to Genetic Resources in Uganda: Strategic Policy Options and Actions to Facilitate Exchange of Germplasm*, Paper prepared for the 2004 Bellagio Meeting on Plant Genetic Resources in East and Central Africa: Protecting and Enhancing the Ability of the Public Sector Scientists to Freely Access Germplasm, Frati Meeting on Genetic Resources, Bellagio, Italy, online: www2.merid.org/bellagio/reports.php (last accessed 1 July 2011).

Van den Hurk, A. (2011) 'The Seed Industry: Plant Breeding and the International Treaty on Plant Genetic Resources for Food and Agriculture', in C. Frison, F. Lopez and J. Esquinas (eds), *Plant Genetic Resources and Food Security: Stakeholder Perspectives on the International Treaty on Plant Genetic Resources for Food and Agriculture*, Earthscan, London.

Wambugu, P., and Z. Muthamia (forthcoming) *Incentives and Disincentives for Kenya's Participation in the Global System for Conservation and Use of Plant Genetic Resources for Food and Agriculture and Proposals to Increase Participation*, Bioversity International, Rome.

Tilander, Y. (2009) *Implementing the Treaty at the National Level: What is the Impact on the Private Sector?* Presentation at the second World Seed Conference, Rome, 8–10 September 2009, Rome.

12

FROM NEGOTIATIONS TO IMPLEMENTATION

Global review of achievements, bottlenecks and opportunities for the Treaty in general and for the multilateral system in particular

Godfrey Mwila

Introduction

Negotiations leading to the creation of the International Treaty on Plant Genetic Resources for Food and Agriculture (ITPGRFA), which had protracted over a period of seven years, were intense and at times frustrating.[1] In fact, it was feared that, after many years of negotiations, with so much invested in terms of financial resources and time, the process was headed for collapse. However, governments and other stakeholders ultimately marshalled adequate political will to overcome all odds and adopt the final text of the ITPGRFA in November 2001 through the Food and Agriculture Organization's (FAO) Conference Resolution 3/2001.

The implementation of the Treaty at the global level was signalled with the convening of the first session of the Governing Body in June 2006, two years following its coming into force. The desire on the part of the Governing Body to accelerate the ITPGRFA's implementation process led to the decision to quickly follow up with the second session of the Governing Body in October the following year, which was considerably sooner than the official rules of procedure dictated. The third session was held in June 2009. Most of what the Governing Body did during these meetings was to put in place procedures and mechanisms to implement the Treaty, with an emphasis on the multilateral system on access and benefit sharing. While the Governing Body has made significant progress on some fronts, it has remained bogged down on others. Ultimately, I argue that this lack of progress can be attributed to the continuing differences of views among contracting parties concerning issues that were never fully resolved during the negotiations of the Treaty from 1994 to 2001.

As a prerequisite to the effective implementation of the ITPGRFA, one would expect the contracting parties and other stakeholders to be more focused on the implementation aspects in their subsequent engagements with the process, putting aside their pre-Treaty negotiation positions. Indeed, one could reasonably expect that in adopting and ratifying the Treaty, countries have basically reconciled with respect to the fundamental issues and could approach the implementation of the Treaty in a more or less refreshed, unified manner. However, the tone and the focus of the discussions concerning implementation during the sessions of the Governing Body clearly show that major divisions and positions in terms of principle remain largely unchanged between developing and developed countries, and they continue to be frequently echoed and affect the pace of discussions related to the Treaty's implementation. This problem was carried over from the negotiation process since there was a clear lack of balance between the topics that were discussed and advanced during the negotiations. For instance, there was financial support for several meetings to discuss the multilateral system (a topic of the utmost importance for developed countries), but there was no financial support for topics such as financial strategy or farmers' rights (which were cornerstones for developing countries). Moreover, in the first session of the Governing Body, the multilateral system was agreed upon, but the financial strategy did not contain much substance and was far from being agreed upon.

It has become apparent that, despite signing on and becoming parties to the ITPGRFA, most countries have still remained skeptical about the potential for the Treaty to serve their interests. As a result, the contracting parties within the Governing Body are continuing to revisit and correct their perceptions of the shortcomings of the key provisions of the Treaty. Not surprisingly, these trends in behaviour are contributing to major bottlenecks in the Treaty's implementation. Some commentators have written about the controversies and mistrust that have historically characterized discussions among these governments, going as far back as 1983 when the International Undertaking on Plant Genetic Resources was being debated (Chapter 6 by José Esquinas-Alcazar, Angela Hilmi and Isabel López Noriega in this volume; Bragdon, 2000).[2]

In the first part of this chapter, I highlight those issues that were unresolved in the ITPGRFA negotiations. In the second part, I analyse how these issues are impacting on the progress of actually implementing the Treaty. In the third part of the chapter, I will focus on future possibilities, highlighting the challenges and opportunities that are available to advance the effective implementation of the Treaty, taking into account the need to finally address some of the fundamental issues that were not resolved during the earlier negotiations.

Before proceeding, it is important to note that my focus in this chapter will be on global-level implementation, primarily involving the Governing Body of the ITPGRFA, and not so much on country-level implementation involving national governments. However, it must be appreciated that compliance in terms of national-level implementation is not possible if financial responsibility is left up to the countries, since developing countries are unlikely to have enough resources to effectively implement the Treaty. Other chapters in this book address the challenges of national-level implementation.

Reflections on the negotiations of the ITPGRFA

The negotiation of the ITPGRFA was to a large extent an act of balancing the interests of developed countries with respect to facilitated access to globally important plant genetic resources for food and agriculture (PGRFA) with the interests of developing countries with respect to the equitable sharing of the benefits arising from the use of PGRFA. Although it was required that the Treaty be in harmony with the objectives of the Convention on Biological Diversity (CBD), governments involved in the negotiations recognized the special nature of PGRFA and used their sovereign rights to establish a multilateral system of access and benefit sharing for a selected number of crop and forage species (Annex 1).[3]

Discussions on what crop species needed to be included in the multilateral system formed a critical part of the negotiations of the ITPGRFA. The two main criteria used to develop this list of crops were their importance for global food security and the interdependence of countries on each other for the genetic resources of these species. As a member of the African group involved in the Treaty negotiations, I recall that although the African region ultimately opted for the smallest number of crop species to be included in the final stages of negotiations for Annex 1, the group had originally preferred a wider inclusion covering as many crop species as possible, providing that no intellectual property rights would be applied. In the end, Africa and other developing countries took a more cautious approach and tended to favour a shorter list because developed countries would not agree to exclude intellectual property rights from the provisions. In contrast, developed countries favoured a more encompassing list. The European region, in particular, wanted to include all species that were relevant for food and agriculture, which is not a surprise since most advanced breeding institutions and programmes are located in developed countries.

It is clear that developing countries wanted to see a treaty and a multilateral system that was balanced between access and benefit sharing, which did not previously exist, and they therefore pushed to use the negotiations on Annex 1 to secure more concrete commitments from developed countries for the provision of funds in order to enhance benefit sharing under the ITPGRFA. Although the majority of countries participating in the negotiation process did adopt the Treaty in the end, individual countries and regions were often not satisfied with all of the aspects of the final text. As in any negotiations, the final text was somewhat of a compromise, which, it was hoped, would benefit all of the countries that would become party to the Treaty.

General expectations

The general expectations concerning the ITPGRFA's implementation are clearly influenced by the positions that were maintained during the negotiations by both developing and developed countries as well as by the enthusiasm that each country has demonstrated for becoming a contracting party. Most countries, both developed and developing, became party to the Treaty on the understanding and appreciation that no country, including their own, had enough crop diversity within its own territories and, therefore, had to rely on diversity from other regions of the world to meet its crop development and improvement needs. It is for this reason that most developed countries, in particular, consider access to PGRFA under the multilateral system to be the major benefit of the Treaty. Although it may seem logical that other benefits, especially monetary gain, would depend on prior grants of access and, therefore, that access needs to be provided before there can be benefit sharing, most developing countries did not share in this logic, tending to expect or favour a situation where access and benefit sharing would be implemented more or less in parallel.

The general feeling among developing countries is that developed countries have had access to PGRFA originating from developing countries for hundreds of years without providing any compensation in most cases and, even worse, have returned these resources to the developing countries as protected varieties for which huge amounts of money must be paid. Ethiopia, for instance, was very clear on this viewpoint in their second country report on the state of PGRFA to the FAO (Ethiopia, 2007, Chapter 4, Section 4.1), which was prepared for the *Second Report on the State of World's Plant Genetic Resources for Food and Agriculture* (FAO, 2010). In this sense, the expectations of developing countries has been to prioritize the measures that will ensure that the benefit-sharing fund is quickly operationalized under the funding strategy of the ITPGRFA. Since it is expected to take a considerable amount of time (no less than ten years) before benefit sharing under the multilateral system could start yielding results, developing countries have also had to push for other measures that could serve this purpose.

Implementation arrangements for some provisions of the Treaty, such as those under Articles 5 and 6, were also not fully clarified in terms of funding sources, although it now seems that these provisions can be implemented using the funds provided under the benefit-sharing fund.

ITPGRFA's implementation experience

Role of the Treaty's Governing Body

It has generally been viewed that a lot has been done, in terms of global efforts, to put the implementation of the ITPGRFA into motion. Much of this effort has been made through meetings of the Governing Body. The initial efforts of this group tackled a number of key tasks aimed at putting in place a mechanism to facilitate the Treaty's implementation as reflected in the report of the first session of the Governing Body (FAO, 2006). These tasks included finalizing the agreement and adoption of the Standard Material Transfer Agreement (SMTA), adopting its rules of procedure and developing the financial rules and funding strategy for the Treaty.[4] Although the role of the Governing Body in the Treaty's implementation seems to be clear from these mechanisms, there seems to be some feeling among developing country parties and civil society stakeholders that it is biased towards the implementation of the multilateral system. This feeling may have been reinforced by the positions that have been taken by some developed country parties regarding the role of the Governing Body in implementing Article 9 on farmers' rights, for instance, in which it stated that discussions on this topic should not be brought into the Governing Body as it is the responsibility of national governments, as clearly stated in one of the provisions in the ITPGRFA. Developing country parties have, however, insisted that the Governing

Body has a role to play in the implementation of the Treaty as a whole, and they have managed to keep the farmers' rights on the agenda of the meetings of the Governing Body.

Significant achievements in Treaty implementation

The adoption of the SMTA as well as the rules of procedures and the financial rules for the Governing Body during the first session of the Governing Body were seen as major breakthroughs that would facilitate the implementation of the ITPGRFA's multilateral system and enable the Governing Body to discharge its functions and play its role in the implementation of the Treaty as a whole. The other achievements included signing agreements between the gene banks under the Consultative Group on International Agricultural Research (CGIAR), thereby bringing the Annex 1 crops in their collections under the multilateral system (Article 15) as well as the relationship agreement between the Governing Body and the Executive Board of the Global Crop Diversity Trust (GCDT).

One of the major steps taken by the Governing Body in fulfilling its functions under its rules of procedure was to establish the ITPGRFA's Secretariat in 2007, replacing the interim secretariat that was provided by the Commission on Genetic Resources. The initial step that was taken was the recruitment of the secretary to the Treaty. The Governing Body mandated its bureau, which was elected for its second session, to undertake this task. By the time the second session was being held towards the end of 2007, a functional, though not complete, ITPGRFA Secretariat was in place and was working hard to prepare and organize this meeting. A major challenge for the Treaty's implementation has related to the functioning of its Secretariat, which has been limited by financial resources, owing to the low level of contributions to the core administrative budget that it receives from its contracting parties.

Implementation of the multilateral system and adoption and use of the SMTA

The SMTA that was adopted at the first session of the Governing Body was seen as a key tool in the implementation of the multilateral system. More significant was the adoption of the SMTA by all of the CGIAR gene banks for use in facilitating transactions for both Annex 1 and non-Annex 1 germplasm collections. During the second session of the Governing Body, the operation of the SMTA, as part of the implementation of the multilateral system, was reviewed. This review also included the inclusion of PGRFA materials into the multilateral system by the contracting parties and by other natural and legal persons pursuant to the ITPGRFA. The positive feedback concerning the operation of the SMTA is to a large extent misleading because the participation of the contracting parties in these efforts is non-existent or insignificant. As could be expected, the CGIAR gene banks have played a leading role in taking clear and concrete steps to include PGRFA in the multilateral system and to distribute it using the SMTA, while the contracting parties have made limited or no progress. The report of the second session of the Governing Body, which noted with appreciation that the collections of the CGIAR gene banks were already being distributed under the terms of the SMTA, testifies to this fact (FAO, 2007).

Role of international organizations: agreements entered into by the Governing Body

From a technical point of view, global-level implementation of the ITPGRFA's provisions and the multilateral system, in particular, is associated with a number of international players. These include international organizations such as the International Agricultural Research Centres (IARCs) of the CGIAR, Bioversity International, the GCDT and the technical units of the FAO.

The PGRFA collections held by the genetic resources centres or gene banks run by the IARCs are considered to be an important global asset. These collections were in existence prior to the ITPGRFA coming into force and were held in-trust on behalf of the international community through agreements signed between these centres and the FAO. The Treaty itself has recognized the importance of these collections and has provided for their inclusion under the multilateral system through agreements signed between the Governing Body of the Treaty and each IARC holding *ex situ* collections of the crops covered by Annex 1 (Article 15 of the Treaty). Reports presented to the meetings of the Governing Body show that the IARCs have made good progress in implementing the multilateral system, particularly in terms of designating material under the multilateral system and actually making it available under the terms and conditions of the Treaty and the SMTA (ITPGRFA, 2007, 2009, 2011).

The IARC collections form by far the largest group of resources currently known to be in the multilateral system and being exchanged through the SMTA. The centres manage the collections under what are generally accepted to be international standards of management, which cover all key aspects of *ex situ* conservation, including procedures for distributing the germplasm materials to users. Since the PGRFA collections maintained by the IARCs are relatively well documented (information about the material is even publicly available through web-based systems), they have been easier to designate under the multilateral system. The IARCs have also been the first to adopt the use of the SMTA, thereby providing an early opportunity to demonstrate the practical operation of the multilateral system and some visibility for the Treaty's implementation. Their experiences have provided a basis upon which the implementation of the multilateral system can be reviewed. The expertise available at these centres in the management of the PGRFA collections is of great value in supporting the provision of technical support in the implementation of the Treaty, in general, and of the multilateral system, in particular.

Another important event was the signing of a relationship agreement between the Governing Body and the GCDT. In this agreement, the Governing Body recognized the GCDT as an essential element of the funding strategy under the ITPGRFA and provided for the provision of overall policy guidance by the Governing Body to the Trust as well as the submission of the reports on its activities to the sessions of the

Governing Body. The GCDT was established almost at the same time the ITPGRFA came into force. The reports provided to the Governing Body indicate that the GCDT has made good progress in fulfilling its mandate of supporting the long-term conservation and availability of *ex situ* collections important for global food security. The main component of the GCDT's activities is the establishment of an endowment fund that will support in the long-term priority *ex situ* collections. Although it has not yet reached its target, the successes scored by the GCDT in terms of fundraising both for endowment and project work have been acknowledged and appreciated by the Governing Body (FAO, 2009). The goal of supporting a number of key crop *ex situ* collections, which will be maintained at various IARCs, has been a first priority for obvious reasons.

These events, together with a four-year project to regenerate global priority crop collections considered unique and threatened around the world, constitute a significant contribution to strengthening the global system of *ex situ* collections and thereby facilitating the implementation of the ITPGRFA, in general, and the multilateral system, in particular. Initially, some contracting parties found it difficult to appreciate the role of the GCDT, viewing it more as a competitor to the benefit-sharing fund under the ITPGRFA's Secretariat than as a complementary entity.

There seem to be mixed feelings among the contracting parties regarding the agreements signed by the Governing Body and the IARCs that hold *ex situ* crop germplasm collections and the GCDT. While most developed countries consider these global instruments to be important for facilitating the implementation of the ITPGRFA, most developing countries are skeptical, viewing them as mechanisms designed to enhance facilitated access from the multilateral system while doing little for benefit sharing.

Delayed issues in Treaty implementation

Although the funding strategy was adopted at the first session of the Governing Body, a lot remained to be done in order to achieve an implementable strategy. Key decisions that accompanied the adoption of the funding strategy at the first session included the establishment of a trust account for receiving and utilizing financial resources that accrue to it for the implementation of the Treaty, in accordance with Article 19.3(f) of the Treaty. This same decision provided for the establishment of an Ad Hoc Advisory Committee, which would, among other things, draft priorities, eligibility criteria and operational procedures for the allocation of funds. The delay in finalizing the funding strategy created a lot of anxiety among developing countries, which saw it as being key to the ITPGRFA's implementation. This anxiety was also amplified by the lack of progress in the implementation of Article 18 – in particular, Article 18.4(a) and (b), which states: 'Contracting Parties shall take the necessary and appropriate measures within the Governing Bodies of relevant international mechanisms, funds and bodies to ensure due priority and attention to the effective allocation of predictable and agreed resources for the implementation of plans and programmes under this Treaty' and calls contracting parties to accord due priority in their own plans and programmes to building capacity in PGRFA.

The Ad Hoc Advisory Committee was later established, according to the recommendations of the Governing Body, and it worked to complete the tasks assigned to it. These efforts led to the development of a strategic action plan, which specified the funding target of US $116 million over a ten-year period. The implementation of this plan in terms of fund mobilization began in 2010 with a target of US $10 million, which has since been achieved and surpassed. The fact that the fund has achieved what it set out to do has brought relief to most contracting parties from the developing world, which had been worried about the lack of financial resources for the implementation of the ITPGRFA. It is hoped that the majority of this money will be contributed by developed country contracting parties pursuant to the expectations of the strategic plan. Although progress has been made in the implementation of the financial strategy, it is still far from reaching its goal as stated in the strategic plan.

The other elements that are considered key in terms of implementing the ITPGRFA, which has been delayed in terms of agreement and adoption by the Governing Body, are the procedures and mechanisms to promote compliance and deal with issues of non-compliance. Substantive discussions on this topic have been postponed since the first session of the Governing Body. Developed country contracting parties consider these procedures and mechanisms to be more or less a prerequisite for the Treaty's effective implementation by the contracting parties. However, the progress anticipated in moving towards an agreed procedure and mechanism was not forthcoming due to the fact that the developing country delegations did not prioritize this agenda item during either of the sessions of the Governing Body since they wanted to invest more time on the funding strategy. This delay frustrated most of the contracting parties from the developed countries who saw this element as being critical in ensuring the effective implementation of the Treaty, in general, and the multilateral system, in particular. It is hoped that this issue will now get additional attention since work on the funding strategy has been finalized. Once full agreement is reached and the procedures and operational mechanism on compliance has been adopted, a Compliance Committee, established pursuant to Articles 19.3(e) and 21 of the ITPGRFA, is expected to begin its work. According to a decision of the Governing Body, a contracting party is allowed to raise any matter concerning its compliance with the Treaty, including potential non-compliance, before the actual procedures and operational mechanisms are finalized.

Concerns, mainly from developing countries, brought the issue of the third party beneficiary into the negotiations of the ITPGRFA – in particular, the operation of the multilateral system. Although it had been emphasized that access to PGRFA under the multilateral system should be provided expeditiously, without the necessity of tracking individual accessions, it was strongly felt that there was a need for some kind of monitoring system that would safeguard the interests of the beneficiaries of the multilateral system in terms of securing benefits. From the point of view of developing countries, the critical element was the need to ensure that the system would deliver benefits at the end of the day. Discussions were protracted, and agreement mainly centred on who should play this role and what it should entail. In the end, it was agreed that the FAO would take the role, which would be limited to basic monitoring and mediation.

[handwritten at top: Lot of overlap could have been avoided]

There were some delays in making the third party beneficiary functional due to a lack of operational procedures. The second session of the Governing Body acknowledged the director-general's acceptance, in principle, of its invitation for the FAO to carry out, as the third party beneficiary, the roles and responsibilities, as identified and prescribed in the SMTA, subject to formal approval upon review of the procedures to be established by the Governing Body. From the perspective of the developing country contracting parties, therefore, the delay in making the third party beneficiary operational was seen as a bottleneck in the full implementation of the multilateral system. The developed countries are to some extent still skeptical about the relevance of the third party beneficiary and would like its role to be as simple as possible and avoid the tendency for it to develop into a bureaucratic entity.

Major challenges and opportunities

There is no doubt that implementation of the ITPGRFA, in general, and the multilateral system, in particular, has faced some challenges. Most of these stem from the negotiation process and are inherent in some of its provisions. There are also developments, both actual and still in the works, that present opportunities to enhance the Treaty's implementation. Some of these challenges and opportunities are discussed in the following sections.

Challenges

Balancing Treaty implementation

The first major challenge confronting the Governing Body is how to deal with the desire, mainly by developing country contracting parties, for a balanced approach in implementing the ITPGRFA. The adoption of the SMTA, the signing of agreements between the Governing Body and the CGIAR gene banks and the creation of the GCDT were all actions that were seen to be biased towards facilitating access to PGRFA in the multilateral system. Delays in reaching agreement and adopting the funding strategy, through which financial resources would be provided for a benefit-sharing fund, were seen as a sign of reluctance on the part of developed country contracting parties to prioritize the implementation of benefit sharing under the multilateral system and other parts of the ITPGRFA. The rationale behind the thoughts of the developing countries was that although the multilateral system encompassed both facilitated access and benefit sharing, the former was not new and was already taking place, while the latter was new and there were no mechanism in place to regulate it. In terms of prioritizing between these two processes, developing countries felt it made sense to prioritize benefit sharing in order to move towards a balanced scenario in terms of the Treaty's implementation. The argument for balancing the implementation of the multilateral system, and in particular for putting into place mechanisms to guarantee the flow of monetary benefits, stems from the understanding that access to PGRFA has been provided in the past, and continues to be provided, without corresponding benefit sharing from the commercial use of PGRFA accruing to the providers. It is generally expected that the contracting parties would view facilitated access as a major benefit of the multilateral system, as reflected in Article 13.1 of the Treaty, since it would lead to a better realization of benefit sharing. Such arguments were echoed in the discussions of the Governing Body.

Under the ITPGRFA, monetary benefits are derived from a royalty payment triggered by the commercialization of products incorporating material received from the multilateral system. It is generally appreciated that the process of developing a variety, which may then be commercialized, takes a long time. A related challenge to realizing such monetary benefits within the multilateral system is knowing when the materials have actually been developed. Can the stakeholders within the multilateral system, who are often a considerable distance away, depend on the honesty and good will of the recipients and users of the accessed PGRFA material to inform them when the materials have been created? It is for this reason, perhaps, that these parties are proposing a mechanism – the third party beneficiary – which would attempt to play the role of a monitoring and mediation mechanism for the multilateral system in order to, among other things, ensure that benefits flow to the multilateral system.

Balancing the implementation of the ITPGRFA is also considered from the point of view of enabling the contracting parties from developing countries to meet their obligations in terms of facilitating the implementation not only of the multilateral system but also of the other parts of the Treaty. There are, for instance, several measures that the contracting parties are required to take under Articles 5 and 6, which deal with conservation and sustainable use. Although access under the multilateral system is important, most developing country contracting parties, particularly the African countries, consider it to be even more important. The expectations are that, through the Treaty, additional financial resources would become available to implement these measures, which would enable these countries to build their national capacity for conservation and sustainable use of PGRFA and, ultimately, help them to tackle problems of food security at the farming, community and national levels.

In pushing for a balanced Treaty implementation, the contracting parties from developing countries were united and consistent in their emphasis on the need to have a funding strategy that delivers the required results, including, among others, the mobilization of financial resources under the direct control of the ITPGRFA's Governing Body. In the opinion of most developing countries, this is the only way that the Treaty can be effectively implemented. While the Governing Body has made progress in finalizing the funding strategy, including the preparation of a strategic action plan and the development of funding targets, one remaining challenge in developing the fund is whether or not there will be adequate incentives or attraction for donors to buy in and contribute the necessary funds. One of the disincentives for donors may be that the administration of such funds would be less efficient since it will be under the FAO, which is considered to be fairly bureaucratic. Furthermore, it is an organization to which they are already contributing funds. Although nobody seems to be overly optimistic about solving this issue, most of the contracting parties from developing countries, a few from developed countries, and several non-governmental organizations are continuing to lobby to mobilize the fund and are hoping that it will be successful. Proponents of a fund that would be under the direct control of the Governing Body cite the FAO's poor implementation of programmes and activities under the Global Plan of Action for the Conservation and Sustainable Use of PGRFA. They fear that these setbacks may also occur with the funding mechanism. Most developed countries, however, are of the opinion that financial resources under existing arrangements, which are not necessarily under the direct control of the Governing Body, could be made available for the Treaty's implementation.

The funding strategy under the ITPGRFA requires that countries prioritize their PGRFA programmes and activities and provide funds for them as an important counterpart to the Treaty's implementation (Article 18.4). The realization of such arrangements, however, faces a considerable number of challenges. If these issues are not addressed, implementation of the Treaty at the national level in most developing countries will be very difficult.

Increasing availability and visibility of PGRFA materials under the multilateral system

The other major challenge in my view, which I think is shared by many stakeholders with interest in the ITPGRFA's implementation process, is how to clearly define what material is included in the multilateral system and is actually available for exchange under the terms and conditions of the SMTA. From what has been reported to date, it is clear that much effort needs to be made in order to provide documentation on the material available in the multilateral system. The Treaty is clear about what material should be included. The difficulty arises from the fact that there is no actual central gene bank identified under the multilateral system into which the contracting parties and other legal and natural persons should deposit their eligible collections. In a sense, the multilateral system is a virtual gene bank from the point of view of germplasm material that is available for exchange. In practice, the PGRFA materials identified under the multilateral system are held by different entities in different countries.

Another related problem concerns knowing and being able to communicate what is actually held in the collections under the multilateral system. Relevant information needs to be provided including passport data, characterization and other descriptive information. Most developing countries do not have proper documentation on their PGRFA collections. It is therefore no wonder that most of the material that has been exchanged in the multilateral system involves collections of the CGIAR gene banks. There is no doubt that in order to facilitate the exchange of materials in the multilateral system some sort of global information system, as envisaged by Article 17 of the Treaty, will need to be developed.

Expanding the crop coverage of the multilateral system

There are a number of stakeholders, including contracting parties from developing countries, who would want to see an expansion of Annex 1 in terms of number of

[Handwritten annotation at top: Annex 1 - why so limited - IPR / lack of B-Sharing]

crops that are included in the multilateral system. Some people find it difficult to understand the logic behind the limited number of crops included under Annex 1, especially if one recalls the aspirations of the key stakeholders in the early stages of the Treaty's negotiations. The African group, for instance, originally advocated a more comprehensive list of crops than that currently included under Annex 1, but it changed its position to a much smaller list since the developed countries were insisting on the use of intellectual property rights on PGRFA within the multilateral system.

Despite the fact that developing countries, for the most part, have appreciated the principle of being interdependent for the PGRFA of most crop species and have recognized the importance of such varieties for global food security, developing countries chose to act cautiously on including a broader list of crops into the multilateral system due largely to their disappointment with the lack of a clear commitment from developed countries to provide financial resources to implement the Treaty and, therefore, their uncertainty over the functionality of the benefit-sharing system under the multilateral system. Those parties that would like to see more crop species included under Annex 1 hope that the Governing Body will review it in the future. To some extent, the *ex situ* collections of crop species that are not contained in Annex 1 and therefore are not under the multilateral system are being left out due to a lack of funding support from some of the funding initiatives, such as those under the GCDT. To some extent, such a shortfall highlights the need to have more crop species included so that they will not be left without financial support. It can be expected that if good progress is not made with respect to the mobilization of financial resources and the functionality of the benefit-sharing arrangements, the prospects for reaching an agreement on the expansion of Annex 1, if it comes up for review, may be bleak.

Opportunities

Regional plant genetic resources (PGR) networks

PGR networks have been instrumental in strengthening national programmes and international collaboration in many parts of the world. These networks also facilitate increased cooperation between developed and developing countries. A number of existing regional and sub-regional PGR networks have many similar objectives, which include enhancing the national capacity for managing PGRFA by providing platforms for the development of common policies, harmonizing regional approaches, promoting the exchange of germplasm and sharing information and technologies. It is therefore envisaged that these networks would provide opportunities to enhance the implementation of the ITPGRFA. In return, the Treaty recognizes the potential that these networks have and calls for them to be strengthened and further established.

Such networks are in existence, or in the process of being established, in all regions of the world. In Africa, they include the Southern Africa Development Community Plant Genetic Resources Programme (SPGRC), which was started in 1989, the East African Plant Genetic Resources Network and the Genetic Resources Network for West and Central Africa (GRENEWECA).

During a workshop organized under the Joint Programme on Capacity Building for Treaty Implementation in Lusaka, Zambia, in 2009, strategies that would enable the SPGRC to enhance its role within the region in implementing the ITPGRFA were discussed. It was felt that these strategies should include capacity building, including policy and legal capacity as well as technical capacity for plant breeding and seed production; the regional harmonization of policies and legislations; and the exchange of information, awareness raising and reporting under the multilateral system. GRENEWECA, which is in the process of being established, is seeking to create a regional system of *ex situ* conservation for priority PGRFA in West and Central Africa and to enhance sub-regional cooperation. Its strategies would incorporate elements that are supportive of the Treaty's implementation, including encouraging member countries to be contracting parties to the Treaty and to use the SMTA for the exchange of Annex 1 and non-Annex 1 crop species.

In the Americas, there are six networks, each consisting of groups of several countries. These are the Caribbean Plant Genetic Resources Network, the Andean Network on Plant Genetic Resources, the Plant Genetic Resource Network for South American Tropics, the Mesoamerican Network of Plant Genetic Resources, the Plant Genetic Resources Network of the Southern Cone Countries and the Plant Genetic Resources Network for North America.

Countries of western Asia and northern Africa are establishing a new network for this combined region within the framework of the Association of Agricultural Research Institutes in the Near East and North Africa. The network would be country driven and would be primarily concerned with the conservation and sustainable use of PGRFA, in line with the ITPGRFA. The regional strategy would also emphasize the harmonization of national legislation and procedures, including legislation necessary for the implementation of the Treaty.

In Asia and the Pacific, there are three main networks. These are the South Asia Network on Plant Genetic Resources, the Regional Co-operation Network in Southeast Asia for Plant Genetic Resources and the Pacific Agricultural Genetic Resources Network.

The European Cooperation on Plant Genetic Resources has established a European Gene Bank Integrated System (AEGIS), which is a European collection consisting of selected accessions designated by the individual countries, through which it hopes to more effectively contribute to the Treaty's implementation. Material designated as part of the European collection would continue to be conserved in the various individual gene banks, but they would be maintained in accordance with the agreed quality standards and they would be made freely available, both within Europe and beyond, in accordance with the terms and conditions set out in the ITPGRFA.[5]

Existing information systems

The various existing information systems provide the necessary documentation tools to facilitate implementation of the multilateral system. These tools include the information systems of the CGIAR's gene banks, both individually and collectively, such as the

System-Wide Information Network for Genetic Resources. Others include the Germplasm Information Network under the Agricultural Research Services of the US Department of Agriculture, DBGERMO, which is a software developed by the Instituto Nacional de Tecnologia Agropecuaria in Argentina and used in several Latin American countries and regional institutions, such as the Tropical Agricultural Research and Higher Education Centre and *EURISCO*, a web-based catalogue that provides information about *ex situ* plant collections in Europe and is hosted and maintained by Bioversity International on behalf of the European Cooperative Programmes on Plant Genetic Resources. Efforts to develop one-stop global information systems to enhance the accessibility of information about existing *ex situ* collections within the framework of the Treaty's implementation are being undertaken and are drawing upon the progress and experiences of these earlier systems.

Improved documentation and information systems will positively impact on the availability and accessibility of germplasm collections in the multilateral system. Major international collections and those of developed countries usually have web-based documentation and information systems that provide such information. Many smaller developing countries, on the other hand, face substantial financial, technical and institutional problems in providing online information to the public on their germplasm collections.

Increased media publicity

There has been a dramatic increase in media publicity regarding the conservation and use of PGRFA over the last two years, chiefly due to the establishment of the Global Seed Vault in Svalbard, Norway. The facility has been built by the Norwegian government to serve as safety backup of last resort for the world's unique PGRFA *ex situ* seed sample collections and is acknowledged and appreciated by the Governing Body of the Treaty as a worthwhile contribution towards the implementation of the Treaty and security of the global system of *ex situ* conservation. Its official launch in February 2008 attracted unprecedented worldwide media attention, which was expected to significantly increase the level of awareness of the ITPGRFA and the value of PGRFA and, therefore, the need for concerted efforts to support its long-term conservation and sustainable use. The attention that the seed vault has attracted provides a unique opportunity to raise the status of the Treaty and enlist high-level policy support and increase options for fundraising.

Increased training in PGRFA

Globally, the increased existence of learning institutions and materials related to PGRFA provide greater opportunity for upgrading human resource capacity that may be required in the implementation of the Treaty, in general, and of the multilateral system, in particular. Formal Bachelor of Science, Master of Science and Doctorate of Philosophy degree programmes that have a special emphasis on biodiversity and genetic resources have been established in several countries as a response to calls for action under the CBD. Support directed at special training programmes designed to cater to the needs of the ITPGRFA, and of PGRFA in general, could serve to accelerate the implementation of the Treaty. In my view, the Treaty's implementation process requires a critical mass of technically trained personnel.

Focus on climate change

The challenges posed by climate change and the role that can be played by using PGRFA to adapt to a changing climate or to play a mitigating role provide an opportunity to further highlight the importance of PGRFA and therefore of the ITPGRFA itself. There has been increased debate and awareness regarding climate change and its actual and potential adverse impact on agriculture and the livelihoods of people living especially in developing countries. Efforts that are being made to demonstrate the role of agriculture and PGRFA in mitigating climate change challenges and preparing farmers to adapt their production systems present a great opportunity to improve or broaden the means for fundraising for the Treaty's implementation.

Policy makers in national governments and the international community need to be convinced that without crop diversity, agriculture, and crop production in particular, cannot adapt to climate change. It should also be emphasized that bringing this information to the public requires that there is a functioning international system (provided by the Treaty) for sharing crop genetic material.

Climate models have predicted dramatic changes in climate in most parts of the world. Most countries in the northern and southern parts of the African continent, for instance, are expected to experience increased warming and drying, which is likely to substantially reduce the yields of staple food crops such as maize (Bellagio Meeting, 2007). One feasible approach to reduce the impact of these shifts in climate will require the promotion of substantial breeding efforts, which will depend on the collection, conservation and distribution of appropriate crop genetic material among plant breeders. Traits of interest for this purpose will include abiotic stress tolerance (drought, salinity and so on) as well as pest and disease resistance.

The implications of climate change on plant genetic resources include the genetic erosion of crop wild relatives *in situ*. To mitigate the possible impact on PGRFA existing *in situ*, including crop wild relatives, priority actions now and in the near future have to include further collection, regeneration, safety duplication and conservation.

Conclusion

The lack of funding and commitment by the contracting parties has slowed the pace of implementing the ITPGRFA, which has led to skepticism among the stakeholders about whether the Treaty can serve their interests or provide them with benefits. For the multilateral system to be fully functional, it has to deliver both of these goals in terms of providing facilitated access to the globally important PGRFA and realizing the flow of both monetary and non-monetary benefits, especially to developing countries and countries with economies in transition. Judging from the Treaty's slow progress as well as the challenges that it is facing, it is unlikely that a review of Annex 1 will be undertaken soon and that agreement on expanding the list will be easy. Unless funding is provided to the contracting parties of developing countries, it is unlikely that the role of developing country parties in the Treaty's implementation, particularly in the exchange of PGRFA under the multilateral system, will change in any significant way in the near future.

Although the funds that have been mobilized and made available through the ITPGRFA's benefit-sharing fund could be seen to be critical in making the benefit-sharing mechanism functional, it is necessary to look at the Treaty's funding strategy more broadly and recognize that other sources of funds, such as those under the GCDT, are contributors to the Treaty's implementation and the realization of benefit sharing.

In addition, there is a need for renewed efforts to strengthen regional PGR networks as platforms to enhance and promote the Treaty's implementation. Currently, support for some of these networks is not encouraging, and unless the trend is reversed it could have adverse implications for implementation, especially among developing country parties. There is need to take advantage of the advances in information technology to develop robust global information systems that enhance the sharing of information and knowledge and that will positively impact on the availability and accessibility of germplasm collections in the multilateral system.

Finally, there is no doubt that the dramatic increase in media publicity regarding the conservation and use of PGRFA over the last few years has increased awareness among policy makers, donors and the general public of the importance of PGRFA and the need for its conservation, especially in view of the anticipated negative impacts of climate change. Linking the challenge posed by climate change and the role that can be played by using PGRFA collections to adapt agriculture to cope with the new conditions created by these changes and to sustain production has the potential of attracting increased political attention at all levels to provide financial resources required to support PGRFA conservation and to implement the Treaty.

Notes

1 International Treaty on Plant Genetic Resources for Food and Agriculture, 29 June 2004, www.planttreaty.org/texts_en.htm (last accessed 8 June 2010).
2 International Undertaking on Plant Genetic Resources, 1983, www.fao.org/waicent/faoinfo/agricult/cgrfa/IU.htm (last accessed 8 June 2010).
3 Convention on Biological Diversity, 31 ILM 818 (1992).
4 Standard Material Transfer Agreement, 16 June 2006, ftp://ftp.fao.org/ag/agp/plant-treaty/agreements/smta/SMTAe.pdf (last accessed 8 June 2010).
5 For more information on the European Gene Bank Integrated System, see www.aegis.cgiar.org (last accessed 8 June 2010).

References

Bellagio Meeting (2007) *The Conservation of Global Crop Genetic Resources in the Face of Climate Change*, Summary Statement from a Bellagio Meeting, 3–7 September, online: http://iis-db.stanford.edu/pubs/22065/Bellagio_final1.pdf (last accessed 15 May 2011).

Bragdon, Susan H. (2000) 'Moving Forward with the International Undertaking: Legal Mechanism to Alleviate Mistrust', *Issues in Genetic Resources*, No. 9, International Plant Genetic Resources institute, Rome.

Ethiopia (2007) 'Ethiopia: Second Country Report on the State of PGRFA to FAO, August 2007', online: http://www.fao.org/docrep/013/i1500e/Ethiopia.pdf (last accessed 21 May 2012).

Food and Agriculture Organization (FAO) (2006) *Report of the First Session of the Governing Body of the International Treaty on Plant Genetic Resources for Food and Agriculture*, Doc. IT/GB-1/06/Report, Madrid, Spain.

——(2007) *Report of the Second Session of the Governing Body of the International Treaty on Plant Genetic Resources for Food and Agriculture*, Doc. IT/GB-2/07/Report, FAO, Rome.

——(2009) *Report of the Third Session of the Governing Body of the International Treaty on Plant Genetic Resources for Food and Agriculture*, Doc. IT/GB-3/09/Report, Tunis, Tunisia.

——(2010) *Second Report on the State of the World's Plant Genetic Resources for Food and Agriculture*, Commission on Genetic Resources for Food and Agriculture, FAO, Rome.

International Treaty on Plant Genetic Resources for Food and Agriculture (ITPGRFA) (2007) *Experience of the Centres of the Consultative Group on International Agricultural Research (CGIAR) with the implementation of the Agreements with the Governing Body, with particular reference to the use of the Standard Material Transfer Agreement*. Information document prepared for the Second Session of the Governing Body of the International Treaty on Plant Genetic Resources for Food and Agriculture. Rome, Italy, 29 October–2 November 2007, Doc. IT/GB-2/07/Inf.11, FAO, Rome, Italy.

——(2009) *Experience of the Centres of the Consultative Group on International Agricultural Research (CGIAR) with the implementation of the Agreements with the Governing Body, with particular reference to the use of the Standard Material Transfer Agreement for Annex I and Non-Annex I crops*. Information document prepared for the Third Session of the Governing Body of the International Treaty on Plant Genetic Resources for Food and Agriculture, Tunis, Tunisia, 1–5 June 2009, Doc. IT/GB-3/09/Inf, 15, FAO, Rome, Italy.

——(2011) *Experience of International Institutions with the implementation of the Agreements with the Governing Body, with particular reference to the use of the Standard Material Transfer Agreement for Annex I and Non-Annex I crops*. Information document prepared for the Fourth Session of the Governing Body of the International Treaty on Plant Genetic Resources for Food and Agriculture. Bali, Indonesia, 14–18 March 2011. IT/GB-4/11/Inf. 10. FAO, Rome, Italy.

Khoury, C., B. Laliberte and L. Gaurino (2010) *Trends in Ex Situ Conservation of Plant Genetic Resources: A Review of Global Crop and Regional Conservation Strategies*, Springer Science and Business Media, New York.

13

THE MULTILATERAL SYSTEM OF ACCESS AND BENEFIT SHARING

Could it have been constructed another way?

Clive Stannard

Introduction

The multilateral system of access and benefit sharing (multilateral system) did not spring whole and complete from the minds of the negotiators but, rather, was built upon a complex history of attempts to resolve a number of long-standing technical, institutional and political tensions between governments in regard to plant genetic resources for food and agriculture (PGRFA). The negotiations were the product of this history. The negotiations of the International Treaty on Plant Genetic Resources for Food and Agriculture (ITPGRFA) drew together two conceptual frameworks for how access to plant genetic resources might be managed and how the benefits might be shared.[1] The first framework – the International Undertaking on Plant Genetic Resources for Food and Agriculture (International Undertaking) – had been developed in the agricultural sector (CPGR, 1989).[2] The second – the Convention on Biological Diversity (CBD) – was developed in the environmental sector.[3] The focus of these two instruments was different, as were their provisions. The negotiations of the ITPGRFA, therefore, inherited a mixed set of conceptual tools that could be drawn upon in making provisions to regulate access and benefit sharing. The most significant of these came from the International Undertaking, which had a complex negotiating history behind it. The political pressure to conform the ITPGRFA as closely as possible to the CBD, however, was very strong.

I shall first consider the focuses that had developed in the negotiating history of the International Undertaking, and how these were addressed in its provisions. I shall then look at the concepts coming from the CBD, and how these related to the concepts that were embodied in the International Undertaking. I shall then isolate a number of key negotiating focuses at the time the negotiations of the ITPGRFA commenced and discuss how these defined and constrained the options that were open to the negotiators. In the course of the negotiations, as the structure of the ITPGRFA was developed, choices had to be made among these options, and new concepts were developed and embodied in the multilateral system as it was finally adopted. It would be a mistake to assume that the current structure of the multilateral system was the only possible one or even the optimal one. With the benefit of hindsight, I shall consider whether the multilateral system could have been constructed in a more effective and less complicated manner.

International Undertaking

The International Undertaking arose from tensions between developed and developing countries over PGRFA and specifically around the establishment of the *ex situ* collections of the Consultative Group on International Agricultural Research (CGIAR). This focus on *ex situ* collections remains a major focus of the ITPGRFA and is reflected in the provisions of Article 15. Moreover, the CBD specifically recognized – in Resolution 3 of the Nairobi Final Act – that 'solutions to [the] outstanding matter' of 'access to *ex situ* collections not acquired in accordance with this Convention' were to be addressed within the Global System for the Conservation and Sustainable Use of Plant Genetic Resources for Food and Sustainable Agriculture, which was based on the International Undertaking.

The establishment in 1974 of the International Board for Plant Genetic Resources (IBPGR) – as a project within the Food and Agriculture Organization (FAO) but, in practical terms, independent, to promote and assist in the worldwide effort to collect and preserve the germplasm needed for future research and production – had been very effective, but it had, at the same time, created a perception in developing countries that germplasm collected in their territories was passing outside their control. This sense was aggravated by the extension of intellectual property rights over plant varieties in developed countries – in particular, through the establishment in 1961 of the International Union for the Protection of New Varieties of Plants (UPOV). Such developments, developing countries felt, could block their access to the genetic resources in international collections that had been collected in their territories as well as to the products derived from them.

The matter came to a head in the twentieth conference of the Food and Agriculture Organization of the United Nations in 1979. The concepts in play – all of which remain central focuses of the ITPGRFA – were the ownership of plant genetic resources, particularly those that had been collected by international institutions from developing countries; the nature and security of *ex situ* collections; guarantees that free exchange would continue; and the benefits that developing countries should receive from these materials. The twenty-first conference in 1981 was also dominated by continued and very tense discussions between the opposed developing and developed countries. Its Resolution 6/81 stated:

> Considering that there is no international agreement for ensuring the conservation, maintenance and free exchange of the genetic resources of agricultural interest contained in existing germplasm banks,

Convinced of the need for such an agreement ...
Request[ed] the Director-General to examine and prepare the elements of a draft international convention, including legal provisions designed to ensure that global plant genetic resources of agricultural interest will be conserved and used for the benefit of all human beings, of this and future generations, without restrictive practices that limit their availability of exchange, whatever the source of such practices.

The twenty-second conference in 1983 adopted the International Undertaking, in accordance with 'the basic principles ... according to which plant genetic resources should be considered as a common heritage of mankind and be available without restrictions for plant breeding, scientific and development purposes to all countries and institutions concerned'. It was the first comprehensive international agreement on any aspect of genetic resources for food and agriculture. The conference also established the Commission on Plant Genetic Resources (CPGR), which in 1995 became the Commission on Genetic Resources for Food and Agriculture (CGRFA). It met for the first time in 1985.

The International Undertaking was a non-binding instrument, to which states could adhere. However, eight developed countries expressed reservations on its adoption, and the CPGR then negotiated a series of agreed interpretations to overcome these objections and seek universality. The concepts in the International Undertaking itself, and those developed through the agreed interpretations, formed the major negotiating tools for the ITPGRFA and the subsequent construction of the multilateral system.

The objective of the International Undertaking was:

> to ensure that plant genetic resources of economic and/or social interest, particularly for agriculture, will be explored, preserved, evaluated and made available for plant breeding and scientific purposes. This Undertaking is based on the universally accepted principle that plant genetic resources are a heritage of mankind and consequently should be available without restriction.

It promoted the exploration of plant genetic resources (Article 3); their preservation, evaluation and documentation (Article 4); access to them for the purposes of scientific research, plant breeding or conservation (Article 5); and the mobilization of financial resources 'to place activities relevant to the objective of this Undertaking on a firmer financial basis, with special consideration for the need[s] of developing countries'. The mobilization of resources was to become a central focus of further negotiations involving the International Undertaking and a formal part of the ITPGRFA – in Article 13.2(d)(ii), as monetary benefit sharing deriving from the use of PGRFA under the multilateral system and in the more general context of Article 18 as 'funding for priority activities, plans and programmes, in particular in developing countries'.

Tensions over the role of the IBPGR continued as:

> [s]ome members considered that the present scientific and technical activities of plant genetic resources conservation and exchange as promoted by the IBPGR in collaboration with FAO were satisfactory, and that possible improvements should be sought within the existing system. The majority of members, however, considered that present activities were not sufficient and that they should be complemented in order to develop a global system on plant genetic resources.
> (FAO, 1983)

The report of the first meeting of the CPGR identified some of these tensions, which were to continue into the negotiations of the ITPGRFA and influence the construction of the multilateral system:

- There is an ambivalence between 'plant genetic resources as a world heritage for the well-being of present and future generations of mankind', as the assistant director-general opening the meeting put it, and national sovereignty: 'The Commission agreed that the sovereignty of governments over their plant genetic resources should be respected and that reciprocity in the exchange process was included in the substance of the Undertaking.'
- The legal status of *ex situ* collections, and the importance of clarifying this status, was a key point: 'The Commission noted that the present informal, bona fide system in germplasm exchange generally worked satisfactorily, but did not provide all legal guarantees many considered necessary to ensure unrestricted exchange of material from base collections.' Countries disagreed as to whether or not a legal framework was needed. The commission recommended 'ascertaining the right of ownership of plant genetic resources held by the organizations and institutions in the IBPGR network'.
- The need for assistance to developing countries was repeatedly stressed.

The search for universal support for the International Undertaking continued in the negotiations in the CPGR, and resulted in three conference resolutions embodying agreed interpretations of the International Undertaking. In 1989, the question of intellectual property rights – specifically plant varietal protection, under UPOV – was addressed. Resolution 4/89 recognized that '[p]lant Breeders' Rights, as provided for under UPOV ... are not incompatible with the International Undertaking'. It simultaneously recognized 'farmers' rights', which it linked to an 'International Fund for Plant Genetic Resources' that the FAO had established but which had drawn only a few tiny contributions, and suggested more might be coming:

To reflect the responsibility of those countries that have benefited most from the use of germplasm, the Fund would benefit from being supplemented by further contributions from adhering governments, on a basis to be agreed upon, in order to ensure for the Fund a sound and long-term basis.

A further resolution in the same year, Resolution 5/89, defined farmers' rights. In 1991, Resolution 3/91 clarified a number of points regarding sovereignty, intellectual property, access and funding:

- 'Nations have sovereign rights over their plant genetic resources';
- 'Breeders' lines and farmers' breeding material should only be available at the discretion of their developers during the period of development'; and
- 'Farmers' Rights will be implemented through an international fund on plant genetic resources which will support plant genetic conservation and utilization programmes, particularly, but not exclusively, in the developing countries.'

The pressure from developing countries for the international mobilization of financial resources continued to mount, and in the context of the International Undertaking, unlike in the ITPGRFA, this pressure was directly linked to farmers' rights. In response, the position of developed countries – that it was necessary to know the needs before being able to mobilize funds – was the prime reason for the preparation of the report on *The State of the World's Plant Genetic Resources for Food and Agriculture* (FAO, 1997a) as well as the adoption of the Global Plan of Action for the Conservation and Sustainable Utilization of Plant Genetic Resources for Food and Agriculture and the Leipzig Declaration (FAO, 1997b), in the context of the International Technical Conference on Plant Genetic Resources, which was held in Leipzig, Germany, on 17–23 June 1996. Such was the state of play at the beginning of the negotiations of the ITPGRFA.

Developments regarding the IBGPR and the *ex situ* collections of the CGIAR took another track, which also influenced the construction of the multilateral system. The matter came to a head in the third session of the CPGR in April 1989. Under continued pressure from developing countries in the Commission, the IBPGR attempted to assert its independence from the FAO by changing its formal status from an FAO project to an independent body and by moving from Rome to Denmark:

The Commission expressed concern with the Board's decision to separate from FAO, particularly as a very large number of related matters appeared to have been inadequately studied. It considered that there had not been proper consultation with FAO ... Many donors of funds, as well as donors of germplasm, expressed surprise and disappointment at not having been consulted with respect to the decision of the Board.

At stake was the ownership and control of the *ex situ* collections:

> The question was raised as to whether [the members of the IBPGR's Board of Trustees], who served in their personal capacity, might change cooperative arrangements that had been established by Governments in a matter that concerned the common heritage of mankind.
>
> *(CPGR, 1989)*

Faced with the opposition of both developed and developing countries, a compromise was found whereby the IBGPR separated from the FAO and became the International

Plant Genetic Resources Institute (IPGRI), within the CGIAR, but remained in Rome. The IBPGR and the FAO agreed on a Memorandum of Programme Cooperation on 21 September 1990. The IPGRI was formally established on 9 October 1991, with its status being formally ratified by the Italian Parliament in March 1994. From this period on, tensions with the IPGRI progressively eased, and, during the negotiation of the ITPGRFA, the assistance of the IPGRI on a number of technical matters was increasingly sought and appreciated by the negotiators.

The question of authority over the *ex situ* collections remained. In accordance with Article 7 of the International Undertaking, the CPGR sought to establish an international network of *ex situ* collections under the auspices of the FAO. A number of countries expressed their willingness to bring their materials into this network, but nothing practical ensued. At its fifth session in 1993, the CPGR 'welcomed the offer made by the International Agricultural Research Centres (IARCs) of the CGIAR to place their base and active collections under the auspices of FAO, and to receive policy guidance from the Commission on these collections', and it 'requested the Director-General to negotiate and, if satisfied, to conclude agreements with the CGIAR Centres' (CPGR, 1993).

After some initial difficulties, these agreements were signed in October 1994. They committed the IARCs to 'hold the designated germplasm in trust for the benefit of the international community, in particular the developing countries in accordance with the International Undertaking' and 'not [to] claim legal ownership over the designated germplasm, nor ... seek any intellectual property rights over that germplasm or related information'. The IARCs 'recognized the intergovernmental authority of FAO and its Commission in setting policies for the International Network'. The agreements also required the centres to ensure that the subsequent recipients of germplasm did not claim legal ownership or intellectual property rights over the germplasm (CGIAR, 2003). A material transfer agreement for use by the IARCs in releasing their materials was accordingly agreed with FAO, which resolved a matter that had been in discussion since the beginning: it stipulated in contractual form that intellectual property rights might not be claimed over any materials that had been provided by a country to the IARCs – in particular, landraces and farmers' materials. The use of a material transfer agreement to regulate the exchange of materials under the International Undertaking was the template upon which the ITPGRFA's multilateral system was subsequently developed.

CBD

The adoption of the text of the legally binding CBD on 22 May 1992, and its entry into force on 29 December 1993, posed a fundamental challenge to the processes underway in the FAO and added further elements to the mix from which the multilateral system was created. In April 1991, the fourth session of the CPGR hopefully 'observed that FAO's experience and its own, in developing various elements of the Global System on Plant Genetic Resources, might be of great value in the formulation and negotiation of a global legal instrument on biological diversity'. It 'noted FAO's active participation in the discussions at UNEP on the draft convention, including the contribution of the FAO Working Group on Biological Diversity in drafting articles for inclusion in the legal instrument' (CPGR, 1991). However, the CBD was negotiated in a non-agricultural forum, which, in fact, took no account of the input of the CPGR and the FAO. It was also a highly political process, reflecting tensions between developing and developed countries that were similar to those in the FAO over access and benefit sharing for biological diversity but with players from the environment and foreign ministries, not agriculture.

The agriculture sector suddenly found itself in a situation where its non-legally binding processes had been trumped by a legally binding instrument that had a scope that included all genetic resources, including agricultural genetic resources. Moreover, because the CBD was seen by many as a victory for developing countries, through the formal assertion of their sovereignty over their biological diversity – which, it was hoped, would bring them considerable financial benefits – a number of the leading countries in the context of the CBD initially expressed hesitation over supporting any further work on plant genetic resources in the FAO's CPGR, when it had to do with access and benefit sharing. A fractioning of the portmanteau concept of 'biological diversity', in order to take into account the specificities of food and agriculture, could in this context be perceived as weakening the CBD. There was initial uncertainty as to whether further, separate work on access and benefit sharing on PGRFA would go ahead. If it did, where would it occur – in the FAO or the CBD – and would an instrument that resulted remain in the agriculture sector or become a protocol of the CBD?

There were two mandates available to those countries that wished to make progress in the agricultural sector: a recommendation in Agenda 21, which was adopted at the United Nations Conference on Environment and Development in Rio de Janeiro on 14 June 1992, to strengthen the FAO's Global System for the Conservation and Sustainable Use of Plant Genetic Resources for Food and Sustainable Agriculture (of which the International Undertaking was the keystone);[4] and Resolution 3 of the Nairobi Conference for the Adoption of the Agreed Text of the CBD, which countries wishing to preserve a distinct role for the agricultural sector had obtained. This resolution:

> Recalling that broadly-based consultations in international organizations and forums ha[d] studied, debated and achieved consensus on urgent action for the security and sustainable use of plant genetic resources for food and agriculture ...
>
> Urge[d] that ways and means should be explored to develop complementarity and cooperation between the Convention on Biological Diversity and the Global System for the Conservation and Sustainable Use of Plant Genetic Resources for Food and Sustainable Agriculture ... [and] further recognize[d] the need to seek solutions to outstanding matters concerning plant genetic resources within the Global System for the Conservation and Sustainable Use of Plant Genetic Resources for Food and Sustainable Agriculture, in particular:
>
> (a) Access to *ex-situ* collections not acquired in accordance with this Convention; and
> (b) The question of farmers' rights.

On this basis, and at the request of the CPGR, the twenty-seventh FAO Conference adopted Resolution 3/93, requesting the director-general:

> to provide a forum for negotiations among governments:
>
> (a) for the adaptation of the International Undertaking on Plant Genetic Resources, in harmony with the Convention on Biological Diversity;
> (b) for consideration of the issue of access on mutually agreed terms to plant genetic resources, including *ex situ* collections not addressed by the Convention;
> (c) as well as for the issue of realization of Farmers' Rights.

The long and complex process that would lead to the creation of the multilateral system, as a cornerstone of the ITPGRFA, had begun. Would something like the legally binding ITPGRFA have come about without the shock administered by the adoption of the CBD? Answering this question raises the impossibility of predicting the past, but it is probable that the negotiations would have continued towards a non-binding and more technical set of arrangements for access and benefit sharing for food and agriculture. The existence of the CBD, and the possibility that the agricultural sector would lose all say in the management of genetic resources which are the basis of arable agriculture and of world food security, kick-started the negotiations and gave new impetus to processes at that stage underway in the CPGR – particularly for the development of *The State of the World's Plant Genetic Resources* and the Global Plan of Action.

The focuses of the negotiation of the multilateral system

Major areas of controversy, which needed to be resolved in constructing the multilateral system, reflected both the prior negotiations in the CPGR, and the CBD and its implementation.

Sovereignty, public goods and pooled goods

From the very beginning of the FAO process, the sovereignty of countries over their plant genetic resources was recognized, but it was not felt to contradict their nature as a common heritage of mankind. With the coming of the CBD, many countries began to see sovereignty much more restrictively and often implemented sovereignty in national law as ownership, usually state ownership. Sovereignty itself – having been recognized in both processes – was never in dispute but, rather, the structure of the access provisions and the uses to which countries would allow the resources over which they had sovereignty to be put. The structures by which plant genetic resources were to be regulated were now seen to derive from a 'common concern of mankind' rather than from the 'common heritage of mankind'.

There was a basic contradiction between the two sets of ideas that needed to be resolved. The ITPGRFA negotiations, like those of the International Undertaking, stressed the value of exchange as well as the huge lost opportunity costs that would be incurred if exchange were bilateralized. From such a perspective, a reduction in exchange and use translated to a lower global agricultural production, greater crop vulnerability and increased global food insecurity. The difficulty of identifying individual agricultural genetic resources of unique value, in the context of repeated exchange, crossing and recrossing, suggested that there was little to be gained from impeding wide exchange. The reality of agriculture, and the thousands of years of exchange and re-exchange, made multilateral solutions an imperative. In the CBD, on the other hand, the value of genetic resources was seen as being derived from their uniqueness or, at least, from being able to market the unique access to, and use of, specific resources. Public, non-appropriable values resulting from exchange were outside such reasoning.

The ITPGRFA resolved this issue in Article 10.2, where it provides that, 'in the exercise of their sovereign rights', the contracting parties will establish the multilateral system and pool the resources to which this applies. This pooled good may then be managed in much the same way as an international public good over which individual sovereignty has not been exercised, nor national property created. Both the International Undertaking and the CBD addressed the *de facto* enclosure of genetic resources, which had previously been seen as a public good or a *res nullus*, of the physical good by systematic collecting and property in collections, and of products derived from these, under intellectual property systems. A key difference, however, was that the products of collecting, under the International Undertaking, remained available for use by others and continued to have the public good characteristic of non-exclusivity. The focus on ensuring the continuing availability and optimal use of public collections therefore remained a central part of the negotiations of the ITPGRFA, whereas in the context of the CBD the very existence of these collections was seen almost as a problem.

Under both instruments, the threat of the loss of genetic resources was recognized. The CBD, in the context of access and benefit sharing, assumes that ensuring appropriable value, by denying general access to genetic resources in order to merchandise unique access and use rights, would both provide an incentive to conserve and provide a flow of funds for conservation. There is no evidence that this process has been particularly successful. The ITPGRFA has addressed the threat of loss by providing an agreed legal framework for international collections and stressing the importance of conservation. It has, however, not made provision for any direct multilateral support for conservation, leaving such efforts to the contracting parties in their individual capacity and to the IARCs that hold the resources and their donors. Nor have the benefits of the multilateral system been applied to conservation. Despite this lack of direct financial support, a major achievement of the ITPGRFA has been the establishment of the Global Crop Diversity Trust, which is a new international organization and an essential element of the ITPGRFA's funding strategy.

There were also crucial differences in regard to intellectual property. Access and benefit-sharing agreements under the CBD invariably assume that the products that are derived from access to the resources are brought under strong intellectual property protection in order to be able to exclude non-licensees. The treatment of intellectual property in the International Undertaking, and subsequently in the ITPGRFA, has followed a different trajectory. Developing countries were initially concerned that materials that they had contributed to international collections could be subject to intellectual property claims. The initial aim was to prevent such claims from limiting access to these materials and, therefore, to maintain them as a public resource. For this reason, the material transfer agreement used by the IARCs following the October 1994 agreements required a recipient 'not to claim ownership over the germplasm to be received, nor to seek IPR over that germplasm or related information' (CGIAR, 2003). In fact, while materials provided by the IARCs should not be eligible for plant varietal protection, applications were on occasion filed (RAFI, 1998). Pursuant to this provision, the International Centre for Tropical Agriculture, supported by the FAO, successfully challenged a US patent over the 'Enola bean' on the grounds that no inventive step had been involved (Correa, 2009).

Benefit sharing in general; monetary benefit sharing in particular

The benefits foreseen by the International Undertaking were general and not linked to the exchange of any specific resource. The benefits would be the result of the free availability of plant genetic resources and the fact that the 'plant genetic resources of economic and/or social interest, particularly for agriculture, will be explored, preserved, evaluated and made available for plant breeding and scientific purposes'. This concept has continued through to the ITPGRFA and is stated in Article 13.1: 'The Contracting Parties recognize that facilitated access to plant genetic resources for food and agriculture which are included in the Multilateral System constitutes itself a major benefit of the Multilateral System.' However, the scope of such benefit sharing has now been limited only to the crops listed in Annex 1 to the ITPGRFA.

Attempts to agree on the mechanisms of benefit sharing in the FAO forums had not come to fruition. The failure of the negotiators within the context of the International Undertaking to find ways to mobilize additional funds for the conservation and sustainable use of PGRFA, particularly in support of the needs of developing countries, predisposed the negotiations of the ITPGRFA to go down the more limited, contractual road opened by the CBD. This direction inevitably led to a single-minded focus on benefit sharing, in general, to the detriment of international arrangements for the sound management of the genetic resources themselves, and to a specific focus on the mandatory sharing of monetary benefits derived from commercial use, which became the almost exclusive topic of the negotiations and culminated in the negotiation of the Standard Material Transfer Agreement (SMTA) at the end of the first session of the Governing Body.[5] This was despite the realization by most of the negotiators that monetary benefits were of less value in facing food security, hunger and climate change than the effective facilitation of wide exchange. It was also despite the tacit recognition that models of mandatory monetary benefit sharing were unlikely to yield much in the way of financial benefits. Its effects on the multilateral system as finally agreed are many. It has led to an enclosure of agriculture by the CBD to the detriment of the public good benefits of wider access.

The scope of the multilateral system

The scope of the ITPGRFA, as it is stated in Article 3, is 'plant genetic resources for food and agriculture', which is usually understood to mean *all* such resources. The scope of its major instrument, the multilateral system, is a sub-set of these resources. Whereas the International Undertaking had addressed all plant genetic resources of interest to food and agriculture, the adoption of the CBD as a binding international instrument that addressed genetic resources in general meant that the negotiation of the ITPGRFA had the task of identifying one by one the crops to which the multilateral system would apply and identifying the uses to which they could be put. Potential monetary benefits were balanced against access, with much horse-trading, and some countries manoeuvring to exclude crops for which they believed they had a national advantage. The limited set of crops in Annex 1 is the result. Although the possibility exists for the list to expand, the provisions of Article 19.2, which require decision by consensus, will be a major obstacle. This is clearly one of the areas in which different solutions could have been sought, either in terms of the structure of access or in terms of the contents of Annex 1.

The status of the *ex situ* collections

Although ownership of, and authority over, the IARCs' *ex situ* collections had been one of the original bones of contention in the CPGR, the tensions around this question had abated by the time the negotiations began. (It remained, however, a sore point to the end with a very small number of countries.) The CGIAR had already accepted inter-governmental authority over these resources. A functioning international system would provide the IARCs with the freedom to operate and – it is still to be hoped – would open greater opportunities of collecting material than had been possible for some time. Since the CBD was not retroactive, these collections were outside its authority, which provided the basis for Resolution 3 of the Nairobi Final Act. FAO Conference Resolution 3/93 went well beyond these collections, in providing a mandate for negotiations on 'access on mutually agreed terms to plant genetic resources, *including* ex situ collections not addressed by the Convention' [emphasis added]. This resolution allowed the ITPGRFA to address, as well, the plant genetic resources held by governments and natural and legal persons, but the IARCs' collections were the key focus throughout, and their mandate crops were therefore the backbone of Annex 1, as it was finally agreed.

The International Undertaking, and in particular the 1993 FAO-CGIAR in-trust agreements that were concluded within its framework, were key factors in the construction and scope of access and benefit sharing under the ITPGRFA. As a result of these agreements, it was from the beginning clear that the ITPGRFA needed to address *all* the plant genetic resources held by the IARCs. Since the IARCs also held limited collections of non-Annex 1 crops, provision was made for both in Article 15, with Article 15.1(a) applying to Annex 1 crops, as part of the multilateral system, and Article 15.1(b) applying to their holdings of other crops, with the obligation that the non-Annex 1 materials held in trust by the IARCs should be distributed under a material transfer agreement that would contain substantially the same benefit-sharing provisions as the SMTA.

Privatization

The fact that there had been increasing privatization of genetic resources was a major influence in many different ways. In many countries, plant breeding had moved, or was moving, from the public sector to the private sector. Intellectual property – whether through plant variety protection or patents – was increasing, and the provisions of Article 27.3(b) of the Agreement on Trade Related Aspects of Intellectual Property Rights specifically required intellectual property protection for plant varieties.[6] In addition, breeders were very jealous of their right to hold genetic material privately and under conditions of trade secrecy.

The privatization of genetic resources – earlier considered a *res omnium* or *res nullus* – was also at the heart of the CBD, which both recognized that biodiversity was a 'common concern of humankind' and put the onus for the conservation of genetic resources on the governments who had sovereignty over them. Most of the necessary funds were expected to come from the commercial use of these resources, and, for this to work, governments had to construct systems through which market forces could play a benign role and create virtuous cycles of use and reinvestment in conservation. The approach attempted to address the problem of what has been called 'the tragedy of the commons', namely that when a resource is a common property the tendency is for individuals to use it without re-investing in its maintenance. In some ways, the aim of the CBD in this regard could be described as an attempt 'to privatize conservation'. The risk is a failure to create an effective market, while entrenching property rights for those few who have an interest in them. Within the context of the CBD, there is little evidence of important resources for conservation having been mobilized in this way.

Within the context of the ITPGRFA's multilateral system, the resources that are expected to derive from monetary benefit sharing 'should flow primarily, directly and indirectly, to farmers in all countries, especially in developing countries, and countries with economies in transition, who conserve and sustainably utilize plant genetic resources for food and agriculture' (Article 13.3). However, the reworking of the International Undertaking's material transfer agreement (which was intended to prevent privatization) to become the ITPGRFA's SMTA (a contractual instrument intended to generate revenues from the private products of the use of PGRFA) is not likely to provide the resources necessary to safeguard and use the world's plant genetic resources for food security and sustainable development.

Nor did the ITPGRFA make financial provision from these resources to support the crucial *ex situ* collections held by international institutions. The Global Crop Diversity Trust, which is an 'essential element' of the ITPGRFA's funding strategy, as well as the Svalbard Global Seed Vault were developed through the initiative of individual donors, and are legally separate from the ITPGRFA but undoubtedly in support of its objectives.

The privatizing focus of the CBD was a major influence on the negotiations of the ITPGRFA as well as on the construction of the multilateral system. It effectively foreclosed the possibility of a more truly multilateral management of PGRFA by focusing on the commercial benefits only, through a contractual instrument (the SMTA) of private contractual law. The obsession with capturing private market values meant that much time and energy was dedicated to working out mandatory monetary benefit-sharing provisions, which were linked to the use of specific samples of plant genetic resources, incorporated into specific new commercial products protected by intellectual property rights, through the value of the actual sales of those products. The multilateral system is therefore an administratively complex, legalistic structure, geared towards capturing private market values to encourage the conservation and sustainable uses of these resources, without clear indication as to how this can be achieved. Since it seeks to draw funds from only a small proportion of the plant genetic resources used in commercial agriculture, it has not internalized the costs of managing plant genetic resources in its market prices, either to prevent their irreparable loss or to facilitate their use today. This failure has been a major impediment to creating a simpler system that is better suited to the situation on the ground.

Weak negotiators

A factor that is extraneous to the influences of both the International Undertaking and the CBD, but which should not be underestimated, is the relative weakness of the agricultural sector and, hence, the generally low level of authority of its negotiators. The lowly position of the Ministries of Agriculture in industrialized countries and the fact that many negotiators were primarily technical experts limited their mandates and negotiating authority. In many national circumstances, Ministries of Agriculture have been hard pressed to advance the distinct needs of agriculture in dealing with other ministries, particularly the Ministries of the Environment and of Trade, because of the importance accorded to the CBD and the higher status of other ministries and of intellectual property authorities in the national pecking order. These problems were thoroughly documented in the study *Why Governments Can't Make Policy: The Case of Plant Genetic Resources in the International Arena* (Petit et al., 2001). The upshot has been that the negotiators have had a very limited ability to envisage solutions that would entail state obligations of any sort or that would require ongoing financial commitments. This situation has predisposed them to solutions constructed around private contracts that require no state obligations and promise benefits that do not come from state coffers, even if more parsimonious – and, in the long run, more economical and effective – solutions could have been found. The negotiators themselves were aware of this weakness, with for example the seventh session of the CGRFA drawing attention in 1997 to 'the need for high-level political involvement in the negotiating process'.

Farmers' rights

The concept of farmers' rights was developed to recognize the specific nature of innovation within traditional farming systems, and they were defined in

Resolutions 4/89 and 5/89 of the FAO Conference, as agreed interpretations of the International Undertaking. They preceded the CBD, which does not consider agriculture in any specific way. The CBD, instead, puts the accent on the holders of traditional knowledge and the role of indigenous and local communities.

In the context of the International Undertaking, farmers' rights were not linked to individual samples of genetic material nor were they construed as property rights. FAO Conference Resolution 5/89 defined them as:

> rights arising from the past, present and future contributions of farmers in conserving, improving, and making available plant genetic resources, particularly those in the centres of origin/diversity. These rights are vested in the International Community, as trustee for present and future generations of farmers, for the purpose of ensuring full benefits to farmers, and supporting the continuation of their contributions.

There have been different currents of opinion. In the case of India, and other countries that have enacted provisions based on its model, farmers' rights have created a form of individual intellectual property over individual traditional resources. Moreover, the discussion of access and benefit sharing in the CBD forums has been increasingly linked to the provisions of Article 8(j), which provides that:

> Each contracting Party shall, as far as possible and as appropriate: Subject to national legislation, respect, preserve and maintain knowledge, innovations and practices of indigenous and local communities embodying traditional lifestyles relevant for the conservation and sustainable use of biological diversity and promote their wider application with the approval and involvement of the holders of such knowledge, innovations and practices and encourage the equitable sharing of the benefits arising from the utilization of such knowledge innovations and practices.

This is a tough enough nut to crack in the context of access and benefit sharing under the CBD. In the context of the negotiations of the ITPGRFA, governments decided relatively early on to separate the realization of farmers' rights from the construction of the multilateral system and to handle farmers' rights in a separate Article 9, which does not foresee the establishment of intellectual property or quasi-intellectual property rights. Benefits to traditional farming communities are therefore construed as generalized, multilateral benefits, which are foreseen primarily under Articles 13.3 and 18.5 but also under Articles 4(d), 5(c), 6.2(b) and (c), and 13.2(b)(iii). This development of farmers' rights on a multilateral basis was probably inevitable, if a truly multilateral system was to be established. It is now a major challenge for the ITPGRFA to ensure that multilateral benefits accrue to traditional farmers in the development of the ITPGRFA's funding strategy.

Private contracts

A key and distinctive structural characteristic of the multilateral system is the management of access and benefit sharing under a private contract, the SMTA, whereby the provider of the goods that the contract regulates obtains no benefit from the contract. This is a complex way to manage a public or pooled good, and it has already piled innovation on top of innovation, including the SMTA itself, in particular the recognition within it of the right of the third party beneficiary – a concept that derives from common law – to enforce its contractual obligations.

As an element of the multilateral system, the third party beneficiary arose in the very late stages of the negotiations and is without any antecedent in either the International Undertaking or the CBD. It was a logical necessity once the decision had been taken to regulate access to, and the use of, a public or pooled good by a private contract between two parties, neither of whom has any beneficial interest in that contract since the benefits devolve on the ITPGRFA itself. Through the SMTA, the parties to it confer on the ITPGRFA a benefit that is distinct from legal ownership or control over the plant genetic resources that it supplies, and recognize the right of the third party beneficiary to take action to enforce the contract (see Chapter 8 by Gerald Moore in this volume).

A further element of the multilateral system is the fact that disputes arising from the SMTA are to be resolved, at the last instance, through international arbitration outside any national jurisdiction. This element, too, was innovative. The material transfer agreement used by the IARCs, in accordance with their 1994 agreements with the FAO within the context of the International Undertaking, similarly provided that disputes were to be settled by international arbitration outside any national jurisdiction. In the case of the agreements, the possible disputes to which this applied were between the IARCs and the FAO. The material transfer agreement under which plant genetic resources were distributed contained no dispute settlement provisions.

The many ways in which the SMTA has had to be tailored in order to serve as the keystone of the multilateral system have resulted in an instrument that, not surprisingly, has created considerable conceptual problems for the private sector, even for industry groups that strongly support the objectives of the ITPGRFA. Whether or not it will be effective in the long term in generating a flow of resources that are able to meet the expectations of the contracting parties remains to be seen.

Concepts from both the International Undertaking and the CBD lie behind the decision to base the most central functions of the ITPGRFA on such a contractual instrument. The existence of a material transfer agreement within the context of the International Undertaking (which was used by the IARCs in the context of the 1994 agreements with the FAO) predisposed negotiators to contractual solutions, although, in fact, this material transfer agreement had a rather different purpose – namely to prevent the misappropriation of farmers' varieties that were being contributed to the international *ex situ* collections. Much more influential was the contractual approach used by many countries in the implementation of the CBD. While national sovereignty allows governments to make multilateral arrangements if they so wish, it is a fact that where governments have moved to regulate access to their national genetic resources, in the application of the CBD, they have almost exclusively done so through arrangements (a combination of permits and contracts) that give the recipient rights to exclusive access and use in exchange for being able to share in the profits from products marketed under patent rights. Biological diversity (often meaning genetic resources) is seen as a permanent and inalienable property of national governments, access to which is regulated by use-licenses.

The model for such arrangements was not taken from the agriculture sector but, rather, from pharmaceutical bioprospecting and the assumption that biodiversity was 'green gold'. In this context, the CBD is a privatizing and enclosing instrument, and its application therefore turns on private law instruments and contracts between provider and bioprospector. What is usually envisaged is the discovery of a bioactive compound in an individual plant species, followed by the isolation, synthesis and patenting of the molecule and the generation of potentially huge profits. This activity has little to do with the realities of food and agriculture: the value in agriculture genetic resources lies at the intra-specific level, not at the inter-specific level – that is, value lies in the genetic diversity within a crop as the basis for the slow process of plant breeding and not in the bioactive molecules found within a particular species.

Individually negotiated contracts between provider and recipient on the pharmaceutical model are very difficult to combine with the realities of plant breeding. Recipients usually seek exclusivity, require providers not to make the material available to other recipients and hold and investigate materials under trade secrecy. Research results are secret, greatly reducing their contribution to the national and international good. The implications are grave for public agricultural research, which is crucial in meeting the needs of developing countries. Potential providers are loath to provide the public sector with access, and receivers starve the public sector of scientific information. Since the pharmaceutical model assumes a patented product, it pushes agricultural innovation towards patents and skews the research agenda away from traditional breeding, plant variety protection, smaller crops and the needs of the poor and developing countries. The result has been to accelerate the tendency that was already present within the agricultural sector of major countries to seek ever-stronger patent protection over seeds and planting materials.

The high transaction costs associated with negotiating, monitoring and enforcing individual contractual access and benefit sharing are a real disincentive to the wide and effective use of agricultural genetic resources. There are symptoms of a market failure in the pharmaceutical sector – that is, an unwillingness to make such contracts in situations of uncertainty, high transaction costs, alternative sources of the same materials and years of research before the potential value can be identified. In the agricultural sector, total or partial market failure is inevitable.

The adoption by the ITPGRFA of an SMTA, and the non-exclusivity of access to these materials, goes a long way towards resolving these problems. The multilateral system has been based on an individual, private sector, contractual instrument. It may not have been the simplest, or most efficient structure, but it is now necessary to create a coherent system around it.

A simpler solution

Could something better have come out of the long and tortured negotiations? The multilateral concepts developed in the negotiating history of the International Undertaking and the privatizing pressures that led to the CBD were at odds throughout the negotiations. Attempts to create a more generalized and open regime of access and benefit sharing for food and agriculture were seen by many developing countries to weaken their negotiating position in the context of the CBD. The failure of the International Undertaking's process to have constructed a realistic system of benefit sharing, despite the intensive process of preparing and agreeing to *The State of the World's Plant Genetic Resources* and the Global Plan of Action for the Conservation and Sustainable Use of PGRFA, substantially weakened the food and agriculture sector. Weak representatives of a relatively weak sector, therefore, had difficulty in thinking afresh and building a new agreement from the ground upwards. Put crudely, the nexus that needed to be addressed was how to conjugate individual property in seeds and crops with access to the genes they contain. What would trigger benefit sharing and who should pay? The conceptual tools coming from both the International Undertaking and the CBD – in particular, the contractual and privatizing elements – linked benefit sharing with individual accessions: a really simple system would have had to break the link between individual accessions and individual benefits.

Clearly, no regime could force all of the holders of genetic material of a particular crop to make it available to anyone who wanted it or to confer a right on someone to be able to demand such resources from anybody who had them. Property in a plant or seed cannot simply be overridden. Moreover, farmers and breeders have materials, which are both unimproved and improved, that they wish to keep for themselves. Without overriding property – physical and intellectual – it is, however, possible to stipulate the rights and duties of classes of persons and bodies in relation to access. The ITPGRFA has done so in a very limited way. Contracting parties are to make available all of the resources that are under their management and control and in the public domain. International institutions can agree under Article 15 to do so, and private persons are encouraged to do so. A more systematic approach would have required national legislation to implement an effective international framework that went beyond a private contractual instrument.

A more fundamental structural question is what triggers the obligation to share benefits: certain specified uses of any genetic resource of the crop in question, the fact that a specific sample or samples has been accessed, or a combination of the two? If a multilateral system is constructed in such a way that the sole trigger for benefit sharing is a specified use, then access becomes a practical matter of seeking agreements that make the largest amount of material available in the simplest way at the lowest cost. The fact of having accessed one specific sample or another of the crop would have no implication for benefit sharing. Since obligations would devolve from use only, there would be no legal need to know where or how such resources were accessed or exchanged after being accessed. The uses that trigger benefit sharing may then be disassociated from access, and the promotion of wide access can be approached separately.

In fact, the experts of two countries at the Montreux Expert Meeting in January 1999 proposed such a system – that is, that access to all plant genetic resources be free, without contractual obligations, but that obligations to share benefits be triggered by the marketing of products that are not freely available to others for further research and breeding. Another country's expert opposed this suggestion, on the grounds that 'our plant breeders would never pay for resources they already have', and the contractual basis of the multilateral system was the result.

The actual trigger for benefit sharing had to be decided. The simplest thing to do would have been commercialization of any product that is a PGRFA of a crop covered by the system. Benefits could also have been crafted to differ in differing circumstances, as is the case in the ITPGRFA, where a distinction is made as to whether or not commercialized products are available to others for further research and breeding. There are various possible ways to establish who should pay, and states could themselves make contributions to a benefit-sharing fund. They might raise the money in any way they wish. The simplest solution would have been to allow each contracting party to decide how to do so. Direct contributions by states were considered at various stages in the negotiations. For example, the draft text of what was then Article 14.2(d)(i) (bracketed), which resulted from the fourth intersessional meeting of the contact group in Neuchâtel, Switzerland, on 12–17 November 2000, read as follows (the details of the draft text are not important; the principle is):

> Each Party to this Undertaking undertakes to pay, in accordance to the agreed Funding Strategy to be established under Article 16, an annual contribution representing *** percent of the value of the crops produced in its territory through the use of plant genetic resources for food and agriculture listed in Article 12 to this Undertaking where such crops are produced from or through plant genetic materials or related processes in respect of which IPR protection has been sought under its national legislation. For this purpose, the value of the crops shall be calculated on the basis of the hectarage of the crops harvested multiplied by the average national yield per hectare for those crops and the average ex-farm price for the current year.

Alternatively, users could be required to make payment directly, or the contracting parties could collect payments from the commercializers.

In structural terms, a system that delinks access from benefit sharing cannot be implemented through a private contractual instrument. It would require the contracting parties to legislate and create the obligation to share benefits by public, rather than private, law. It would require the state to enforce the execution of obligations. Many negotiators may simply have been unable to envisage state obligations because they were unable to obtain adequate negotiating authority due to the relative weakness of their ministries at home and were under pressure from the national authorities committed to the CBD to explore privatizing solutions in a context where the use of intellectual property rights in agriculture was expanding rapidly.

Two matters that would need to be considered are the perceived risk of 'free-loading' by states that are not contracting parties and the questions of non-food and agriculture uses. Again, there are a number of possible solutions. Upon entry into the markets of contracting parties, similar commercial products of non-contracting parties could be required to pay a levy corresponding to the payments originating from the commercializers of a contracting party. An SMTA that imposes the benefit-sharing obligations of the ITPGRFA could be foreseen, for use only when the gene banks in the contracting parties have supplied materials to users in non-contracting parties. The question of non-food and agriculture uses could be viewed in the same way that the ITPGRFA views them, as being outside its ambit. A wide reading of what is meant by agriculture – and agriculture traditionally includes the cultivation of crops for food, fibre, fuel and energy products, building material and a wide variety of other uses – would prevent the benefits of the use of crop plants of all types draining out of the benefit-sharing mechanisms of the ITPGRFA. There is anyway very little evidence of large profits being generated from non-food and agriculture uses of PGRFA.

The uses to which benefits should be put is a separate question. One of the failures of the ITPGRFA is that it has not sought, in any way, to internalize the costs of the conservation of plant genetic resources in the price of products within the multilateral system.

Solutions based on the delinking of access and benefit sharing, and of individual accessions to individual benefits, have the advantage of simplicity, transparency and low transaction costs, with no registration and no tracking of materials. They are also fully compatible with property rights. However, their greatest advantage is that they address in a generic manner the genetic resources of a crop rather than a specific sample of resources. The multilateral system, as it was developed in the negotiations, addresses only specific physical samples of genetic resources, not the genetic resources themselves. This fact creates a complex set of differing property rights over what is essentially the same material. Given that essentially the same material may have been acquired under any one of the sets of conditions, and that individual samples under different conditions may have been interbred, the implementation of the multilateral system is fraught with complexities and uncertainties. Box 13.1 illustrates a long-lived field containing an Annex 1 crop, from which samples have been collected at various times.

Box 13.1

Collected before the entry into force of the CBD (29 December 1993):

- held privately, unencumbered by any obligations;
- held in an IARC;
- distributed unencumbered before the October 1994 agreements with the FAO; and
- distributed after October 1994, but before the entry into force of the ITPGRFA, subject to the material transfer agreement implementing these agreements.

> Collected after the entry into force of the CBD (29 December 1993) and before the entry into force of the ITPGRFA (29 June 2004):
>
> - held privately, possibly under conditions imposed by the provider;
> - held in an IARC under special conditions imposed by the provider:
>   - may only be distributed under the special conditions imposed by provider:
>   - distributed before October 1994: distributed unencumbered;
>   - distributed after October 1994, but before the entry into force of the ITPGRFA: subject to the material transfer agreement implementing the FAO/IARC agreements and may now be distributed under the SMTA.
>
> After the entry into force of the ITPGRFA (29 June 2004):
>
> - may only be distributed by contracting parties and their institutions, as well as international institutions, under the SMTA.
>
> Materials held privately by legal and natural persons:
>
> - any natural or legal person may decide to distribute materials under the SMTA and
> - any natural or legal person may also decide to distribute their materials under individual contracts outside the multilateral system.

Could the multilateral system have been constructed in another way?

There is no teleology in evolution. It does not function to achieve an end. Genes are recycled and reworked by the chance events of selection for functions different to those they originally played. Evolution is a '*bricholeur*', making and patching and re-patching with the tools that come to hand. For very understandable reasons, the multilateral system was similarly cobbled together from earlier concepts, but there was no inevitability in the solutions that were finally adopted and there was a lost opportunity to construct a simpler and more efficient system.

Like an organism created by the hazards of evolution, the multilateral system nonetheless lives and works and has many defects of design due to the re-use of older organs for purposes for which they were not designed. The imposition of the SMTA, which removes individual negotiations from the equation and functions across national boundaries, is a real achievement. The contractual obligation it contains to resolve disputes regarding benefits by binding international arbitration strengthens it as a truly international instrument. However, the most innovative element of the multilateral system is undoubtedly the third party beneficiary, which for the first time in the management of an international public or pooled good provides the international community with a right to intervene in private contracts. The system would not otherwise be workable. If the fashion of managing international public goods through private instruments continues, the third party beneficiary is likely to become the fountainhead of a whole new form of international law. The complexities of the multilateral system will require substantial and sustained political will and commitment from the contracting parties to make the system work. We cannot, like Pangloss, simply assume that all is for the best in the best of all possible worlds. *Il faut cultiver notre jardin.*

Notes

1 International Treaty on Plant Genetic Resources for Food and Agriculture, 29 June 2004, www.planttreaty.org/texts_en.htm (last accessed 10 May 2011).
2 International Undertaking on Plant Genetic Resources for Food and Agriculture, 1983, www.fao.org/waicent/faoinfo/agricult/cgrfa/IU.htm (last accessed 10 May 2011).
3 Convention on Biological Diversity, 31 ILM 818 (1992).
4 Agenda 21, 13 June 1992, UN Doc. A/CONF. 151/26 (1992).
5 Standard Material Transfer Agreement, 16 June 2006, ftp://ftp.fao.org/ag/agp/planttreaty/agreements/smta/SMTAe.pdf (last accessed 10 May 2011).
6 Agreement on Trade Related Aspects of Intellectual Property Rights, Annex 1C of the Marrakech Agreement Establishing the World Trade Organization, 15 April 1994, 33 ILM 15 (1994).

References

Commission on Plant Genetic Resources (CPGR) (1989) *Report of the Third Session of the Commission on Plant Genetic Resources*, Rome, Italy, 17–21 April, online: ftp://ftp.fao.org/docrep/fao/meeting/015/aj401e.pdf (last accessed 16 May 2011).
——(1991) *Report of the Fourth Session of the Commission on Plant Genetic Resources*, Rome, Italy, 15–19 April, online: ftp://ftp.fao.org/docrep/fao/meeting/015/aj419e.pdf (last accessed 17 May 2011).
——(1993) *Report of the Fifth Session of the Commission on Plant Genetic Resources*, Rome, Italy, 19–23 April, online: ftp://ftp.fao.org/docrep/fao/meeting/015/aj632e.pdf (last accessed 17 May 2011).
Consultative Group on International Agricultural Research (CGIAR) (2003) *Booklet of CGIAR Centre Policy Instruments, Guidelines and Statements on Genetic Resources, Biotechnology and Intellectual Property Rights*, Version II, System-wide Genetic Resources Programme, Rome, online: www.cgiar.org/corecollection/docs/sgrp_policy_booklet_2003.pdf (last accessed 17 May 2011).
Convention on Biological Diversity (CBD) (2005) *Handbook of the Convention on Biological Diversity*, 3rd edition, Section IX: Nairobi Final Act of the Conference for the Adoption of the Agreed Text of the Convention on Biological Diversity, Montreal, online: www.cbd.int/doc/handbook/cbd-hb-09-en.pdf (last accessed 17 May 2011).
Correa, Carlos M. (2009) *Trends in Intellectual Property Rights Relating to Genetic Resources for Food and Agriculture*, Background Study Paper no. 49, FAO Commission on Genetic Resources for Food and Agriculture, Rome, online: ftp://ftp.fao.org/docrep/fao/meeting/017/k533e.pdf (last accessed 9 January 2011).
Food and Agriculture Organization (FAO) (1983) *Report of the Twenty-Second Session of the Conference of FAO*, Rome, Italy, 5–23 November, FAO, Rome.

——(1997a) *The State of the World's Plant Genetic Resources for Food and Agriculture*, FAO, Rome, online: ftp://ftp.fao.org/docrep/fao/meeting/015/w7324e.pdf (last accessed 10 January 2011).
——(1997b) *Global Plan of Action for the Conservation and Sustainable Utilization of Plant Genetic Resources for Food and Agriculture and the Leipzig Declaration adopted by the International Technical Conference on Plant Genetic Resources Leipzig*, Germany, 17–23 June 1996, FAO, Rome, online: http://typo3.fao.org/fileadmin/templates/agphome/documents/PGR/GPA/gpaeng.pdf (last accessed 22 May 2011).
Petit, M. et al. (2001) *Why Governments Cannot Make Policy: The Case of Plant Genetic Resources in the International Arena*, Centro Internacional de la Papa, Lima, Peru.
Rural Advancement Foundation International (RAFI) (1998) *Recent Australian Claims to Indian and Iranian Chickpeas Countered by NGOs and ICRISAT*, News Item, online: www.etcgroup.org/en/node/428 (last accessed 9 January 2011).

14

THE MOVING SCOPE OF ANNEX 1

The list of crops covered under the multilateral system

Bert Visser

Introduction

The multilateral system of access and benefit sharing (multilateral system) of the International Treaty on Plant Genetic Resources for Food and Agriculture (ITPGRFA) is currently limited in scope to the list of 64 food and forage crops included in Annex 1 of the Treaty.[1] On the one hand, the existence of a finite list of crops logically follows from the agreement to establish a multilateral system and was reached rather early in the negotiations on the Treaty. This multilateral system needed to be defined and limited, and the list helped provide a solution to this problem. On the other hand, the negotiations of the contents of the list itself reflect how perspectives on access and benefit sharing differed between regions and how these different perspectives resulted in divergent views on the composition of the list. In this sense, the negotiations of the content of the list represented a microcosm of the larger political and legal debates that were so central to the negotiations of the ITPGRFA.

Whereas many developing countries were inclined to regard the multilateral system as an experiment that had to show its effectiveness and its value in terms of monetary benefit sharing, developed countries in general pointed to the access to genetic resources as a major benefit in itself. The former view resulted in cautious positions regarding the length of the list, whereas the latter position favoured the inclusion of all plant genetic resources for food and agriculture (PGRFA) by default. The fact that the International Agricultural Research Centres (IARCs) of the Consultative Group on International Agricultural Research (CGIAR) already hosted international, publicly available collections that contained a major share of the genetic resources of staple crops of essential importance to all countries – in particular, developing countries – also influenced the debate and the final decisions on the composition of the list. The 35 food crops and 29 forage crops in Annex 1 of the

ITPGRFA represent a compromise, which was arrived at over the course of several negotiating meetings. This compromise blended a proposal made by the African region for only nine food crops with a proposal made by the European region to include 287 crops.

This chapter analyses the negotiating process that led to the current composition of the Annex 1 list, highlighting how differing perspectives on more fundamental issues – with the largest differences lying between developed and developing countries – influenced the final outcome of the list. The chapter also considers how the perspectives on the importance of biodiversity for global survival have changed in the nine years since the Treaty was adopted, and whether or not expansion of the Annex 1 list might be possible in light of those changes.

A chronology of the negotiations on the list

Resolution 3 of the Nairobi Conference for the Adoption of the Agreed Text of the Convention on Biological Diversity (CBD), which was agreed in 1992, identified the Food and Agriculture Organization's (FAO) Global System on Plant Genetic Resources as the context within which issues left outstanding by the CBD (access to existing *ex situ* collections and farmers' rights) should be addressed.[2] The FAO and its Commission on Genetic Resources for Food and Agriculture (CGRFA) took up this challenge.[3]

Between November 1994 and November 2001, the CGRFA, which was the FAO body in which the ITPGRFA was negotiated, convened nine times to develop the Treaty, in three regular and six extraordinary sessions. To support its work, the working group of the CGRFA convened a number of times early in the negotiating process, while a specially established contact group also convened six times between September 1999 and June 2001. An expert meeting in Montreux in 1999 played a pivotal role in establishing the principles for this internationally binding agreement. Whereas all of the countries that were members of the CGRFA (more than 150) took part in the negotiations of the commission, the informal consultations in the contact group were only attended by delegations from 40 countries and the European Union, and the Montreux meeting was only attended by 24 participants.

Over time, the negotiations resulted in four consecutive negotiating drafts, a consolidated negotiating text, the chairman's elements derived from the Montreux meeting, and a composite draft text incorporating the chairman's elements. However, the composition of a list of crops was not discussed until late in the negotiations. In fact, in the regular sixth session of the CGRFA in June 1995, a 'list by way of example' was presented for the first time, containing a large number of genera and crops – a total of 231! Consecutively, in May 1997, a vast majority of participants in the seventh regular session of the CGRFA agreed in principle to establish a multilateral system to facilitate access in an efficient, effective and transparent way. However, only in the fourth extraordinary session of the CGRFA, which took place in December 1997, did the notion take shape that the multilateral system would have to facilitate access to PGRFA through a list of major crops, which was yet to be determined. After the initial submission of a list 'by way of example', it took until the third meeting of the contact group in August 2000 before formal discussions on the list really began to come together. Less than one year later, the final composition of the list was reached in June 2001, although negotiations on the list carried on until the meeting of the FAO Council in November 2001.

The list of crops took shape in several phases, and, over this period of time, it was variously referred to as an 'example proposed during the sixth session of the CGRFA of a list of genera important for food and agriculture', a 'Tentative List of Crops (Article 11.3.3)', and a 'List of crops covered by/under the Multilateral System'. During a late phase of the negotiations, additional lists circulated under the heading 'Lists of crops discussed and pending acceptance by some regions/countries, during the meetings of the Working Group for the list of crops at the sixth extraordinary session', namely 'Crops under discussion (Spoleto – Table II)' and 'Additional crops under discussion' (CGRFA, 2001). Country and region contributions also referred to 'listed crops essential to food security' (United States), 'crops/genera of basic importance for human world food consumption' (Brazil), and 'proposed list of crops of the Latin American and Caribean Region'.

Whereas the early negotiations on the list are characterized by individual country contributions, later negotiations show mainly regional positions and contributions. Such regional positions could sometimes only be reached after lengthy and complex consultations at the regional level. In the case of the meetings of the G-77 and China, reaching effective conclusions necessitated even more complex and time-consuming consultations.

The negotiations on the composition of the list formed a late phase in the development of the multilateral system, and Annex 1 proved to be one of the major missing pieces in the final agreement of the ITPGRFA. Finalizing the list of crops in Annex 1, using regional approaches that ranged from nine to 287 crops, remained the largest obstacle in the final negotiation process, and there were good reasons for this part of the negotiations to be late and complex.

Annex 1 and the CBD

To suggest a direct linkage between the various perspectives on the list of crops contained in Annex 1 under the multilateral system, on the one hand, and the CBD, on the other hand, may seem a little far-fetched to some, but this was certainly not the case. The adoption of the CBD in 1992 caused a definitive paradigm shift, away from the concept of biodiversity (including genetic resources) as the heritage of humankind and towards the concept of national sovereignty over genetic resources. According to the CBD, states can decide the conditions for access to their genetic resources, and, as a result, the introduction of the term 'country of origin' is very significant.[4] It was the mandate of the CGRFA to revise the International Undertaking on Plant Genetic Resources for Food and Agriculture (International Undertaking) to bring it into harmony with the CBD.[5] The FAO and its commission were requested to do so in order to resolve issues regarding the status of existing *ex situ* collections – in particular, the collections held by the IARCs.[6]

Many delegates participating in the negotiations of the ITPGRFA operated from the perspective of national sovereignty, taking the principles on which the CBD was based as the default for access and benefit-sharing arrangements and being prepared to consider the need and justification for exceptions to the rule for PGRFA.[7] In this same perspective, the Brazilian delegation, during the second meeting of the Contact Group in April 2000, stated that the negotiations on the multilateral system and its Annex 1 should be considered a 'window' to be opened on the bilateral benefit-sharing arrangements of the CBD, but that such action was only justifiable if the 'window' was small, had clearly defined limits, and would be an important contribution for food security, represented by the basic staple food crops, to the benefit of all countries (CGRFA, 2000a).

In the final days of the negotiations on the Treaty in the FAO Council in November 2001, a number of developing countries also presented their view that, in effect, 'crops on the list represent gifts by countries to the rest of the world' (IISI, 2001a). Such wording was clearly motivated by the principle of national sovereignty enshrined in the CBD. However, other perspectives prevailed among the participants from developed countries, where any limitation to access to PGRFA was seen as a threat to the interests of plant breeders and, thus, to major breeding advances and world food security. Naturally, the interests of plant breeders also included the need to limit additional costs stemming from new policy instruments as much as possible. Hence, throughout the negotiations, the notions of 'window' and 'gift' often came up against the notion of the need to promote plant breeding. Both perspectives recognized the importance of global food security. However, other elements also played a role in the discussions and negotiations, as will be discussed in the following section.

The linkage between Annex 1 and other major components of the ITPGRFA

The negotiations on Annex 1 – that is, the list of crops covered by the multilateral system – cannot be properly understood without reference to the negotiations on other major elements of the Treaty, in particular:

- the scope of the multilateral system (what is included and what is not);
- the agreed types of use for which access is provided;
- the provisions on benefit sharing, including the notion that access to plant genetic resources for food and agriculture represents a benefit in itself;
- (other) financial arrangements to facilitate the implementation of the ITPGRFA; and
- the options to obtain intellectual property rights on products derived from materials included in the multilateral system (see, for example, the statements of five regions contained in CGRFA, 2000b; IISI, 2001a).

In discussing the scope of the multilateral system, discussions arose whether or not it would only include material from the public domain, whether the multilateral system would only cover material acquired before the coming into force of the CBD (prior to December 1993), and whether the multilateral system would cover *ex situ* as well as *in situ* materials (CGRFA, 1997a). Developing countries feared that if only material in the public domain would be incorporated in the multilateral system, they would still lack access to the more advanced, and therefore more valuable, breeding materials developed in the private sector. Developed countries felt that it was important to incorporate materials found *in situ* in the multilateral system, since this category encompassed wild crop relatives from which so many useful resistance and tolerance traits could be derived. In addition, most countries were prepared to agree that, in practice, it would be inefficient to try to distinguish materials acquired by a country or collection holder before the coming into force of the CBD from materials obtained thereafter. As long as agreements on all these issues were pending, it was felt to be premature to negotiate the contents of the list of crops covered in Annex 1 under the multilateral system.

Type of use formed a major concern for a number of developing countries that wished to limit the use of material obtained from the multilateral system for purposes of research, breeding and training for food and agriculture and not for other uses, as was ultimately reflected in the final text of the ITPGRFA (Article 12.3(a)). In fact, only during implementation of the Treaty was direct use by small-scale farmers dealt with in particular as a separate issue.

> **Box 14.1 Text of Article 12.3(a)**
>
> 12.3 Such access shall be provided in accordance with the conditions below:
>
> (a) Access shall be provided solely for the purpose of utilization and conservation for research, breeding and training for food and agriculture, provided that such purpose does not include chemical, pharmaceutical and/or other non-food/feed industrial uses. In the case of multiple-use crops (food and non-food), their importance for food security should be the determinant for their inclusion in the Multilateral System and availability for facilitated access.

While the provisions on non-monetary benefit sharing could be derived from the concepts elaborated in the CBD (Articles 16–18), the negotiations on an agreement on monetary benefit sharing heavily impacted on the preparedness of developing countries to include crops in the list. This was also the case for the discussions on financial arrangements. Although framework decisions on mandatory monetary benefit sharing were reached at a rather late stage in the negotiations, there was no time to work out the details for such mandatory benefit sharing, such as the triggers for payment and the level of monetary benefit sharing. Such details had to be worked out in the subsequent negotiations of the Standard Material Transfer Agreement (SMTA), which were finalized in 2006.[8] In addition, the elaboration of the funding strategy that was discussed in the first three sessions of the Governing Body of the

ITPGRFA between 2006 and 2009 provided more clarity on the objectives, targets and expectations of the access and benefit-sharing arrangements. Lack of clarity about the nature and level of benefit sharing made many developing countries hesitant in accepting an expanded list of crops covered under the multilateral system.

> **Box 14.2 Framework agreements on mandatory benefit sharing: Articles 13.2(d) and 18 of the ITPGRFA**
>
> 13.2 The Contracting Parties agree that benefits arising from the use, including commercial, of plant genetic resources for food and agriculture under the Multilateral System shall be shared fairly and equitably through the following mechanisms: the exchange of information, access to and transfer of technology, capacity-building, and the sharing of the benefits arising from commercialization, taking into account the priority activity areas in the rolling Global Plan of Action, under the guidance of the Governing Body:
>
> ...
>
> (d) Sharing of monetary and other benefits of commercialization
> (i) The Contracting Parties agree, under the Multilateral System, to take measures in order to achieve commercial benefit-sharing, through the involvement of the private and public sectors in activities identified under this Article, through partnerships and collaboration, including with the private sector in developing countries and countries with economies in transition, in research and technology development;
> (ii) The Contracting Parties agree that the standard Material Transfer Agreement referred to in Article 12.4 shall include a requirement that a recipient who commercializes a product that is a plant genetic resource for food and agriculture and that incorporates material accessed from the Multilateral System, shall pay to the mechanism referred to in Article 19.3f, an equitable share of the benefits arising from the commercialization of that product, except whenever such a product is available without restriction to others for further research and breeding, in which case the recipient who commercializes shall be encouraged to make such payment.
>
> The Governing Body shall, at its first meeting, determine the level, form and manner of the payment, in line with commercial practice. The Governing Body may decide to establish different levels of payment for various categories of recipients who commercialize such products; it may also decide on the need to exempt from such payments small farmers in developing countries and in countries with economies in transition. The Governing Body may, from time to time, review the levels of payment with a view to achieving fair and equitable sharing of benefits, and it may

also assess, within a period of five years from the entry into force of this Treaty, whether the mandatory payment requirement in the MTA shall apply also in cases where such commercialized products are available without restriction to others for further research and breeding.

Article 18 – Financial Resources

18.1 The Contracting Parties undertake to implement a funding strategy for the implementation of this Treaty in accordance with the provisions of this Article.
18.2 The objectives of the funding strategy shall be to enhance the availability, transparency, efficiency and effectiveness of the provision of financial resources to implement activities under this Treaty.
18.3 In order to mobilize funding for priority activities, plans and programmes, in particular in developing countries and countries with economies in transition, and taking the Global Plan of Action into account, the Governing Body shall periodically establish a target for such funding.
18.4 Pursuant to this funding strategy:

(a) The Contracting Parties shall take the necessary and appropriate measures within the Governing Bodies of relevant international mechanisms, funds and bodies to ensure due priority and attention to the effective allocation of predictable and agreed resources for the implementation of plans and programmes under this Treaty.
(b) The extent to which Contracting Parties that are developing countries and Contracting Parties with economies in transition will effectively implement their commitments under this Treaty will depend on the effective allocation, particularly by the developed country Parties, of the resources referred to in this Article. Contracting Parties that are developing countries and Contracting Parties with economies in transition will accord due priority in their own plans and programmes to building capacity in plant genetic resources for food and agriculture.
(c) The Contracting Parties that are developed countries also provide, and Contracting Parties that are developing countries and Contracting Parties with economies in transition avail themselves of, financial resources for the implementation of this Treaty through bilateral and regional and multilateral channels. Such channels shall include the mechanism referred to in Article 19.3f.
(d) Each Contracting Party agrees to undertake, and provide financial resources for national activities for the conservation and sustainable use of plant genetic resources for food and agriculture in accordance with its national capabilities and financial resources. The financial resources provided shall not be used to ends inconsistent with this Treaty, in particular in areas related to international trade in commodities.

> (e) The Contracting Parties agree that the financial benefits arising from Article 13.2d are part of the funding strategy.
> (f) Voluntary contributions may also be provided by Contracting Parties, the private sector, taking into account the provisions of Article 13, non-governmental organisations and other sources. The Contracting Parties agree that the Governing Body shall consider modalities of a strategy to promote such contributions.
>
> 18.5 The Contracting Parties agree that priority will be given to the implementation of agreed plans and programmes for farmers in developing countries, especially in least developed countries, and in countries with economies in transition, who conserve and sustainably utilize plant genetic resources for food and agriculture.

Developed countries argued that facilitated access to PGRFA that are included in the multilateral system constitutes in itself a major benefit – a notion that has been incorporated in the text of the ITPGRFA (Article 13.1). However, since most plant breeding is concentrated in developed countries and most benefits are felt to accrue to developed countries, this facilitated access was less convincing for many developing countries, particularly because they expected that most varieties developed from genetic materials obtained from the multilateral system for agriculture in developed countries would not be suitable for use in an unmodified way in their own agricultural systems.

Disagreement over the list of crops revealed deep political divides on the issue of trading agricultural resources. Some participants in the negotiations indicated that without a clear consensus on whether PGRFA would include a plant's genetic parts and components (which was eventually agreed upon in Article 12.3(d) of the Treaty), it would be very difficult for a country to know what it was agreeing to in this virtual market exchange of agricultural genetic materials (IISI, 2001c). In fact, lack of clarity in the agreed text resulted in the additional inclusion of the provision 'in the form received', which was essential for developed countries. While it can be assumed that any new product derived from the original plant genetic resources obtained from the multilateral system may be protected by intellectual property rights, it remains to be seen how much the original genetic resource needs to be changed for intellectual property protection to be widely acceptable, and how limitations on the use of PGRFA obtained from the multilateral system by very wide patent applications may be avoided.

Many developing countries feared that various forms of intellectual property rights would limit, in practice, access to, and use of, products derived from genetic materials obtained from the multilateral system. In this context, negotiations on the modalities of access also stimulated a discourse on the list of plant genetic resources to be included in the multilateral system early on in the work of the contact group. Some delegates were of the opinion that the list should be narrowed if obtaining intellectual property rights on the products of materials obtained from the multilateral system would be allowed.

In fact, the African region proposed a long list of crops early during the negotiations under the condition that use of the materials under the multilateral system would not be amenable to intellectual property protection. Only when developed countries maintained their condition that obtaining intellectual property right protection would have to be possible, and would indeed form a basis for benefit sharing, did the African region shift to a position that defended a very short list of crops (CGRFA, 2000b).

In conclusion, the negotiating process on each of these elements directly influenced countries' and regions' perspectives on the composition of the list. Thus, postponing negotiations on the composition of the list made sense since its fate was to be determined by the outcome of the negotiations on these other major elements of the Treaty.

Fundamental perspectives on the functionality of the list

From a more fundamental perspective, the list is an inseparable part of the multilateral system, and this system incorporates two components that, in turn, are also inseparable: (1) benefit sharing necessitates access and (2) without benefit sharing no preparedness to facilitate access can be expected. Many developed countries tended to focus heavily on the access component and regarded benefit sharing other than by making the products of breeding with materials from the multilateral system freely available as a politically unavoidable consequence. However, many developing countries took the opposite stance. For them, benefit sharing represented a final recognition of the use that developed countries had made of the genetic resources maintained and developed in their countries for centuries. When mentioning the term access and benefit sharing, developing country representatives tended to hear only benefit sharing, while developed country representatives heard only access. For example, it was very important for the African region that benefit sharing would contribute to their own capacity building in the area of plant breeding. In general, many developing countries felt that if developed countries thought the conservation of plant genetic resources was essential for plant breeding, it would only be logical that they would have to pay for this conservation. Again, these divergent views complicated discussions and decision making on the list of crops.

In one of its interventions during the final FAO Council meeting in November 2001, Norway protested against this particular sentiment, stating that all countries were both providers and recipients of PGRFA. In a similar stance, the European region had long since argued that a short list of crops would, in particular, damage the position of developing countries since so many flows of plant genetic resources were largely in the direction of developing countries, particularly plant genetic resources originating from the collections of the IARCs. Developing countries were looking for a balance between providing access under the multilateral system and obtaining benefits from the use of their own resources.

Criteria for the establishment of the list

In the report produced by the chairman of the eleventh session of the Working Group of the CGRFA, which was held in December 1996, explicit reference is made to food security. It states that the ITPGRFA (which is then still referred to as the International Undertaking)[9] should apply to PGRFA, with specific reference to food security (CGRFA, 1996a). The same report also makes note of 'the scope of any agreement on access and benefit-sharing', mentioning food security as well as countries' interdependence as two issues to be resolved. Interdependence between countries was addressed in the FAO's report on the *State of the World's Plant Genetic Resources for Food and Agriculture*, which was released in 1996, showing that for their major food crops all regions depended heavily on plant genetic resources originating from other regions. Finally, the report concluded that

> one option would provide for access in accordance with national legislation, and sharing out the benefits derived, on a multilateral basis ... This [option] would apply to a list of genera, covering both *in situ* and *ex situ* material, as well as material collected before and after the entry into force of the CBD. The list could be based on importance for world food security and great worldwide interdependence. There was wide agreement that this proposal might provide a useful starting point, although the disadvantages of limited inclusive lists were also stressed.
>
> *(FAO, 1997)*

Subsequently, the report of the fourth extraordinary session of the CGRFA in December 1997 refers to conclusions of the Contact Group: 'In this context, it was noted that the criteria that had been used to establish the Tentative List of Crops annexed to Article 11 were (i) their importance for food security at local or global levels, and (ii) countries' interdependence with respect to plant genetic resources' (CGRFA, 1997b). Hence, while other criteria for the scope for the multilateral system had been suggested in the course of the negotiations, including the degree to which the material is endangered by genetic erosion, the role of the material in sustainable agricultural production, the social and/or economic importance of the material, and the strategic economic importance of the material to the source country, these two criteria remained the only acceptable criteria for the establishment of the list (*ibid.*). It is fair to say that these two criteria would never be able to be applied in an absolute manner, and so it left substantial room for disagreement and negotiation, as is apparent from the negotiating process.

In the end, the two criteria of food security and mutual interdependence also ruled out the incorporation of forestry crops and medicinal crops in the list of crops covered by the multilateral system. Incorporation of these crops had been considered in the early stages of the negotiations (CGRFA, 1994a, 1995).

The collections held by the IARCs

During the entire negotiating process, the status of the IARC collections formed a major element of the discussions. In a report to the sixth session of the CGRFA in June 1995, the signing of the agreement by the FAO and 12 of the IARCs was mentioned, placing their 'designated germplasm' under the auspices of the FAO and recognizing 'the intergovernmental authority of FAO and its commission in setting policies for the International Network' (CGRFA, 1995). This agreement formed a clear early signal that the IARC collections had to be covered by the Treaty and that the conditions regarding access to these collections and the sharing of benefits derived from their use should be addressed. Moreover, since the body that gave the centres policy advice concerning their international collections was the same body hosting negotiations of the Treaty, it only made sense that the collections of the CGIAR centres would be included within the scope of the ITPGRFA.

In fact, the first negotiating draft of November 1994 refers extensively to the work done, and the collections held, by the IARCs, long before the concept of the multilateral system became established and recognized in the course of 1997 (CGRFA, 1994b). Providing access to the material held in the 'base collections' to parties to the ITPGRFA for purposes of scientific research, plant breeding or genetic resources conservation was already foreseen. In the Montreux expert meeting, which took place in January 1999, participants reconfirmed that the multilateral system would initially cover a limited selection of crops determined on the basis of criteria of food security and interdependence. The collections of the IARCs would also form part of the multilateral system on terms previously agreed with these centres. The composite draft text of the International Undertaking, which was released in April 1999, provided text on the multilateral system and explicitly incorporated 'all material held in ex situ collections by such of the International Agricultural Research Centres of the Consultative Group on International Agricultural Research' (CGRFA, 1999). The final text of the Treaty refers to the IARC collections in Articles 11.5 and 15, clearly placing the *ex situ* collections listed in Annex I and held by the IARCs in the multilateral system.

There are two primary reasons why the incorporation of the IARC collections in the multilateral system is significant for the composition of the list of crops under Annex 1 in the multilateral system. First and foremost, the composition of the IARC collections and, in particular, the crops represented in those collections directly influenced the discussions on the composition of the list. The status of the IARC collections had been questioned for a number of years – in particular, by countries from which genetic resources in those collections had been obtained as well as by some countries hosting the IARCs – and the revision of the International Undertaking and the negotiations on the ITPGRFA offered the opportunity to come to a definite agreement on the status of these international collections. In practice, it was only logical to place these collections in the multilateral system that was under discussion. To a large extent, the crops held by the IARCs had represented global staple crops as well as forages, and there was a strong tendency from many countries to adjust the final list of crops in such a way that the list would encompass all, or nearly all, of the crops held in the IARC collections. In fact, this is what happened, although, in dealing with the IARC collections in Article 15, the final text of the Treaty still distinguishes between PGRFA listed in Annex 1 and PGRFA other than those listed in Annex 1.

Second, experts of the IARCs were often consulted by the members of the CGRFA who were negotiating the Treaty, and these experts made regular pleas to solve the status of the IARC collections, in the interest of their own centres, by including their collections in the multilateral system to the fullest extent possible. In effect, all major crop holdings represented in the IARC collections, with the exception of some millets, soybeans, forages and tree species, have eventually become incorporated into Annex 1.

On the composition of the list of crops

The same report by the chairman of the eleventh session of the Working Group of the CGRFA, which called for the development of a list of crops, also mentioned that 'various ways of developing lists were considered: (i) starting from a comprehensive list (such as that in Annex 1 of the Third Negotiating Draft), and excluding those taxa on which agreement could not be reached, or (ii) beginning from a short agreed list [such as those given in the attachments to this Appendix D] and agreeing on further genera to be included' (CGRFA, 1996a). In fact, the second approach appeared to form the only viable option.

Before an agreement on the composition of the list of crops could be reached, it was necessary to agree on the identification of the crops concerned. The technical advice that was offered resulted in the adoption of the genus as a biological identifier of the crop, since in almost all cases the gene pool – that is, all of the species with which the crop can be hybridized with more or less difficulty – is fully included in the crop genus (CGRFA, 1998a).[10]

The first list that can be found in the documents on the negotiations of the ITPGRFA is the one that was appended to the report of the sixth session of the CGRFA in June 1995 (CGRFA, 1995, Annex 1 to Appendix 1). This list of 231 crops is presented as an 'example proposed during the sixth session of the Commission of a list of genera important for food and agriculture' and is subdivided in 21 crop categories, including major grain crops, minor grain crops, major grain legumes, minor grain legumes, cereals from other families, major starch crops, minor starch crops, oil crops, fruits, shrub fruits, tree fruits, vegetable crops, nuts, spices, herbs, beverages, fibres, sugar crops, industrial crops, forage grasses and forage legumes. This list, however, seems to have only functioned as a reference. It was never seriously discussed. In 1996, during the third extraordinary session of the CGRFA, various submissions were made regarding the composition of the list (CGRFA, 1996b). The submission of the United States proposed inclusion of those plant genetic resources for which there was a global interest in maintaining unrestricted access, and it listed 25 crops plus an unspecified number of forage crops. Brazil submitted a proposal based on the relevance of crops for world food security, human world food consumption and greater interdependence worldwide. This proposal included 25 crops, of which 20 were shared with the list proposed by the United States.[11] A submission made by France made no reference to a list, since France and the European region in general had major objections to a list since they were fearful of limitations to diversification, human dietary needs and global food security. The fourth negotiating draft still contained the original list of 231 species/crops, which was tabled in 1995 by way of example.

In 1998, a new draft text for the Treaty appeared as a consolidated negotiating text. To this text, a new Annex 1 was attached (CGRFA, 1998b). This list contained 37 crops, including all crops earlier proposed by North America and the Latin American group as well as some additional crops. In addition, the list of grass and legume forages had been revised and now reflected 28 grass forages and 33 legume forages. In response to this list, five regions proposed varying draft lists of crops during the third inter-sessional meeting of the Contact Group, which took place in August 2000. The chair of this meeting opened the floor for statements from the regional groups addressing the list of crops covered by the multilateral system (CGRFA, 2000c). The Latin American group stated that consensus on a combined list of crops relevant for food security had been reached at a recent regional meeting. The region noted that their submission of 29 crops to the Contact Group was subject to conditions, including the adoption of a mechanism for benefit sharing, the consensual adoption of the ITPGRFA text, the use of genetic material covered under the multilateral system only for food purposes and acquired access to genetic material relevant to the private sector. The African region announced submission of a limited list of nine crops. To the African region, this smaller list was to be seen as a 'trial sample' that would allow countries to build trust and gauge the levels of success and failure before the list was expanded. In the view of some of the other delegations, this African proposal not only severely restricted the level of access but, at the same time, also limited the potential for benefit sharing. The European regional group recalled its position that the multilateral system should apply to all plant genetic resources for food and agriculture, noting that world food security reached beyond the small number of large crops grown across the world. The Europeans underscored the need for inclusion of a wide list of plant genera, which would encompass 287 crops compiled through consultations with stakeholders at all levels. Among the other delegations, the European proposal raised concerns since this proposal was seen to be an attempt to open access to an expanded range of genetic resources, which were not all directly related to food security. However, according to the European group, an expanded list might in fact be more beneficial to developing countries since they had been the major net recipients of PGRFA over the past decades. Asia submitted a list of 22 crops based on the criteria of global food security and interdependence, noting that their list reflected consensus in the region, although some countries in the region (notably Japan and South Korea) had preferred expanded lists. This region also stated that future additions to their list would depend upon progress in implementing the ITPGRFA – in particular, the components addressing the multilateral system and access and benefit sharing in general. The North American regional group supported the list of 37 crops appearing in Annex 1 of the composite draft text.

The sixth meeting of the Contact Group, which took place in April 2001, took up the issue of the list of crops, and a technical group on the list of crops was established that reported on its work at the end of the session (IISI, 2001b). This technical group made substantial progress. It worked on the premise that all regions could propose the inclusion of crops. Those crops to which no other region objected would stay, otherwise they were removed. This approach, which was probably the only realistic one, caused crops to be removed rather easily. The reservation of only a single region or country was sufficient for the removal of a crop from the consensus list. The working group reported consensus on a list of 30 crops and considerable support for a further 17 crops (crops for which only one country or region had placed reservations). It also reported that much more discussion was needed on the list of forage crops, and it asked for technical advice regarding the genera by which the agreed crops could be properly identified. The list of 30 crops lacked a number of important crops that were featured in the earlier version attached to the composite draft text in August 2000 – namely soybeans, peanuts, sugar cane, tomatoes, onions, melons, pumpkins, flax and cotton and oil palm – while a few new crops had been added, including apples and triticale (a hybrid between wheat and rye). This list of 30 crops was quite similar to the final list, which was established in the sixth extraordinary session of the CGRFA, wherein the majority of the text of the ITPGRFA was agreed upon. Late in the proceedings of this meeting, breadfruit, asparagus, strawberries, grass peas and eggplant were added, and so the final list comprised 35 crops in total.

The final discussions on the expansion of the list of crops took place within the G-77 and China between developing countries (Latin America, Africa and Asia). Although several attempts to make changes were made in the final hours of the session of the FAO Council in October and November 2001, no further changes were agreed upon. The final list contained 35 food crops. Much less discussion was spent on the list of forage crops, although a large number of the originally proposed forage crops were also objected to and removed from the list. Of the 28 proposed forage grasses and the 33 originally proposed forage legumes, 12 forage grasses and 15 forage legumes, as well as two other forage crops, remain.

In the last stages of the negotiations on the list of crops, many delegations changed gears. Negotiations did not so much centre on which crops countries and regions had to offer to the multilateral system but, rather, which crops countries and regions wanted to obtain from the multilateral system for their own agricultural development. This outcome led to the removal of a number of crops in the last stage of the negotiations. Attempts by the European region to agree on a regular and scheduled revision of the list of crops also were not successful. In line with the draft version of the ITPGRFA's article on the coverage of the multilateral system negotiated during the eighth regular session of the CGRFA in 1999, which made reference to a periodic review of Annex 1 as well as of the provisions on access, benefit sharing and financial arrangements, the region proposed adding an article stating that the multilateral system would be including the crops listed in an additional annex of crops (Annex E) after five years following the treaty's entry into force, provided that the funding strategy and the SMTA had been adopted by then. A number of developing countries stated that the successful review of the treaty's financial mechanism, benefit-sharing provisions and SMTA would be necessary before the list could be expanded. Another proposal by the European region for a biannual review of the list was not accepted either (IISI, 2001a).

When the European region asked for consideration to be given to expanding the list, developing countries responded by calling into question the need to review the mechanisms such as the funding strategy and commercial benefit sharing, in order to ensure an appropriate balance in securing their perceived gains under the Treaty. The rather limited composition of the list served to anger several developed countries, which felt that the developing countries were not really dedicated to the cause of the Treaty and the issue of global food security. Many developing countries responded by saying that trust was needed before the list could be revisited and possibly expanded. This response was a reference to the outcome of the negotiations on other components of the ITPGRFA – in particular, the provisions on benefit sharing and financial arrangements and the interpretations of the text associated with intellectual property rights.

Some reflections on the future

The contents of the list were one of the last things negotiators struggled over. In the lead up to the adoption of the Treaty's text, countries made last-minute deals about the inclusion of some species and unilaterally withdrew their consent to the inclusion of others. Over the course of the negotiations and across regions, the potential scope of the list expanded and contracted dramatically, shifting from possibly including all agricultural plants to only comprising a narrow list of just nine species or genera.

The negotiations on the list of crops were characterized by two extreme positions taken by Africa, which favoured a very short list on the one hand, and Europe, which favoured a very comprehensive list on the other hand. Other regions kept to the middle ground or remained sitting on the fence for a long time. Regions objected to the inclusion of certain crops not because these crops were considered to be unimportant for food security but, rather, because the genetic resources of these crops were considered to be, first and foremost, economic assets for those particular regions. This is how soybean and groundnut, sugar cane and oil palm disappeared from the list, even though they had been featured for a considerable period. It also explains why some crops that no country would seriously consider to be of utmost importance to food security are included on the list, such as asparagus and strawberries. Ultimately, no region objected to a proposal to include these crops, and so they stayed on the list.

This chapter has examined the relationship between the evolution of the negotiators' thinking about the scope of materials covered in the multilateral system and the way that other, more fundamental issues were resolved over the course of the negotiations. Looking back at the negotiations of Annex 1, one can attempt to consider the possibility of adding crops to the Annex 1 list. However, no specific provisions to regulate such a revision of the list have been incorporated in the text of the ITPGRFA, although text to that effect was proposed in the final stages of the negotiations. While changes in the composition of the list of crops are possible, such changes will require the consensus of the parties to the Treaty. Given the sharp divides between regions based on fundamentally different perspectives and expectations, such revisions are not likely in the short term. Only if the implementation of the ITPGRFA in general, and the benefit-sharing arrangements and the funding strategy in particular, appear to be successful and no negative effects result from the execution of the intellectual property rights, might the occasion arise for an expansion of the list of crops. Growing food insecurity might add to the willingness of countries to re-examine the composition of the list.

One could argue that the adoption of the draft Strategic Plan for the Implementation of the Benefit-Sharing Fund of the Funding Strategy has built the basis for improved benefit sharing between regions, which will benefit developing countries, but the five-year target of US $116 remains to be reached in the next few years. Only when this target is realized might there be sufficient trust between regions and confidence in the acquisition of future benefits to move towards an extension of the list.

While the need to adapt to climate change is likely to dramatically increase interdependence between countries and regions for their plant genetic resources for food and agriculture, it is likely that policy makers will need several years at least to analyse the full effects of climate change on agricultural production in their countries and to realize that it will further increase their interdependence on plant genetic resources from other countries and regions in an unprecedented way. Although these two developments might foster a re-evaluation of the list of crops, another development of importance will have a less predictable outcome – namely, the volume and scope of intellectual property rights in general and patents in particular as well as the degree by which these issues will influence the access of developing countries to modern breeding materials. Ultimately, these factors will have an enormous influence on the willingness of the parties to re-examine Annex 1. As a result, no early and easy re-examination of the list of crops in Annex 1 is to be expected on the basis of changed circumstances. In short, expansion can and will come hand in hand only with increased trust in the implementation of the ITPGRFA and with a realization of all of the benefits of facilitated access.

Notes

1 International Treaty on Plant Genetic Resources for Food and Agriculture, 29 June 2004, www.planttreaty.org/texts_en.htm (last accessed 30 March 2011).
2 Convention on Biological Diversity, 31 ILM 818 (1992) [CBD].
3 The Commission on Genetic Resources for Food and Agriculture was known as the Commission on Plant Genetic Resources until 1996.
4 CBD, *supra* note 2, Article 15 and Article 2 respectively.
5 International Undertaking on Plant Genetic Resources for Food and Agriculture, 1983, www.fao.org/waicent/faoinfo/agricult/cgrfa/IU.htm (last accessed 30 March 2011).
6 Also known as Centres of the Consultative Group on International Agricultural Research or CGIAR centres.
7 The recently adopted Nagoya Protocol on Access and Benefit Sharing and the Fair and Equitable Sharing of Benefits Arising from Their Utilization, 29 October 2012, www.cbd.int/abs/text/ (last accessed 30 March 2011).
8 Standard Material Transfer Agreement, 16 June 2006, ftp://ftp.fao.org/ag/agp/planttreaty/agreements/smta/SMTAe.pdf (last accessed 30 March 2011).
9 This document was the non-legally binding precursor of the ITPGRFA.
10 In the case of Annex 1, notable exceptions are formed by the *Brassica* complex and wheat, where other genera than the crop genus also provide for hybridization, and *Citrus* where related genera provide the root stocks.
11 Differences between the lists proposed by the United States and Brazil regarded tanier, taro, chickpeas, pigeon peas, oranges, sugar beets, sugar cane, pumpkins and tomatoes.

References

Commission on Genetic Resources for Food and Agriculture (CGRFA) (1994a) *Revision of the International Undertaking, Mandate, Context, Background and Proposed Process*, First Extraordinary Session, Doc. CPGR-Ex1/94/REP, FAO, Rome, online: www.fao.org/nr/cgrfa/cgrfa-meetings/cgrfa-comm/first-extra/en/ (last accessed 30 March 2011).

——(1994b) *Revision of the International Undertaking*, Doc. CPGR-Ex1/94/4, FAO, Rome, online: ftp://ftp.fao.org/docrep/fao/meeting/016/aj684e.pdf (last accessed 30 March 2011).

——(1995) *Report of the Commission on Plant Genetic Resources, Sixth Session*, Doc. CPGR-6/95/REP, FAO, Rome, online: ftp://ftp.fao.org/docrep/fao/meeting/015/aj595e.pdf (last accessed 30 March 2011).

——(1996a) *Report by the Chairman of the Eleventh Session of the Working Group of the CGRFA*, Doc. CGRFA-Ex3/96/Rep, Appendix D, FAO, Rome, online: ftp://ftp.fao.org/docrep/fao/meeting/016/aj704e.pdf (last accessed 30 March 2011).

——(1996b) *Report of the Commission on Genetic Resources for Food and Agriculture*, Third Extraordinary Session, Doc. CGRFA-Ex3/96/Rep, Appendix D, Attachments 1–4, FAO, Rome, online: ftp://ftp.fao.org/docrep/fao/meeting/016/aj704e.pdf (last accessed 30 March 2011).

——(1997a) *Characterization of Various Options for Scope and Access, and a Notional Assessment of the Pros and Cons*, submitted by the International Plant Genetic Resources Institute and the Food and Agriculture Organization, Doc. CGRFA-7/97/9, FAO, Rome, online: ftp://ftp.fao.org/docrep/fao/meeting/015/aj611e.pdf (last accessed 30 March 2011).

——(1997b) *Report of the Fourth Extraordinary Session of the Commission on Genetic Resources for Food and Agriculture*, Doc. CGRFA-Ex4/97/Report, FAO, Rome, online: ftp://ftp.fao.org/docrep/fao/meeting/014/aj593e.pdf (last accessed 30 March 2011).

——(1998a) *Technical Aspects Involved in Developing a List of Crops for the MLS within the Revised International Undertaking: Relevant Characteristics of the Crops and Genera in the Tentative List of Crops Annexed to Article 11 of the Consolidated Negotiating Text*, Docs. CGRFA-Ex5/98/Inf.1 and CGRFA-Ex5/98/Inf.1/Annex, FAO, Rome, online: ftp://ftp.fao.org/ag/cgrfa/ex5/E5I1E.pdf (last accessed 30 March 2011).

——(1998b) *Consolidated Negotiating Text resulting from the Fifth Extraordinary Session of the CGRFA*, Doc. CGRFA-Ex5/98/REPORT, Appendix C, FAO, Rome, online: ftp://ftp.fao.org/docrep/fao/meeting/014/aj587e.pdf (last accessed 30 March 2011).

——(1999) *Composite Draft Text of the International Undertaking on Plant Genetic Resources, Incorporating the Chairman's Elements*, Doc. CGRFA-8/99/13, Annex, FAO, Rome.

——(2000a) *Statement by the Brazilian Delegation*, Doc. CGRFA/CG-2/00/TXT, Appendix B, FAO, Rome, online: ftp://ftp.fao.org/docrep/fao/meeting/014/aj538e.pdf (last accessed 30 March 2011).

——(2000b) *Information Provided by the Regions on the List, during the Third Intersessional Meeting of the Contact Group*, Doc. CGRFA/CG-4/00/Inf.4, FAO, Rome, online: ftp://ftp.fao.org/docrep/fao/meeting/014/aj530e.pdf (last accessed 30 March 2011).

——(2000c) Composite Draft Text of the International Undertaking on Plant Genetic Resources for Food and Agriculture, incorporating the Texts of Articles 11, 12 and 15, Negotiated during the Commission's Eight Regular Sessions and the Texts of Articles 13, 14 and 16 Negotiated during the First and Second Intersessional Meetings of the Contact Group. Doc. CGRFA/CG-3/00/2, FAO, Rome, online: ftp://ftp.fao.org/docrep/fao/meeting/014/aj536e.pdf (last accessed 30 March 2011).

——(2001) *Report of the Commission on Genetic Resources for Food and Agriculture*, Sixth Extraordinary Session, Doc. CGRFA-Ex6/01/REP, Appendix E, FAO, Rome, online: ftp://ftp.fao.org/docrep/fao/meeting/014/aj357e.pdf (last accessed 30 March 2011).

Food and Agriculture Organization (FAO) (1997) *State of the World's Plant Genetic Resources for Food and Agriculture*, FAO, Rome, online: ftp://ftp.fao.org/docrep/fao/meeting/015/w7324e.pdf (last accessed 10 January 2011).

International Institute for Sustainable Development (IISI) (2001a) 'Negotiations on the ITPGRFA', *Earth Negotiations Bulletin* 9(213), online: www.iisd.ca/vol09/enb09213e.html (last accessed 30 March 2011).

——(2001b) 'Sixth Inter-sessional Contact Group on the revision of the International Undertaking on PGRFA, in harmony with the CBD', *Earth Negotiations Bulletin* 9(191), online: www.iisd.ca/vol09/enb09191e.html (last accessed 30 March 2011).

——(2001c) 'Summary of the sixth extraordinary session of the CGRFA', *Earth Negotiations Bulletin* 9(197), online: www.iisd.ca/vol09/enb09197e.html (last accessed 30 March 2011).

15

BUILDING A GLOBAL INFORMATION SYSTEM IN SUPPORT OF THE INTERNATIONAL TREATY ON PLANT GENETIC RESOURCES FOR FOOD AND AGRICULTURE

Caroline Ker, Selim Louafi and Myriam Sanou

Introduction

The International Treaty on Plant Genetic Resources for Food and Agriculture (ITPGRFA) establishes a multilateral system of access and benefit sharing (multilateral system), which pools together material from national, regional and international gene banks.[1] It consists of a global pool of genes from 64 major crops and forages, which may be obtained for utilization and conservation in research, breeding and training, on the basis of a standard contract agreed multilaterally. The multilateral system can be seen as the most integrated and collective way of managing globally and in a coordinated way a distributed pool of genetic resources. In this regard, the multilateral system could be seen as the most advanced expression of the willingness and need for global cooperation on the management, conservation and distribution of plant genetic resources for food and agriculture (PGRFA).

The ITPGRFA has allowed its contracting parties to move from a common interest in a global resource (similar to the approach of the Convention on Biological Diversity) to a shared and assumed responsibility for their global management.[2] Information about PGRFA is an essential element to enable the multilateral approach followed by the Treaty to efficiently function. Indeed, information plays a key role in facilitating the exchange of multilateral system material by providing the precise details of what each gene bank holds and where it holds it. As a non-monetary benefit-sharing component, the exchange of information also plays an important political role by increasing knowledge about plant genetic resources worldwide. Information exchange, which is contained in Article 17 of the ITPGRFA, is also a supporting component of the Treaty's implementation and covers a wide range of needs and expectations that the Treaty has managed to pool together. The global information system mandated in Article 17 constitutes a means of cooperation that exists in addition to the multilateral system, which is the central element of the

Treaty. If properly designed and implemented, it could constitute a dynamic platform for evidence-based decision making with respect to policy and practice in PGRFA conservation and use. The objective of this chapter is to shed some light on the challenges that face the global information system in addressing the fragmented sources of data, users and knowledge domains of PGRFA.

We argue in this chapter that the foundation of the global information system in the multilateral framework of the ITPGRFA is both an opportunity and a justification for building an informational commons for plant-related information. This chapter provides an overview of the technical and legal challenges in making the global information system functional. The first part describes the possible range of information that could be part of the global information system, its intended users and beneficiaries and its technical constraints and challenges. The second part covers the legal challenges of establishing a fully integrated information commons.

An informational platform on 'scientific, technical and environmental matters'

The global information system was established in Part V of the ITPGRFA, as a supporting component of the Treaty. Article 17 of the Treaty is drafted in very general terms stating that the global information system will facilitate the exchange of information on scientific, technical and environmental matters related to PGRFA. This description is arguably much wider in scope than it would need to be if it was only dealing with the multilateral system itself.

Content of the global information system

Information related to the multilateral system

The multilateral system is the core element of the ITPGRFA, around which all of the other substantive elements are more or less related. The multilateral system provides rules that coherently articulate into one single global legal framework the physical genetic material (that is, the germplasm) and the information about it. In addition, it provides the legal status for the innovations that are based on, and embodied in, this genetic material. The multilateral system also has several features that provide specific and essential roles for a variety of information. First of all, the multilateral system is not a physical centralization of material in one single global Annex 1 gene bank. Rather, it is a distributed system of gene banks gathered together according to a set of common agreed rules of access and benefit sharing. It consequently depends on numerous actors around the world who put the material that they physically hold into the common global gene pool that the multilateral system constitutes. In addition, this material – and the information about it – is constantly evolving as a result of exchange, evaluation, characterization and plant breeding. The multilateral system is consequently not a static and finite set of elements but, rather, a dynamic and infinite entity.

These two qualities of the multilateral system (distributed and dynamic) ensure that it is only as useful as the information that is contained in it at any time. Three categories of information can be distinguished:

- information about plant genetic resources (documentation of genetic resources in the multilateral system and their characteristics, including where they can be found physically);
- information about agreements that have been entered into and the material that has been transferred (by way of the Standard Material Transfer Agreement (SMTA)),[3] which includes the minimum information legally needed to initiate a dispute settlement procedure; and
- feedback information about the use of the material (non-commercial benefit sharing).[4]

Category 1 and 3 might cover in substance the same items of information. However, the difference would be that Category 3 provides the dynamic part of the multilateral system since it describes the information added about the use of the material and associated information that is derived from Category 1. Category 1 and 3 are covered by Article 13.2(a), which requires that information is made available through the global information system, 'which shall, *inter alia*, encompass catalogues and inventories, information on technologies, results of technical, scientific and socio-economic research, including characterization, evaluation and utilization, regarding those plant genetic resources for food and agriculture under the Multilateral System'. Category 3 is also covered by Article 6.9 of the SMTA, which obliges the recipient of a given PGRFA to 'make available to the Multilateral System, through the information system provided for in Article 17 of the Treaty, all non-confidential information that results from research and development carried out on the Material'. This obligation serves to increase the number of requests as well as the supply of resources that are in the system, and it also increases the PGRFA's value thanks to the input of new available information related to such resources. This process is fully in line with the benefit-sharing principle that mandates that access to the resource must benefit all of the players in the system.

Consequently, Category 3 information serves to enrich Category 1 information by including more qualitative details that serve the various Treaty objectives in support of sustainable agricultural development, food security, conservation and sustainable use of PGRFA, including information about organic alternatives to chemical inputs, the results of genetic analysis that have been performed on varieties and so on.

Role as a knowledge base on PGRFA and an early warning system

Since the scope of the global information system is clearly not only limited to the PGRFA under the multilateral system, it also comprises the information on all PGRFA, pursuant to Article 5.1.[5] As a supporting component of the ITPGRFA and bearing in mind the objectives of promoting *in situ* conservation (Article 5.1(d)), the conservation and sustainable use of PGRFA (Articles 1 and 6.2) and farmers' rights (Article 9), the constitution of the global information system may provide an ideal tool for enabling the exchange of information on all of the dimensions and types of knowledge associated with PGRFA, from conservation to product development.[6] The provision of information management systems lies at the heart of all conservation and use activities. It supports effective and sustainable complementary PGRFA conservation and encourages and facilitates the use of PGRFA diversity for crop improvement. For instance, for the purposes of conservation and sustainable exploitation, data on the spatial patterns of biodiversity (species distribution and infra-specific or ecosystem variability in space) are desirable (Harrer et al., 2002).

Beyond this accession-level information, which is useful for the exchange of PGRFA, the global information system could also contribute to providing information on the various dimensions of agricultural biodiversity in order to stimulate communication within the global agro-biodiversity community (farmers and indigenous communities, researchers, maintainers, donors and decision makers). Its contribution to such an endeavour could be made by improving the accessibility to information about various PGRFA crosscutting issues, which are addressed in structured (databases) and unstructured systems (publications, web portal and so on). Examples of useful information that the global information system could help to provide are, *inter alia*, the identification of hotspots of diversity for crop wild relatives (CWR) and landraces; the management and monitoring techniques for CWR and landraces in protected areas and in on-farm systems; the experiences with neglected and under-utilized species valuation and the tools and practices relevant to using agro-biodiversity for coping with climate change.

In addition to collating or making accessible the existing information, the global information system could also provide a support in the longer term for value-added information by providing a synthesis of available information, assessing the major knowledge gaps and identifying new research areas that are more targeted to the needs of farmers and indigenous communities. Unlike the *ex situ* approach, the conceptual basis on which the *in situ* approach is based is still very fragmented and context specific. Supplying farmers with information on sustainable methods of cultivation would contribute to the preservation and development of *in situ* conservation knowledge and, more generally, to the realization of farmers' rights since such rights notably aim at enabling farmers to maintain and develop crop genetic resources (Fridtjof Nansen Institute, 2009).[7] More generally, it has been determined that it is necessary to encourage cooperation between the different stakeholders in order to implement successfully *in situ* conservation (Brush, 1994). In this way, farmers, scientists, national agricultural research services, international centres such as Bioversity International and non-governmental organizations might advantageously share information on *in situ* conservation topics.[8] By providing farmers with access to, and the ability to share, such knowledge, the global information system could be considered in itself an element of the benefit-sharing scheme. It would also be in line with Article 13.3, which states that the benefits of the multilateral system should flow primarily to farmers who conserve and sustainably utilize PGRFA, at least as far as non-monetary benefits are concerned.

More generally, the global information system could also function as a broad capacity development mechanism by providing training material and information services to support the establishment and/or development of national programmes for PGRFA through institutional strengthening and human resources development. This support might include, for instance, PGRFA projects databases or any tools that could help to identify needs in relation to PGRFA. Policy makers could potentially draw a substantial profit from a database that inventories the policies that other states have experimented with in implementing Articles 5 and 6 on conservation and sustainable use or Article 9 on farmers' rights.

Finally, a global information system for PGRFA could contribute to such an early warning mechanism by enabling and facilitating the monitoring of bio-indicators relating to the loss of landraces and monitoring qualitative and quantitative indicators in sensitive regions of wild habitat (FAO, 2002).

Article 17.2 of the ITPGRFA specifically mentions the provision of early warning in the case of threats to the efficient maintenance of plant genetic resources. Examples of such risks might include:

- the displacement of farmers' varieties by modern varieties and the loss of landraces;
- the destruction of habitats of wild relatives and of agricultural landscapes and traditional production systems; and
- the destruction of *in* or *ex situ* collections and on-farm conservation and sustainable use practices.

Institutional arrangements

Principles

Two principles stated in Article 17 of the ITPGRFA are worth recalling: (1) the necessity to base the global information system on existing systems and (2) the need for inclusiveness. The global information system does not operate in a vacuum, it is clearly important for it to avoid adding another layer of information that could compete with, or duplicate, existing ones. The informal working group that was brought together by the Food and Agriculture Organization (FAO) in May 2002 to discuss the global information system recommended bringing together 'relevant information systems/services in a manner and/or by means of a mechanism, to allow access the corporate information sources (that is, the joint information holdings of these information systems/service) and enable and facilitate the exchange of information between these systems/services, and others' (FAO, 2002). There are already several initiatives existing at the global (both international – that is, state-driven – and transnational – for example, crops networks), regional and national levels (we will come back to some of these initiatives in the second part of this chapter). The experience accumulated, both at the conceptual (design) and technical levels, is a clear advantage that the global information system could benefit from. It should be noted, however, that such a plethora of initiatives is also a factor in the duplication of efforts and overlaps, which, in turn, increases the cost of coordination at the global level. Coordination challenges involve activities such as technical inter-operability (through standards and common format) and technological capacity building to address the huge discrepancies between countries regarding a lack of technical and personal capacity. The challenges posed by the existence of private rights on the information have to be taken into account as well.

As a supporting component of the ITPGRFA, the global information system should also adequately cover all dimensions of PGRFA, thereby administering to the information needs of the widest range of users. The Treaty has indeed managed to pool together, in one single global framework, the interests of a wide range of stakeholders involved in the use and conservation of PGRFA. The global information system that it aims to create could be seen as the glue that holds the different substantive issues/topics together and forms the link between them. By using the opportunities that are offered by the new technology of information, the global information system could indeed offer the possibility of addressing some of the coordination challenges that are linked to the distributed nature of the different information providers and users (horizontal integration) at their various levels of governance (vertical integration).[9]

Horizontal integration refers to the coordination needs between crops, diversity levels (from gene to habitats), policy issues and knowledge domains. The ITPGRFA has much less to say about vertical integration (integration of the different levels of governance from local (crop specific, sector specific or territorial) to global). Nevertheless, from the existing information science literature as well as from prior experiences within the global information system, one can already infer the following basic principles, namely that the information should be (1) managed as close as possible to its source; (2) readily available and easy accessible; (3) structured and searchable in such a way that it is useable by a broad range of users in different countries; and (4) supported through common, free and open software standards.

Users and beneficiaries

It is commonly understood in the literature of information science that the distinction between production, distribution and consumption of knowledge is more and more blurred. The technical possibilities offered by the Internet have accentuated this tendency. The success of the global information system will depend on the direct participation of the four corresponding stakeholder groups:

- the research and scientific community – in particular, the public agricultural research sector;
- the users, including commercial users, of genetic resources such as industry, plant breeders and so on;
- the government decision makers involved in PGRFA policy activities; and
- the conservers and developers of genetic resources on-farm – that is, the farmers and farming communities.

TABLE 15.1 Potential users and beneficiaries of the global information system

	Beneficiary	Main purpose
Global level	International research networks, including the International Agricultural Research Centres	Knowledge sharing
	Multilateral partners, donors and bilateral aid agencies	More coherent and better coordinated technical and financial support to country-led plans and programmes
	Governing Body	Monitoring informed decisions
National and local levels	Governments and their associated institutions involved in genetic resource policies	Better national plans regarding the enrichment of gene pools of crops that are essential for national food security
	Gene bank managers and researchers/breeders in the public sector and breeders in the private sector	Facilitated access to material and access and sharing of knowledge, information and data
	Representatives of civil society	Better participation in, and contribution to, policy debates
	Farmers	Informed decisions, increased participation and facilitated access and sharing of benefits and information
	Organizations and other development partners that participate in country-led processes	Improved coordination and better project design

Several groups could potentially benefit from a PGRFA global information system at the different levels of governance. Table 15.1 summarizes these different beneficiaries according to their level of governance.

Technical challenges and opportunities

Considering the diversity of actors involved at the different levels of governance (local, national, regional and global), it would be foolish to think that the global information system could meet the enormous need for information in a comprehensive and organized way. Prior international experience with establishing and managing a global information system shows that several challenges have to be addressed: (1) the lack of equal technical capacity between actors and countries and a scarcity of human resources; (2) the lack of incentives to share information and/or fragmented reporting responsibilities; (3) the existence of different sensitivities (sovereignty, ownership, secrecy and confidentiality); (4) a fragmented knowledge domain and lack of common formats, protocols and standards to ensure inter-operability; (5) scattered data and information and shortcomings in relation to timeliness, availability, reliability and the relevance of the information; (6) inadequate coordination and initiatives that overlap and lead to a duplication of efforts; and (7) shortcomings in the ability to turn data into policy-relevant information.

These challenges have been well documented in the literature, and none of them can be easily addressed in a one-size-fits-all approach (Alavi and Leidner, 2001; Brown and Duguid, 2001; Pan and Leidner, 2003). Indeed, even if none of these issues can be entirely ignored, they do not all apply to the same degree in the case of the global information system for PGRFA. Some of the specificities of the ITPGRFA make some of these challenges probably less acute than others. First of all, the Treaty includes some legally binding obligations in relation to reporting requirements and information sharing. These reporting requirements regard taxon-related information with respect to the use of material from the multilateral system and are derived from the SMTA.[10] Much of this information is of prime importance to the third party beneficiary in initiating dispute settlement, should that become necessary. However, beyond this legal purpose, part of these reporting requirements is also a 'natural' component of non-confidential information sharing. The Treaty very much emphasizes this aspect by making exchange of information an essential element of benefit sharing. Indeed, it distinguishes three non-monetary benefit-sharing components. They are essential elements of equity in the access and benefit-sharing 'grand bargain', especially in the case of the Treaty's multilateral system, whereby the monetary element is expected not to be as important as it is in the case of wild genetic resources. In addition to the legal obligation mentioned earlier, one can then also rely on a kind of political or moral obligation in relation to the sharing of information.

In addition, the entire set of rules governing the multilateral system is based around the concept of 'accession' used by gene bank managers worldwide. Accession can include a lot of seeds, *in vitro* plantlets or tissues. The germplasm is what is actually exchanged, and it corresponds to a portion of the accession. The accession bears a unique identifier that is attached to all of the derived samples, enabling anyone to identify the relationship of the sample to the accession. The accession consequently constitutes the common 'language' and common smallest unit in *ex situ* conservation: (1) identifying the gene bank entry from which the derivate samples are created and exchanged physically across borders and (2) facilitating the ability for much information to be attached to it. In terms of information management, it is undoubtedly an extremely important asset to have such a pre-established 'standard' to which the various levels of data can be attached. It constitutes, with minimum transaction costs, the first level of technical inter-operability necessary in the attempt to establish a global information system.

To conclude, the global information system on PGRFA, as envisaged under Article 17 of the ITPGRFA, could consist of a decentralized network of systems rather than one single system, with equally decentralized management. It would be a 'community of practice', including its information providers and partners, which are diverse and disparate in the same way as its users. This structured networking and broker approach incorporates information on the integration of other supply sources at the search level, using the decentralized capabilities of the Internet. The global information system could establish quality criteria for the selection of cooperating sites and for the standards of information and data that is provided. However, this work is limited in the case of PGRFA information by the fact that international standards already exist on PGRFA. In addition, the information requirements decided by the Governing Body provide a basis for harmonization.

However, technical opportunities and solutions have never been sufficient on their own, and it should be acknowledged that the information needs are very heterogeneous and, while the integration of several layers of information is always sought, it is not always easy to combine them in a way that is useable by the different stakeholders. It is also worth noting that, on the supply side, data put into the public domain are scattered across an increasing number of knowledge domains, which make their interoperability quite challenging to concretize. Furthermore, the existence of political, institutional and individual sensitivities on the way some data are managed and their flow organized have often served to impede the full possibilities offered by information and communication technologies. By providing specific examples, the following section discusses how, from a legal perspective, the 'commons' approach could allow parties to overcome these sensitivities. Some lessons are drawn for the global information system in the conclusion.

Legal challenges in building an information commons

While the ambition of the global information system is to ease access to a very wide range of useful information for research and the sustainable use of PGRFA, the availability of such information may be limited by various proprietary rights. Both private and public entities are entitled to such rights.

In this section, we will first present the different proprietary rights that might influence the realization of the global information system. We will then present some legal responses to address the challenges posed by such rights, looking for inspiration in the existing provisions that will encourage the publication of the results of public-funded research. We will also demonstrate that the protection of traditional knowledge and farmers' rights under the ITPGRFA provides a resource in accommodating the protection of such knowledge and its availability under the global information system. Finally, examples of an 'informational commons' will provide an interesting mechanism for the rationalization and sharing of scattered informational resources. A focus on the characteristics of these commons will enable us to appraise their relevance with the objectives of the Treaty.

A general overview

Information on PGRFA is under the control of its holders, which include states, public or private gene banks, public or private research centres, private breeders, farmers and so on. Such control is mandated by different legal sources. The availability of information through the global information system will be possible under these constraints of ownership.[11] The control of information might come from the exclusive rights that holders have on such information. Intellectual property rights comprise such exclusive rights. They grant the rights owner a monopoly on the exploitation of the protected subject matter – be it an invention, a work, a plant variety, software or whatever. From an economic point of view, the policy justification for such a monopoly is the encouragement of investment in the production of goods and inventions. When information on PGRFA is protected by intellectual property rights, the authorization of its rights holder is necessary to allow access and use of the information through the global information system.[12] The exclusivity that characterizes intellectual property rights implies that such authorization by the rights holder is necessary even if the information is available elsewhere – for instance, in an existing database.

Copyright might protect information relevant for the global information system. Authorization of the copyright holder, for instance, will be necessary to provide access to scientific literature through the global information system. It must be noted that while copyright protects a great deal of material, it does not extend to taxonomic data and biological nomenclature (Agosti and Egloff, 2009, 6).[13]

Databases might as well be protected by intellectual property rights. Depending on the countries, databases might be protected by copyright, and in the European Union databases are protected both by copyright[14] and a *sui generis* right.[15] When applicable, intellectual property rights in databases require the authorization of the rights holders of the database to make it available through the global information system, whether for the selection and use of some data it contains or for pooling it with another database. The integration or connection of databases into the global information system will therefore require the authorization of their original rights holders, even if such databases are already openly accessible.[16]

Control on information might also result from a non-exclusive right. One example of non-exclusive right comes out of the factual control that a holder has on a piece of information (assuming that she or he has no intellectual property rights on this information). It gives the individual the discretion to authorize access and use of information that is in its possession only (and because it is in its possession) and to stipulate conditions for such access through contractual means (by stipulating a fee, personal authorization, a confidentiality clause, a commercial use exclusion and so on). Such non-exclusive rights are thus exercised through contractual rights. A secrecy policy can, for instance, be implemented through such contractual means. The conclusion of non-disclosure or confidentiality agreements with third parties will enable the holder of the information to prevent third parties from disclosing the information although she or he has no intellectual property right on it. However, contrary to exclusive rights, non-exclusive rights cannot be asserted *erga omnes*. Contractual rights can only be asserted against parties that have agreed upon the contractual clauses with the information holder. A holder of information that has no exclusive rights on it can therefore not oppose access to, nor condition the use of, this information when it is in possession of another person willing to give access to it or when it is public.

Given such availability, it is not necessary for the user to ask for access from such a holder who has therefore no opportunity to enter into a contractual relation and impose contractual use conditions to the said user. This implies that if non-intellectual property-protected information is available without having to comply with any contractual conditions or terms of use, this information might be made available through the global information system. The collaboration of the holder of the information will be necessary only when the information is not available elsewhere with no, or more compliant, conditions. This constitutes the main distinction between non-intellectual property-protected information and intellectual property-protected information. However, it should be noted that some other legal constraints might marginally limit such breathing space. Tort law, for instance, will limit the boundaries of such freedom. The protection of undisclosed information, as mandated by Article 39 of the Agreement on Trade Related Aspects of Intellectual Property Rights (TRIPS Agreement), might also limit the possibility to lawfully integrate 'undisclosed' information in the global information system.[17] However, the conditions for protection of 'undisclosed' information according to the TRIPS Agreement make it unlikely to apply to the constitution of the global information system (this provision aims to protect undisclosed information, whose secrecy contributes importantly to its commercial value and has been the object of efforts to maintain its secret by its holder).

In a previous section, we suggested that the global information system would provide an opportunity to share and enhance the use of farmers' knowledge with respect to cultivation, conservation and breeding methods. If farmers' knowledge were to be shared through the global information system, this shall be done in compliance with the provisions related to the protection of traditional knowledge, which is enshrined in Article 8j of the Convention on Biological Diversity (CBD), Article 5bis of the Nagoya Protocol on Access and Benefit Sharing and the Fair and Equitable Sharing of Benefits Arising from Their Utilization, the farmers' rights of the ITPGRFA and the applicable national implementations of these international norms.[18] These provisions might condition access and/or use by third parties of traditional knowledge on the prior informed consent of local communities and the respect of mutually agreed benefit-sharing provisions. In such circumstances, and subject to national provisions, the consent of the concerned farmers might be required before giving access to tradition knowledge through the global information system, along with an agreement on a mechanism to share the benefits that might come out of the uses to be made of the knowledge. Yet what if the knowledge is already available elsewhere – for instance, in a public database inventorying traditional knowledge, or if it can be obtained from another entity which has already accessed such knowledge? The answer mainly depend on the national provision implementing the CBD and the Nagoya Protocol, and the mutually agreed terms to be negotiated with the holders of the knowledge (are prior informed consent and benefit sharing for the use of publicly-available traditional knowledge required, is third-party transfer authorized?).[19] In conclusion, it must be underlined that access and benefit-sharing national provisions might be diverse, and this diversity has an important impact on the observance of legal prescriptions that condition both the input of traditional knowledge in the global information system and the access and the use that it will enable.

Legal resources for the sharing of informational resources

In the context of numerous proprietary rights affecting information, what legal resources do we have to ensure that information is widely available through the global information system? Answers to this question involve the specific nature of the information on PGRFA that the global information system aims to pool together as well as the collaborative nature of the ITPGRFA's framework. We will first have a look at the specific case of the information generated through public funds and the legal initiatives that have already been taken in favour of the accessibility of such information. We will then consider the farmers' rights that have been established by the Treaty as a legal means to encourage the sharing of farmers' knowledge. Finally, we will investigate the resource that an informational commons scheme would offer to address property rights.

Public sector and publicly funded information

Following the same logic that applies to the physical plant material under the control of a contracting party and in the public domain that falls automatically under the multilateral system, data belonging to the public sector should be made easily available through the global information system. Article 13.2(b) requires contracting parties to make information on PGRFA available under the multilateral system, notably through the global information system. Although this stipulation is submitted to the applicable legislation, in connection to property rights, for instance, it would be inconsistent for a contracting party, which has ratified the ITPGRFA to assert its own property rights on such information, to deny making them openly available through the global information system. Article 5 and Article 17 are also relevant in this respect, along with the sharing of information as a benefit-sharing mechanism, stipulated in Article 13.2(a).

Large quantities of relevant PGRFA information is held by the international collections built from samples collected in the wild or on farms or donated and maintained by the International Agricultural Research Centres (IARCs) of the Consultative Group on International Agricultural Research (CGIAR). The *ex situ* germplasm collections held by the IARCs were put under the auspices of the Food and Agriculture Organization (FAO), according to the 1994 agreements signed by the centres and the FAO. According to the FAO-CGIAR agreements, the said germplasm is held in trust by the IARCs, which 'should not claim legal ownership over the germplasm, nor shall it seek any intellectual property rights over that germplasm or related information'. In accordance with this principle, there should be no obstacle to the availability of information related to germplasm held in trust by the centres.

However, the global information system could also benefit from a second kind of national or regional legal resource: the adoption of an open access policy for scientific data generated as a result of public funds. These regulations might be applicable to a great amount of data, given that they might be applicable to private entities holding publicly funded data. These regulations have been adopted in consideration of public policy objectives such as the advancement of science, education or transparency goals. Due to the fact that in many cases information on PGRFA in the global information system might be considered public sector information or scientific data, it is worthwhile to look more closely at the underlying principles and legal mechanisms envisaged by these open data texts.

With respect to the public sector information, the European Union has been without doubt one of the most active organizations in promoting its openness. It adopted EC Directive 2003/98 on the Re-Use of Public Sector Information in 2003.[20] This directive introduces a common legislative framework that regulates how public sector bodies must make their information available for re-use.[21] The Organisation for Economic Co-operation and Development (OECD) has also adopted a Recommendation for the Enhanced Access and More Effective Use of Public Sector Information (2009).[22] The recommendation contains some principles that go even further than the obligations of EC Directive 2003/98. In the field of scientific data, a very relevant text is the OECD Recommendation on the Principles and Guidelines for Access to Research Data from Public Funding (2007).[23] The main purpose of this text is to support the development of good practices in the area of access to research data and to foster their global exchange and use (Fukasaku and Pilat, 2007, 2). The recommendation cites the key principles that should be taken into account by the member states when developing their national strategies to promote access and use to research data from public funding. These key principles relate to openness, flexibility, transparency, legal conformity, the protection of intellectual property, formal responsibility, professionalism, inter-operability, quality, security, efficiency, accountability and sustainability.

In relation to the information contained in *ex situ* collections, the OECD Best Practice Guidelines for Biological Resource Centres (BRCs) also encourage the availability of information on the biological resources contained in their collections.[24] In some cases, it might be even considered that in those countries with public sector information laws, public sector information obligations might be applicable to the information managed or generated by the BRCs under the control of public entities. Open access has also been encouraged by the Conference of the Parties (COP) to the CBD. In COP-8 on Decision VIII/11 on Scientific and Technical Cooperation and the Clearing-House Mechanism, it invites the parties and other governments to provide free and open access to all past, present and future public good research results, assessments, maps and databases on biodiversity, in accordance with national and international legislation. And, finally, the Governing Body of the ITPGRFA has chosen a similar approach when deciding that information generated by projects funded through the benefit-sharing fund shall be made publicly available within one year of the completion of the project (FAO, 2009, Appendix A, 16). The adoption by the contracting parties of a regulation advocating the availability of publicly funded information through the global information system would consist of a useful means to implement Article 17 of the ITPGRFA since it would favour the availability of a large amount of information internationally, including information not held entirely by public bodies.

Farmers' rights

We have demonstrated earlier in this chapter that the global information system might also provide an opportunity to encourage the use, dissemination, exchange and development of farmers' knowledge about the sustainable use of PGRFA. However, both the philosophy that lies behind the access and benefit-sharing principle of the CBD and its national implementation must duly be taken into account. Article 9 of the ITPGRFA, which acknowledges farmers' rights, also provides for the protection of farmers' knowledge. Legal resources to address these challenges are also found in Article 9 itself. First, sharing farmers' knowledge within the global information system would provide simultaneously a catalogue of such knowledge. Cataloguing such information would contribute to the protection of such knowledge in accordance with Article 9.2(a), by protecting it from misappropriation and enclosure by third parties (in the form of a patent involving such knowledge, for instance). Second, the voluntary disclosure of farmers' knowledge via the global information system might also provide an opportunity to conceive of a benefit-sharing system for the use of such knowledge. This benefit-sharing mechanism would be similar to the one that falls under the multilateral system established for the use of PGRFA. Benefit sharing would result from the facilitated access to other contributing farmers' knowledge and best practices. Encouragement to share the monetary benefit arising from the use of farmers' knowledge and best practices could also be developed in accordance with the spirit of Article 13(d)(ii) of the Treaty. Finally, the publication of this knowledge would enhance its dissemination, further use and, thus, development, which would certainly contribute to its protection against perdition.

A 'commons' approach

The challenge of making information available for new research in a spirit of collective social welfare has already been faced by other initiatives. Especially since the democratization of database and networking facilities (thanks to the development of information and communication technologies and the Internet itself) and with the opportunities that these technical developments have provided for the enhancement of research, notably through the creation of global, interconnected databases, suggestions have been made to address the legal obstacles that are limiting these technologies from reaching their full potential.

Such strategies have taken place in the scientific research field where the availability of research tools is crucial for downward innovation and where many of these research tools have an informational nature. The following section will focus on the informational commons, which has been extensively implemented in the larger context of biological research and includes examples such as the Human Genome Project,[25] HapMap,[26] Biobricks[27] and BiOS.[28] We will refer to the 'informational commons' as a resource that is subject to common access and/or use. It involves the participation of a multiplicity of entities that contribute to the production of the commons, putting data and information that they hold into the commons to make it available to the other contributors to the commons or beyond. A commons is thus characterized by an open-access policy – that is, free and non-discriminatory access.

The most relevant existing commons for the global information system are those collective open informational databases that pool information in the biological information field. In doing so, these informational commons address the issue of the dissemination of information throughout multiple databases around the world. They acknowledge the fact that the coordination that is entailed in gathering together these scattered informational resources might considerably contribute to their very objective, which is to make the information easily available for research. Examples include the European Plant Genetic Resources Search Catalogue (EURISCO), the System-Wide Information Network for Genetic Resources (SINGER), the US Department of Agriculture's Germplasm Resources Information Network (GRIN), the Global Biodiversity Information Facility (GBIF), Strain Info and GENESYS.

Some informational commons

Hosted, developed and coordinated by Bioversity International, EURISCO is a European network database of information about *ex situ* plant collections maintained in European germplasm collections.[29] Launched in 2003, it addresses the dissemination of information among European countries by providing a web-based catalogue of plant genetic resources data that are in the possession of *ex situ* national inventories. The data are about food crops, forages, wild and weedy species, including cultivars, landraces, farmers' varieties, breeding lines, genetic stocks and research material. Thanks to the efforts to assemble all of this data behind one single portal, EURISCO provides facilitated access to 'passport data on more than 1 million samples of crop diversity representing 5,387 genera and 34,823 species (genus-species combinations including synonyms and spelling variants) from 40 countries'. It thus provides access to 'more than half of the ex situ accessions maintained in Europe'. The goal of such a 'one-stop shop' web portal goes beyond merely assembling the scattered bits of information, as EURISCO works on the development of search facilities and web services 'that are necessary components of an Internet-based information infrastructure for ex situ plant genetic resources'.[30]

The GBIF is an Internet-based distributed network of inter-operable databases on biodiversity data maintained by numerous institutions around the world.[31] This decentralized network contains primary biodiversity data including information associated with specimen occurrences documented in biological collections, records of plants and animals, organism names and metadata, observations of plants and animals in nature, and results from experiments. This government-initiated initiative provides a web portal to more than 10,000 data sources disseminated from around the world. Note that the GBIF also provides capacity-building programmes for enabling relevant institutions to integrate the GBIF.

The Strain Info operates in the field of microbial information where the microbial information is disseminated throughout over 500 biological resource centres.[32] The project integrates information contained in the many biological resource centres on microbial collections, and information on the materials possessed by third parties such as GenBan or Swissprot. It offers a bio-portal that provides the research community 'with the most complete and correct integrated view on microbial information' accessible over the Internet (Van Brabant et al., 2006).

SINGER is a relational database that compiles gene bank data from the databases of 11 gene banks supported by the CGIAR.[33] It provides common access to information about the samples of crops, forages and tree germplasm held by individual and independent gene banks by bringing them together and allowing for their easy access and interrogation. SINGER combines a plant material multiprovider platform with a recently developed ordering system.

GRIN is an initiative of the United States Department of Agriculture's Agricultural Research Service.[34] It provides information on plant germplasm held in the United States and also integrates information from 25 independent gene banks. GRIN is a centralized computer database that, among other objectives, is designed to permit flexibility to users in storing and retrieving information, reduce redundancy of data and relate pertinent information about each accession. All data are publicly available and can be fully downloaded.

GENESYS is a web portal that publishes PGRFA data from Category 1 (as specified in the first section of this chapter) and will use the gene bank information management system (GRIN-Global) as the channel for getting quality data. The ambition of GENESYS is to constitute a global information system that comprises characterization, evaluation data and environmental data on PGRFA and compiles data from EURISCO, SINGER and GRIN, using these characterization and environmental parameters for querying data. A germplasm request facility has also been developed in order to enable material to be ordered from a providing gene bank (see Arnaud et al., 2010).

Box 15.1 GENESYS: A global accession-level information system for crop genetic resources

Data standards and gene bank information management systems give gene banks the tools they need to document their collections – but still leave isolated 'islands' of information. For a plant breeder seeking useful materials, the multilateral system is only as good as the information and the information systems that describe these materials. Providing such information is a 'distributed' function; it is not centralized but, rather, is handled by the gene bank and information system managers throughout the world.

Moreover, information may be available but not readily accessible. Anyone seeking to identify all accessions with a particular trait in collections around the world would be faced with the challenge of first identifying the collections of interest and then making multiple searches to find the information they are looking for.

Several national or regional databases link gene bank catalogues, including accession-level information, but are crop or regionally focused – their mandates limit their scope for expanding to provide a global information system for PGRFA. Other efforts, such as the GBIF (www.gbif.org), are broadly targeted at biodiversity, in general, and not at plant genetic resources, in particular, and, hence, are not set up to handle information about the traits of individual accessions.

What is needed is a platform that links national, regional and international gene bank databases and that will allow users to search across gene bank holdings using combinations of data on the characteristics, environments and other aspects of genetic diversity in order to identify accessions of interest and order samples online. The last function is vital if users are to be encouraged to make use of gene bank holdings. Ideally, the platform should also provide access to complementary information, including links to publications relating to specific accessions or collections, germplasm management guides and directories of crop collections.

There are a number of germplasm portals in existence that provide some of this functionality (see, for example, the Germplasm Clearing House Mechanism prototype, http://chm.grinfo.net/index.php), but these portals mainly focus on passport-level information and not the trait data that will support increased use of the collections.

In 2008, Bioversity International, the Global Crop Diversity Trust, and the Secretariat of the ITPGRFA initiated a project to develop just such an integrated global information resource, bringing gene bank databases and existing crop and regional databases together in order to create a single information portal to facilitate access to, and use of, accessions in *ex situ* gene banks.

In September 2010, the prototype portal, called GENESYS, contained 2.3 million accession records and some 3 million phenotypic records for 22 crops. GENESYS has since been launched. GENESYS offers a new paradigm for access to, and use of, PGRFA data, not only because of the number of accessions it contains but also because any user can create custom queries across passport, phenotypic and environmental data categories to suit their specific requirements.

The quantity and quality of data available in GENESYS is an important measure of its value to the global community. While it is based on three quite substantial information networks, there is still the need to continue to attract additional gene banks and/or other data providers to expand the coverage of both passport and phenotypic information. One way that this will be achieved is by deploying the GRIN-Global gene bank management system throughout a number of gene banks during 2010 and 2011. In addition, any data provider will be able to provide their information to GENESYS after completing a data-sharing agreement to do so. All data providers will then have access to a direct data control (DDC) utility that will allow them to upload, edit or delete their own data via the Internet. In the future, it is anticipated that web services will become available to permit a more automatic uploading mechanism where applicable.

Data providers will be provided with tools to perform primary data quality checks; map to their database and upload their passport data into the portal's data warehouse; document the meta-data associated with their characterization and evaluation data; and then upload this type of data into the portal's data warehouse.

In addition, environmental data have been harvested for accessions with geo-references to enable querying on a range of climatic and/or edaphic parameters, and geo-referenced accessions can be displayed on maps, which will help curators identify geographic gaps in their collections. Search functionality allows users to search for accessions on any combination of data. This might include, for example, searching for chickpea landraces that originate from regions in West Asia with an annual rainfall of between 150 and 350 millimetres, have a specific coloured petal and a specific grain size. The value of this type of information has already been demonstrated in the selection of wheat accessions for mining new functional alleles at a powdery mildew locus (Bhullar et al., 2009).

The GENESYS portal is being developed and tested in collaboration with a wide range of stakeholders, but especially those who are likely to be data providers (gene bank curators) and those who will actually use the PGRFA for crop improvement, including plant breeders.

Characteristics of the informational commons

The existing informational commons enable us to put forth their characteristics and to place them in perspective with the characteristics of the ITPGRFA and the objectives of Article 17 on the global information system.

Enhanced access to informational resources

Addressing the multiplication of databases in the biological field, these informational commons pool disseminated information. They offer a 'one-stop-shop' access portal to the information, exempting researchers from browsing many different informational systems one after the other. In some instances, the informational commons may also undertake to rationalize and synchronize more effectively the information contained in the different informational systems, which can be done thanks to applications working at signalling and/or deleting redundancies or inconsistencies between data originating from different databases.

The informational commons hold the significant ability to enhance access to the informational research resources and are generally motivated by the ambition to foster research on biological resources. The result of such pooling efforts consists of a scientific resource that benefits all of the participants in the pool (and beyond). The global informational system provides research with a new resource with an enhanced value, and consisting in a facilitated practical accessibility of its informational components.

Voluntary participation in the commons

It is noteworthy that information providers contribute to the commons on a voluntary basis. Data providers voluntarily collaborate to pool their individual resources. They bring their databases and information to the system and might even in some cases collaborate to effectively coordinate, integrate and synchronize them. They have chosen this type of management of their intellectual assets, instead of a more 'traditional' management policy consisting in restricting and monetizing access to their intellectual assets. However, thanks to the strategy of the commons, they have gained access to other contributors' resources and to the commons. This alternative value provides the motivation that lies behind the participation in a commons. The motivation that lies behind such 'democratic' management of one's resources, where intellectual property rights or secrecy could have been asserted instead, usually comes from a shared goal or interest. Relevant to the scientific informational commons, the motivation that is shared by the information providers comes out of a common belief in the virtue of the dissemination and availability of informational resources for scientific research. This is not mere philanthropy. In a very similar way to the commons approach of the multilateral system of the ITPGRFA, these informational commons are underlined by the idea that by contributing to the common resource, any participant (and the scientific community in general) will also benefit from the contributions of all of the other participants in the system, thanks to the access that he or she enjoys in the commons and vice versa. This belief in the virtue of the dissemination of scientific results rests on the interdependency of researchers and is a translation of the Mertonian norms that are used to characterize access policy in the scientific world. Open access is therefore appropriate to the commons strategy, as it notably generalizes the reciprocity mechanism, which encourages facilitated mutual access to each other's resources.[35]

Open access to the commons

Conceived to enable, promote and ease access and use of data that are useful to scientific research, the informational commons are usually characterized by an open access policy. These portals permit free and non-discriminatory access to the information they shelter, under conditions that are agreed upon through the acceptance of standardized, non-discriminatory terms of use.[36] This system avoids the cost of transactions that one faces when trying to access informational resources on an individual basis by their holder, which is a further advantage of pooling resources within one single access policy. However, such open access might need to be limited to certain kinds of use, such as EURISCO's limitation for scientific research, breeding, genetic resource conservation, the sustainable management of genetic resources or academic, informational or personal use.

The data holders authorize free access and use of their resources through the commons. This upstream authorization by the contributors empowers the manager of the global information system to allow access under an open access scheme to the information contained. This open access licensing approach, thus, is the legal mechanism throughout which data providers voluntarily make their intellectual assets available via the global information system. The open spirit of such a system has notably found its place in copyright. Here, the principle of permission has been used as a tool to create a commons sphere in the area of copyrighted works. As pointed out by Séverine Dusollier (2007, 1394), licensing is employed to promote a collective access to, and sharing of, intellectual resources produced and distributed through a logic that is opposed to proprietary exclusion.[37] Exclusive rights end up being used to enhance and facilitate access, in comparison to a situation in which users need to search for each potential holder of an intellectual property right on the work or on any component of it in order to negotiate use conditions and related royalties. The open access schemes for scientific publications and software are based on licenses attached to the copyrighted work.

Such an approach, however, has extended beyond the sphere of software and copyright to reach the biotechnology field (see the examples of the BiOS, Hapmap and Biobricks projects) as well as the plant material sector (the multilateral system of the ITPGRFA). In these fields, where innovation relies on prior innovations and where it is said that 'researchers stand on the shoulders of giants', the commons have largely been implemented as an alternative to the proprietary strategy that research centres have adopted for the licensing of their research results. Fearing that such an approach would hamper access to research results and thus discourage downstream innovation, open access projects were built in an attempt to ease access to scientific results.[38] In addition to providing authorization for access and use, the commons also began to encompass a mechanism to ensure that the resources remained free. While authorizing many uses for these resources, the license requires the user not to appropriate the resource (see the HapMap project at its onset, and the BioBricks and the BiOS projects).[39] A similar principle can be found in Article 12.3(b) of the ITPGRFA, which prevents beneficiaries of the multilateral system from 'claim[ing] any right that limit the facilitated access to the plant genetic resources for food and agriculture, or their genetic parts or components, in the form received from the Multilateral System'.

Another mechanism implemented by the commons to grant accessibility to these resources is the release of the information into the public domain. This strategy excludes the resource from patentability, given the patentability condition of novelty (see the Human Genome and HapMap projects). The integration of such copyleft-like mechanisms into the global information system should be considered. For instance, the drafting of a clause that would prevent the misappropriation of farmers' knowledge that has been placed in the system would grant its further use while protecting such knowledge. The drafting of such a clause might be done in such a way that it does not prevent the appropriation of invention or knowledge derived from the original knowledge, as long as the latter remains free to be accessed and used.

Conclusion and perspectives

The enormous success of the ITPGRFA has been to create a biological commons for (some) PGRFAs. The multilateral system establishes an international common pool for certain kinds of plant genetic resources that will guarantee facilitated access to these material resources and the sharing of benefits derived from their use. In return for placing their materials into the pool, contracting parties enjoy access to other contracting parties' PGRFA as well as other benefits such as the exchange of information; access to, and transfer of, technology; capacity building; and the sharing of benefits arising from commercialization.

The use of the commons scheme for the global information system would provide the same advantages and incentive for sharing informational resources. In enabling facilitated access to other contributors' resources, the commons scheme provides an incentive to contribute to the production of the commons by bringing one's own resources to the group. In providing an incentive to produce a new resource (the commons), this approach offers an alternative to strong intellectual property. Remember that from an economic perspective intellectual property protection is grounded on the necessity to foster, to incentivize innovation (that is, the creation of new resources such as works, inventions, plant varieties and databases). While intellectual property rights encourage investment in innovation/creation, thanks to the monopoly it grants the inventor/author on the exploitation of the new, more efficient resource, its regrettable effect, however, is to restrict access to such resources. With the commons approach, the access to the resource produced is preserved (even enhanced), while the incentive of any given contributor to take part in the creation of the commons remains given the access he or she will enjoy to this new, global resource that is enriched with other data providers' inputs. This is an interesting aspect of the commons particularly when it encompasses resources for further research.[40]

The collaborative nature of the commons also proves appropriate given that the projected global information resource will benefit all participants and is consistent with the cumulative, incremental and interdependent nature of scientific progress. Finally, the open access scheme makes legal access to information possible while at the same time respecting property rights and freedom of contract. As a result, one might say that the provision of an open global information system to the scientific community, thanks to the mutual contributions of the members of this community, is a form of 'open and collaborative research' (Rai, 2005).

The rooting of the global information system in the multilateral collaborative framework of the ITPGRFA is crucial since it is both an opportunity and a justification for building an informational commons for plant-related information. On the one hand, it was opportune since the PGRFA users have already been sensitized to their interdependency under the framework of the Treaty. The multilateral scheme of the

Treaty is particularly appropriate for the constitution of a complementary global informational resource, composed of data coming from many providers. On the other hand, the rooting of the global information system in the multilateral collaborative framework is a justification for building an informational commons because it is a very valuable tool in reaching the goals of the Treaty while addressing the legal challenges that the sharing of information faces.

Notes

1 International Treaty on Plant Genetic Resources for Food and Agriculture, 29 June 2004, www.planttreaty.org/texts_en.htm (last accessed 11 May 2011) [ITPGRFA].
2 Convention on Biological Diversity, 31 ILM 818 (1992) [CBD].
3 Standard Material Transfer Agreement, 16 June 2006, ftp://ftp.fao.org/ag/agp/planttreaty/agreements/smta/SMTAe.pdf (last accessed 11 May 2011) [SMTA].
4 It should, however, be immediately noted that these three categories of information are not of equal status. Since the whole functioning of the multilateral system on access and benefit sharing is based on the very idea of reducing transaction costs for the exchange of plant genetic resources for food and agriculture (PGRFA), it is explicitly mentioned in the Standard Material Transfer Agreement (SMTA) that there is no need to track accession. Reporting on information about agreements that have been entered into in the context of the multilateral system is sometimes perceived as an additional burden that could hamper the flow of genetic resources. At the same time, because the system is based on trust building and the enforcement of contracts managed at the global level, there are some minimal reporting obligations attached to the exchange and use of PGRFA, which need to be established. The balance between the need not to track and the ability for the third party beneficiary to fulfill its function is a delicate one that was discussed at the third session of the Governing Body in Tunis in June 2009. It is argued here that, even though it has been decided, for reasons of confidentiality, not to publicly disclose this second layer of information, it constitutes the legal basis around which some behavioural change in terms of information exchange at the global scale could happen, for all non-confidential information.
5 ITPGRFA, *supra* note 1, Article 5.1:

   Each Contracting Party shall … in particular, as appropriate:

   (a) Survey and inventory plant genetic resources for food and agriculture, taking into account the status and degree of variation in existing populations, including those that are of potential use and, as feasible, assess any threats to them;
   (b) Promote the collection of plant genetic resources for food and agriculture and relevant associated information on those plant genetic resources that are under threat or are of potential use;
   (c) Cooperate to promote the development of an efficient and sustainable system of *ex situ* conservation, giving due attention to the need for adequate documentation, characterization, regeneration and evaluation, and promote the development and transfer of appropriate technologies for this purpose with a view to improving the sustainable use of plant genetic resources for food and agriculture.

6 See also the Global Programme of Action's Activity 4 '[p]romoting in situ conservation of wild crop relatives and wild plants for food production' and Activity 12 '[p]romoting development and commercialization of under utilized crops and species'. See Resolution 5/89 of the Food and Agriculture Organization's Conference (Twenty-fifth Session, Rome, 11–29 November 1989, Doc. C 1989/REP).

7 The preservation of farmers' varieties has also been recognized as being critical for the realization of the right to food and the maintenance of agro-biodiversity by the special rapporteur to the United Nations. Accordingly, it was advised that proactive policies be undertaken in order to allow farmers to prosper and be innovative as to the development of varieties (De Schutter, 2009, 15).

8 The importance of information flow between research centres and farmers has been underlined – particularly so that the centres can provide technical back up to the farmers for their conservation methods and so that they can gather useful factual information for their research (Brush, 1994, 16). Cooperation and the exchange of information among researchers is also important given the multidisciplinary character of the scientific research on the erosion phenomenon and on appropriate conservation methods (*ibid.*, 22).

9 Indeed, as a result of data computerization, accessibility of different sources of information has increased considerably, while allowing new linkages to be built between different databases and sources of information.

10 This reporting requirement (see FAO, 2008) includes the provisions of Article 5e (information given by a provider to the Governing Body about material transfer agreements entered into); Article 6.4(b) (notification by a provider that material has been transferred to a subsequent recipient); Article 6.5(c) (the notification of the Governing Body, by a provider, about material transfer agreements entered into, in the specific case of plant genetic resources for food and agriculture under development) and Article 6.11h (notification by a recipient of having opted for payment in accordance with this article), and in Annex 2, para. 3 (reports on payment due); Annex 3, para. 4 (notification by the recipient of a decision to opt out of the alternative payment scheme provided for in Article 6.11); and in Annex 4 (signed declaration of having opted for the alternative payment scheme provided for in Article 6.11).

11 Although we will not deal further with this point, it must be reminded that the challenge that those private rights pose to the global information system will potentially have to be considered at two different levels: at the level of integration of information into the global information system and at the level of subsequent use of this information by users of the global information system

12 The present contribution is limited to general statements as to the potential application of intellectual property rights on the resources that will be covered by the global information system. Further analysis will be necessary once that positive content and the *modus operandi* of the global information system is known.

13 Copyright protects original, literary and artistic works. Article 2 of the Berne Convention for the Protection Literary and Artistic Works refers to 'literary and artistic' works as 'every production in the literary, scientific and artistic domain, whatever may be the mode or form of its expression'. This enumeration includes 'works relative to geography, topography, architecture or science'. This is a mere illustrative and non-exhaustive list. Thus, copyright also protects other original works not cited in Article 2 such as maps, software, databases, films or multimedia productions. According to Article 2.5, compilations or databases may also be protected by copyright: 'Collections of literary or artistic works such as encyclopedias and anthologies which, by reason of the selection and arrangement of their contents, constitute intellectual creations shall be protected as such, without prejudice to the copyright in each of the works forming part of such collections.' See also Article 2–5 of the World Intellectual Property Organization Copyright Treaty, adopted in Geneva on December 20, 1996. Criteria and the degree of originality required for copyright protection is not defined and may vary from one country to another and from one category of work to another (Sterling, 1998, 254). In any case, the author of the work must enjoy some freedom to create – that is, the work should be the author's own intellectual creation.

14 The protection of a database by copyright is established by international treaties. Article 2.5 of the Berne Convention, *supra* note 13, makes the protection of '[c]ollections of literary or artistic works' mandatory. Article 5 of WIPO Copyright Treaty, *supra* note 13,

and Article 10.2 of the Agreement on Trade Related Aspects of Intellectual Property Rights, Annex 1C of the Marrakech Agreement Establishing the World Trade Organization, 15 April 1994, 33 ILM 15 (1994) [TRIPS Agreement], establish the protection of compilations of data or other material, whether in machine readable or other form, which by reason of the selection or arrangement of their contents constitute intellectual creations.
15 Created by EC Directive 96/9 on the Legal Protection of Databases, the *sui generis* right lasts 15 years and prevents the extraction or the re-utilization of substantial parts of the database and the repeated and systematic extraction and/or re-utilization of insubstantial parts of the database.
16 These pre-existing databases are, *inter alia,* SINGER, WIEWS, EURISCO, the GBIF and any private database to be connected to the global information system.
17 TRIPS Agreement, *supra* note 14.
18 The CBD, *supra* note 1, recognizes the sovereignty of States over their natural resources and submits access to the latter to mutually agreed terms on the conditions of access and the sharing of benefit arising from such access. Article 8j calls for the protection of traditional knowledge in these terms:

> Subject to its national legislation, respect, preserve and maintain knowledge, innovations and practices of indigenous and local communities embodying traditional lifestyles relevant for the conservation and sustainable use of biological diversity and promote their wider application with the approval and involvement of the holders of such knowledge, innovations and practices and encourage the equitable sharing of the benefits arising from the utilization of such knowledge, innovations and practice.

Nagoya Protocol on Access and Benefit Sharing and the Fair and Equitable Sharing of Benefits Arising from Their Utilization, 29 October 2012, www.cbd.int/abs/text/ (last accessed 11 May 2011).
19 See, for instance, the Peruvian law introducing a protection regime for the collective knowledge of indigenous peoples derived from biological resources, whose Article 42 stipulates that 'Indigenous peoples possessing collective knowledge shall be protected against the disclosure, acquisition or use of that collective knowledge without their consent and in an improper manner provided that the collective knowledge is not in the public domain.' Traditional knowledge, published in Article 17, Public National Register of Collective Knowledge of Indigenous Peoples is such public domain traditional knowledge, the use of which is thus not submitted to prior consent.
20 EC Directive 2003/98 on the Re-Use of Public Sector Information, OJ L 345, 31 December 2003, 90–96, for more information, see http://ec.europa.eu/information_society/policy/psi/index_en.htm and www.epsiplus.net/ (last accessed 11 May 2011).
21 In the directive review, one area of particular interest was the availability of scientific information paid for by public funds. It was considered, in line with the *Council Conclusions on Scientific Information in the Digital Age: Access, Dissemination and Preservation*, 2832nd Competitiveness (Internal market, Industry and Research) Council meeting (Brussels, 22 and 23 November 2007) and with the *Commission Communication on Scientific Information in the Digital Age* (COM(2007) 56 final) that this information should be widely available and useable by all in order to maximize its usefulness for research and innovation. See the Communication from the Commission to the European Parliament, the Council, the European Economic and Social Committee and the Committee of the Regions, *Re-use of Public Sector Information – Review of Directive 2003/98/EC*, COM(2009) 212 final, 8, and, in particular, the *Commission Staff Working Document accompanying the Communication*, SEC (2009) 59.
22 Recommendation for the Enhanced Access and More Effective Use of Public Sector Information, No. C(2008)36.

23 Recommendation on the Principles and Guidelines for Access to Research Data from Public Funding, www.oecd.org/dataoecd/9/61/38500813.pdf (last accessed 11 May 2011).
24 Best Practice Guidelines for Biological Resource Centres, www.oecd.org/dataoecd/7/13/38777417.pdf (last accessed 11 May 2011).
25 See Human Genome Project, www.ornl.gov/sci/techresources/Human_Genome/home.shtml (last accessed 11 May 2011).
26 See HapMap, http://hapmap.ncbi.nlm.nih.gov/ (last accessed 11 May 2011).
27 See Biobricks, http://bbf.openwetware.org/ (last accessed 11 May 2011).
28 See BiOS, www.bios.net/daisy/bios/home.html (last accessed 11 May 2011).
29 The International Plant Genetic Resources Institute is an autonomous international scientific organization, supported by the Consultative Group on International Agricultural Research.
30 All of the quotations come from the EURISCO website, http://eurisco.ecpgr.org/about/about_eurisco.php (last accessed 11 May 2011).
31 About GBIF, see www.gbif.net (last accessed 11 May 2011).
32 Strain Info, www.straininfo.net/ (last accessed 11 May 2011).
33 SINGER, http://singer.cgiar.org/ (last accessed 11 May 2011).
34 GRIN, www.ars-grin.gov/npgs/index.html (last accessed 11 May 2011).
35 Those commons are built on the belief that collaboration and multi-utilization of each other's resources might be more beneficial to each participant (and the advancement of progress in general) than an access policy conceived individually and on a restrictive basis. Each contributor to the mutual resources enjoys access to a new, global resource, whose value is higher for his or her research activities than his or her own individual resource. The pooling of the efforts of the members of the scientific community to produce the commons appears consistent with the fact that all members will profit from the common resource. Such a pooling mechanism has enabled the production of resources that might not have been produced if they had had to be taken in charge on an individual basis (or at least with more difficulty). Consider the production of an effective, global informational system, an inventory of genetic variation (HapMap project), a toolbox of synthesized bioparts (Biobricks project). The commons have much in common with public goods. A link between the suitability of the commons and the cumulative character of scientific progress and innovation might be made in this case. It is acknowledged that the outcome of a given scientific research usually rests on prior discoveries and progress. Given such externalities characterizing the progress of science, the pooling of efforts to produce common resources appears appropriate.
36 This description of open source licensing is a working definition. It has no normative ambition, and the use of this concept in the present contribution might differ from previous definition efforts. For an in-depth analysis of the characteristics of open source licensing, see Hope (2009).
37 This is clearly reflected on the creative commons web page, http://wiki.creativecommons.org/History (last accessed 11 May 2011): 'We use private rights to create public goods: creative works set free for certain uses.'
38 See, among others, Cukier (2003), Boettinger and Burk (2004) and Kumar and Rai (2007).
39 Note, however, that in such case the authorization might not be given thanks to the intellectual property rights but on a mere contractual basis, as the intellectual property rights may not be applicable.
40 The sustainability of the open source model as a means to incentivize innovation in a for-profit context has been challenged, see Van Overwalle (2001).

References

Agosti, D., and W. Egloff (2009) 'Taxonomic Information Exchange and Copyright: The Plazi Approach', *BMC Research Notes* 2: 53, online: www.biomedcentral.com/1756-0500/2/53 (last accessed 11 May 2011).

Alavi, M., and D. Leidner (2001) 'Knowledge Management and Knowledge Management Systems: Conceptual Foundations and Research Issues', *MIS Quarterly* 1: 107–25.

Arnaud, E. et al. (2010) 'A global portal enabling worldwide access to information on conservation and use of biodiversity for food and agriculture', in Lisa Maurer and Klaus Tochtermann (eds), *Information and Communication Technologies for Biodiversity Conservation and Agriculture*, Shaker Publishers, Aachen.

Bhullar, N., K. Street, M. Mackay, N. Yahiaoui and B. Keller (2009) 'Unlocking Wheat Genetic Resources for the Molecular Identification of Previously Undescribed Functional Alleles at the Pm3 Resistance Locus', *Proceeding of the National Academy of Sciences* 106(23): 9519–24.

Boettinger, S., and D. Burk (2004) 'Open Source Patenting', *Journal of International Biotechnology Law* 1: 221–31.

Brown, J.S., and P. Duguid (2001) 'Knowledge and Organization: A Social Practice Perspective', *Organization Science* 12(20): 198–213.

Brush, S. (1994) *Providing Farmers' Rights through In Situ Conservation of Crop Genetic Resources*, Commission on Plant Genetic Resources, First Extraordinary Session, Rome, 7–11 November, online: ftp://ftp.fao.org/ag/agp/planttreaty/gb1/BSP/bsp3E.pdf (last accessed May 2012).

Cukier, K.N. (2003) 'Community Property: Open-Source Proponents: Plant the Seeds of a New Patent Landscape', *Acumen* 1: 54–60.

De Schutter, O. (2009) *Seed Policies and the Right to Food: Enhancing Agrobiodiversity, Encouraging Innovation*, Special Report on the Right to Food, Sixty-third session of the UN General Assembly, October 2009.

Dusollier, S. (2007) 'Sharing Access to Intellectual Property through Private Ordering', *Chicago-Kent Law Review* 82(3): 1391–1435.

Food and Agriculture Organization (FAO) (2002) *Towards a Global Information System on Plant Genetic Resources for Food and Agriculture*, Informal Consultation on a Global Information System on PGRFA, FAO, 9–10 May.

——(2008) *Information Management in Support of the Global System for the Conservation and Sustainable Use of Plant Genetic Resources for Food and Agriculture: Global Information on Germplasm (GIG)*, Second technical consultation on information technology support for the implementation of the multilateral system of access and benefit sharing of the International Treaty, Rome, 2–3 December 2008, Doc. IT/GB-3/TCIT-2/08/Inf.2.

——(2009) *Report of the Third Session of the Governing Body of the International Treaty on Plant Genetic Resources for Food and Agriculture*, Tunisia, 1–5 June, Doc. GB-3/09/REPORT, online: www.planttreaty.org/meetings/gb3_en.htm (last accessed 11 May 2011).

Fridtjof Nansen Institute (2009) *Information Paper on Farmers' Rights at the Third Session of the Governing Body of the ITPGRFA*, Tunis, Tunisia, 1–5 June, online: www.farmersrights.org/pdf/FNI-infopaper-on-FR-GB3.pdf (last accessed 11 May 2011).

Fukasaku, Y., and D. Pilat (2007) 'OECD Principles and Guidelines for Access to Research Data from Public Funding', *Data Science Journal* 6: OD4–11.

Harrer, S., F. Begemann, J.D. Jiménez Krause and S. Roscher (2002) 'Distributed Databases Retrieval Systems in Germany as a National Approach in an International Context', in *Managing Plant Genetic Diversity*, CABI Publishing, Wallingford, UK.

Hope, J. (2009) 'Open Source Genetics: Conceptual Framework', in Geertrui Van Overwalle (ed.), *Gene Patents and Collaborative Licensing Models: About Patent Pools, Clearing Houses, Open Source Models and Liability Regimes in Human Genetics*, Cambridge University Press, Cambridge.

Kumar, S., and A. Rai (2007) 'Synthetic Biology: The Intellectual Property Puzzle', *Texas Law Review* 85: 1745–68.

Pan, S.L., and D.E. Leidner (2003) 'Bridging Communities of Practice with Information Technology in Pursuit of Global Knowledge Sharing', *Journal of Strategic Information Systems* 12(1): 71–88.

Rai, A. (2005) 'Open and Collaborative Research: A New Model for Biomedicine', in Robert W. Hahn (ed.), *Intellectual Property Rights in Frontier Industries: Software and Biotechnology*, AEI-Brookings Press, Durham, NC.

Sterling, A.L. (1998) *World Copyright Law*, volume 34, no 2, Sweet and Maxwell, London.
Van Brabant, B., P. Dawyndt, B. De Baets and P. De Vos (2006) 'A Knuckles-and-Nodes Approach to the Integration of Microbiological Resource Data', in *On the Move to Meaningful Internet Systems 2006: OTM 2006 Workshops,* volume 4277/2006, Springer, Berlin and Heidelberg, online: www.springerlink.com/content/l8387320170x6rjj/full-text.pdf (last accessed 11 May 2011).
Van Overwalle, G. (2001) 'Individualism, Collectivism and Openness in Patent Law: Promoting Access through Exclusion', in Jan Rósen (ed.), *Individualism and Collectiveness in Intellectual Property*, Edward Elgar, Cheltenham, UK.

16
COLLECTIVE ACTION CHALLENGES IN THE IMPLEMENTATION OF THE MULTILATERAL SYSTEM OF THE INTERNATIONAL TREATY

What roles for the CGIAR centres?

Selim Louafi

Introduction

This chapter aims to identify the role of the International Agricultural Research Centres (IARCs) of the Consultative Group on International Agricultural Research (CGIAR) in overcoming the collective action problems that are associated with the implementation of the multilateral system of access and benefit sharing (multilateral system). These centres constitute a network of 15 institutions located in different parts of the world and hold collections of crop genetic resources from various countries. They are major players in the community of plant genetic resources for food and agriculture (PGRFA), and their role has been acknowledged in the International Treaty on Plant Genetic Resources for Food and Agriculture (ITPGRFA) in several provisions – in particular, in Article 15 as supporting components of the Treaty and its multilateral system.[1] More interestingly, it can be argued that the multilateral system and all of its innovative elements have been made possible thanks, among other things, to the very existence of the IARCs, which were already freely distributing across borders material coming from different regions. In other words, the ITPGRFA has formalized an orientation that was already de facto pre-existing.

Collective action problems occur in situations in which the uncoordinated action of individuals is less beneficial or efficient than coordinated action. Collective action problems often arise in relation to public goods, which attract free riders who do not contribute to the production or conservation of the goods (see Olson, 1971; Ostrom, 1990). The principle of establishing the multilateral system was based on the recognition of the need for international collective action to manage a common resource – PGRFA – in a manner that is more beneficial and efficient to everyone than individual action would be. The generally recognized high degree of interdependence of countries for plant germplasm is the main justification for the collective mechanism that was finally formalized with the establishment of the multilateral system of the

ITPGRFA. In fact, the Treaty recognizes specifically 'that plant genetic resources for food and agriculture are a common concern of all countries, in that all countries depend very largely on plant genetic resources for food and agriculture that originated elsewhere'. In addition, Annex I, which lists all of the crops that fall under the multilateral system has been 'established according to criteria of food security and interdependence' (Article 11.1). The Treaty introduces an element of collective constraint, mainly through contractual obligations that it imposes on the recipients of multilateral system material. The objective of the standardized access and benefit-sharing provisions, as reflected in the Standard Material Transfer Agreement (SMTA), is to reduce transaction costs that would otherwise occur if access to, and the sharing of benefits derived from, PGRFA would be subject to bilateral negotiations rather than to a multilaterally agreed standard agreement.[2]

The ITPGRFA is based on the assumption that collective rules are expected to produce better outcomes than the pursuit of individual self-interest. However, establishing and implementing collective rules entail costs as well, which might, in turn, create a new public good dilemma (see Dedeurwaerdere, 2009). Indeed, decreasing the transaction costs linked to the international exchange of PGRFA under the Treaty's multilateral system (through the establishment of a SMTA agreed multilaterally) would require increased coordination costs at the global level both to establish and to monitor such an exchange. Despite the expected benefits of these collective rules, not everyone can agree on the additional cost that would be necessary to contribute to their creation and maintenance.

This chapter addresses three public good dilemmas associated with the implementation of the multilateral system and identifies how the IARCs could help to address them: (1) the coordination of different governance levels (from local to global); (2) the management of diverging interests and expectations within the PGRFA community; and (3) the reference to several global challenges. Ultimately, the chapter argues that it is essential when dealing with the implementation of the multilateral system to adopt a wide perspective that helps to situate the multilateral system in its broader environment. Being by constitution beyond the interests of the national perspective, the IARCs are indeed uniquely situated to help implement the multilateral system in a way that reflects its real multilateral and commons nature.

Multilateral system, sovereignty and stakeholder participation

Box 16.1

Fact

The ITPGRFA leads to a redefinition of stakeholders' responsibilities regarding PGRFA use and management at the different levels of governance, from the local to global level.

> *Collective action problem*
>
> Public good dilemma: politicization of implementation process
>
> *Solution*
>
> The redefinition of responsibilities (reinforcing the global level at the expense of the national one) increases the relevance of international public good research practices. This gives specific responsibilities to the IARCs and requires more inclusiveness in the decision-making process, which the centres could help to foster.

A new way of exerting sovereignty

The ITPGRFA has led to a new way of exerting national sovereignty. Through the establishment of the multilateral system, contracting parties, 'in the exercise of their sovereign rights', as Article 10.2 states explicitly, have agreed to establish a multilateral system to facilitate access to specific PGRFA and to share the benefits arising from the utilization of these resources, as laid down in the Treaty and as operationalized through the SMTA. Through the establishment of the multilateral system, contracting parties have agreed to defer a part of their responsibility for PGRFA management from the national level to the international level. The underlying rationale is that they will all gain more by having access to all of the resources in the multilateral system than they would have by restricting access to their own. Moreover, the multilateral system will allow for the monitoring and control of the flow and use of PGRFA and thus generate the knowledge needed to operationalize benefit sharing.[3]

The multilateral system can be seen to be the most integrated and collective way of managing globally and in a coordinated and coherent way a distributed, but still common, pool of genetic resources. In this regard, the multilateral system might qualify as the most advanced expression of the need for cooperation in the management, conservation and distribution of PGRFA. As an operational system at the highest level of governance – that is, the global level – the multilateral system has to work in different parts of the world with different legal and administrative systems, cultures and traditions. The implementation of the multilateral system, therefore, poses major challenges to governments at various levels – the global level as much as the technical and national levels. Moreover, for this operational system to be efficiently managed, a high level of shared responsibilities must be maintained.

Usually, international texts simply provide general guidelines, and implementation is a subsequent step that is taken according to the national translation of these guidelines. They are usually interpreted in coordination, and/or in consultation, with the stakeholders who are subject to the regulation. Compared to other provisions of the Treaty, or even to the vast majority of international instruments, the multilateral system is quite unusual in the sense that it breaks from the usual strategy of shared responsibilities, wherein state representatives negotiate international norms in the form of general guidelines, national decision makers elaborate public policies taking into account these guidelines and local or sectoral actors manage the local resources within the given framework.

The multilateral system, in contrast, sets precise rules for the inclusion of materials by the contracting parties. More importantly, the SMTA is a very operational instrument that applies directly (that is, without the necessary involvement of state or administrative representatives) to the stakeholders involved in the exchange of PGRFA. It does not require any administrative steps at the national level, even in the case of reporting procedures. Hence, the direct link that is established between the stakeholders and the Governing Body of the Treaty (through its secretariat) provides an opportunity to reinforce the global commons nature of the resource exchanged and the responsibilities attached to its management, beyond or aside from the national states.

In accordance with Article 15, the IARCs are already experiencing, in a far more acute way, this direct connection with a global policy process. While this has not always been felt positively by some of the constituencies, when viewed from a broader perspective it is hardly questionable that this direct link is increasing the level of relevance of the tools designed and their effectiveness (Dryzek et al., 2005; Koontz and Thomas, 2006).[4] Participation and inclusiveness are consequently two important features of the multilateral system that could ensure its effective implementation.

A need for more inclusiveness in the decision-making process

What are the consequences of this new way of exerting sovereignty? The cooperative logic underlying the functioning of the multilateral system requires moving away from the traditional negotiation mode that relies on a pure interest-based conception of sovereignty. By including material from the Article 15 bodies, the multilateral system goes beyond a simple state membership club. In addition, through its agreed set of rules for PGRFA exchange, universal functional links are established between governmental and non-governmental actors at the global level and across scale (between national or local stakeholders and the Governing Body). Implementing such an integrated and operational system at the global level could only occur if one goes beyond the classical divided, dichotomous and linear sequence from policy to implementation. Increased and continuous interactions between negotiators and stakeholders are needed to ensure that issues will not be dealt with entirely from a political perspective, on the one hand, or from a technical or administrative perspective, on the other hand. Those constant interactions are also required because the multilateral system is a dynamic framework that includes a number of elements, some of which are still being set in place.[5] Indeed, properly monitoring such a dynamic and operational system at the global level depends upon the constant interaction between users and negotiators.

Consequently, the contracting parties need to move from a negotiating mode to an operative and functional attitude that is requisite for the implementation process. Keeping a strong division between decision and implementation opens up the possibility for policy makers to avoid responsibility. For instance, although the negotiations proceeded along the divided line of providers and recipient countries, this divide is no longer able to deal with the rationale and requirements of the multilateral system's implementation. Continuing to refer to this dividing line is a way for some countries to delay the implementation process by waiting for some proof of good faith from other countries before taking action themselves. For example, notifying the Secretariat about the country collections that are included in the multilateral system and providing information about such collections is not an obligation. It is, however, a practical measure that will make the multilateral system work better. Not all countries think that providing such information is a good idea, simply because at this point in time they are not willing to go beyond the legal obligations contained in the Treaty. The same logic applies also to information attached to the samples. The legal obligations regarding the sharing and availability of information about PGRFA remain quite vague. However, on the other hand, since the multilateral system is a distributed and dynamic system of collections worldwide, it is only as useful as the information that it contains and makes available.

This change of behaviour can only take place if the complexity of the implementation process is properly taken into account (Grindle and Thomas, 1990). And for this to happen, it is important that state and administrative representatives take actions not just because they are legally required to do so according to the Treaty but also because these actions will make the multilateral system really operative. The change of behaviour can only take place if those who have a real interest in seeing the multilateral system implemented are more familiar with the functioning of the system at the global level. Such knowledge requires creating spaces of dialogue and a two-way flow of information, beyond national borders, in order to ensure as far as possible a continuum between decision and implementation. For the moment, this brokering role is being played by only a few individuals who have managed, through their skills and position, to speak both the negotiation and technical languages. Yet such efforts are far from being enough, and it is crucial to think about a more institutionalized learning process to cope with the unique implementation challenges posed by the multilateral system.

Being very often situated at the nodes of different stakeholders involved in the exchange and use of PGRFA (national agricultural research systems, gene banks, farmers, universities, breeders and non-governmental organizations), the IARCs are definitely in a position to play the honest broker, provided they manage to go beyond simply reflecting the corporate views of the scientists in this political debate (whether as gene bank managers or as breeders). We consider in detail in the next section what such a brokering role would look like and what the advantages or disadvantages would be for the IARCs to play this role.

Recommendation for the IARCs: make the IARCs brokers between the levels of governance (vertical integration)

It is our assumption that such a multilateral approach has been made possible thanks, among other things, to the very existence of the IARCs, which have been already freely distributing material coming from different regions across borders. The ITPGRFA has reaffirmed and formalized this orientation and expanded it to all member states. In addition, the collective nature of the multilateral system has re-legitimized and given a higher profile to global public good (GPG)-oriented research, which neither one individual state by itself nor the market (through the private sector) could have undertaken (or, if so, under much more restricted conditions). It is not by chance that this concept arose in the IARCs in the context of the Treaty's negotiation and process. Operationalizing the GPG concept at the IARC's level has been already subject to fierce debates since the 2000s (Anderson, 1998; Dalrymple, 2008; Gardner, 2003; Kanbur et al., 1999; Pinstrup-Andersen and Mengistu, 2008; Ryan, 2005; Sandler, 2002). This has impacted IARC's research practices in reorienting research agenda towards the production of GPGs. Indeed, this concept has led, at least theoretically, to research products and outputs that (1) go beyond situation-specific locations or sectors and (2) are made widely available without restrictions (technical or intellectual property-related). The basic rationale is to ensure that research products and outputs benefit the largest range of stakeholders.

The spirit with which the Treaty's multilateral system has been designed draws clear inspiration from this approach and applies it to PGRFA listed in Annex 1 and their related information. However, applying it at the inter-governmental level requires an even bigger change of attitude. As mentioned earlier, responding to this challenge requires, among other things, repeated interactions between stakeholders within and between national boundaries in order to exchange experiences and provide the necessary feedback information needed by the delegates/negotiators to update and monitor the process.

Through their extensive experience as users of PGRFA beyond national boundaries, the IARCs could use the unique position that the ITPGRFA is providing them as Article 15 bodies to: (1) engage with stakeholders in order to be in a better position to manage the system in relation to the negotiators and (2) use the incentive provided by ITPGRFA's implementation to strengthen national PGRFA communities by helping them to better plan and define their national strategies of conservation and use of PGRFA.[6]

The literature on network governance shows that a network, through a decentralized and incremental process of interaction and emulation, could be a source of convergence (Raustiala, 2002). They provide a means for the transfer of regulatory ideas and policies. This role was played successfully by the IARCs during the negotiation process of the ITPGRFA and was undertaken in two manners: first, in a passive way through the *de facto* 'community of practice' that the IARCs and their partners were forming; and, second, in a more active way, which consisted of playing a brokering role between the different stakeholders (Louafi, 2011; Pistorius, 1997).

Examining these mechanisms in more detail is beyond the scope of this chapter. Yet what is interesting to note is that both roles were also relevant during the implementation phase. The experience accumulated by the IARCs and its network in terms of implementation was instrumental in helping to strengthen the capacities of other stakeholders that were less familiar with the provisions of the Treaty. The diffusion of regulatory approaches through international networks and peer-to-peer cooperation is something that is very often overlooked, although it is an essential element for effective implementation, especially when regulatory power is diffuse (Kaniaru, 2002; Raustiala, 2002).

However, more importantly, in accordance with the idea of the 'community of practice', the way in which the IARCs are implementing the Treaty is crucial in helping to move towards an operative mode and away from a negotiating mode (that is, the politicization of the implementation process), in which a lot of state members seem to be stuck. Indeed, the IARCs' implementation practices could potentially help develop best practices and be used as a standard which the Governing Body, from a legal perspective, and the other stakeholders, from a practical perspective, could build upon and/or get some inspiration from. This way of approaching the roles of the IARCs could help to end the division between decision and implementation, which was referred to in the previous section. These roles could also help to introduce a necessary level of flexibility and adaptability in the management of the system by helping to conceive of policies as experiments, the outcomes of which constitute opportunities for learning and improving the subsequent decisions (McNie, 2007).

However, experience shows that those roles, nevertheless, have to be carefully undertaken. Financial, intellectual and technical resources mobilized by actors participating in the collective process are often unequally distributed. The IARCs are indeed far ahead in implementing the multilateral system and using the SMTA compared with most of the state parties. While IARC users are very familiar with the use of contractual agreements in the exchange of PGRFA, others have never used formal agreements to transfer or receive materials. The whole contractual approach is new to them. Precisely because of this discrepancy between the experience of the IARCs and the national stakeholders, the actual or perceived risk of 'elite capture' is difficult to avoid. This risk refers to the fact that the outcomes of any collective action process could be captured by the more informed and capable stakeholders, who are in a better position to impose their views and strategies. This discrepancy has somehow put pressure on the system, as users from the IARCs are insisting on addressing legal issues which are far removed from most countries' practical experience in regard to the multilateral system and the SMTA. This pressure has led to the politicization of some technical issues, which has further delayed the design of the solutions necessary to address these legal issues.

Similarly, early IARC compliance has paradoxically undermined the urgency of others to comply. In other words, by adopting a very generous interpretation of the GPG policy in making available material even to non-parties, it could be argued that the IARCs have, unconsciously, created a kind of free-rider attitude (SGRP, 2010). Consequently, this elite attitude could undermine the legitimacy of the IARCs to be able to efficiently play the role of bridge builder, which is necessary to foster better inclusiveness at the different levels of governance (international, regional and national).

Multilateral system and horizontal cooperation: toward an integrated approach on PGRFA

> **Box 16.2**
>
> *Fact*
>
> Within a highly politically charged environment, the ITPGRFA combines in a single framework several interests and approaches previously presented as irreconcilable.
>
> *Collective action problem*
>
> High transaction costs associated with the establishment of partnerships and collaborations.
>
> *Solution*
>
> By providing equitable 'rules of the game' at the global level, the Treaty's multilateral system potentially decreases these high transaction costs. The IARCs should build upon such an enabling framework to initiate and test ways of working more efficiently in partnership on PGRFA.

Making the ITPGRFA a cooperative framework across disciplines, approaches and actors

The ITPGRFA provides an enabling framework that creates the global conditions for attenuating the effects of centrifugal forces (intellectual property, sovereignty and so on). Indeed, it has managed to create a framework that strengthens the capacities of different actors to cooperate across existing geographical, organizational and intellectual (scientific disciplines) boundaries. The Treaty as a whole recognizes the legitimacy of these different approaches from conservation to use. References are made to collection, conservation, pre-breeding, participatory plant breeding, on-farm management and crop improvement. Consequently, even though the core and most operational element around which all of the functional articulations take place is the multilateral system, the Treaty envisages collaboration along the whole chain of PGRFA management and use, with stakeholders involved at all levels of development.

Rightly or wrongly, the ITPGRFA is nevertheless still perceived as being biased towards formal conservation in *ex situ* conditions. It is true that the Treaty negotiation process was mainly engaging '*ex situ*' experts (gene bank curators) and, indeed, the multilateral system is somehow responding pretty well to this specific category of users and the material they hold. However, in its Articles 5 and 6, the Treaty puts conservation and use clearly on an equal footing. Some dissonant voices are claiming that, if the multilateral system is effectively facilitating access for gene banks to exchange their material, there is little evidence that the material held in these collections is being used extensively in breeding programmes and even less by farmers.

It is important to understand the reasons for such a perception. Some of them relate to the technical solution, the SMTA, the complexity of which makes it ill-adapted to the whole range of situations and stakeholders and which, furthermore, still contains 'unclear' provisions. These shortcomings have been already discussed in the previous section, and the IARCs' experience in using material transfer agreements and the SMTA for their exchanges of material is crucial to help countries and users to address them.

However, more importantly, it should be recognized that there are many objective technical reasons that explain the feeling of many parties that the multilateral system is more of a 'gene bank club' than a powerful mechanism able to cover a wide range of uses and needs. Lack of information about the very availability of the material, lack of characterization and evaluation data, lack of capacity or technology to use the material are the main reasons that have been put forward. This feeling is accentuated by a perceived lack of benefit sharing, both current and in the foreseeable future. In addition, there are other more structural reasons for the lack of linkages between the formal and informal systems of genetic resources research, development and distribution in many developing countries. This situation has increased the uncertainty surrounding the potential benefits of the multilateral system for agricultural production.

Many of these shortcomings are beyond the scope of the ITPGRFA, but it is important to note that some responses have already been built into the Treaty in order to help it address these issues (at least partially) in the longer term. On the one hand, various non-monetary benefit-sharing mechanisms are foreseen and, on the other hand, voluntary contributions to the benefit-sharing fund have been introduced to compensate for the expected lack of monetary benefit arising from the use of material coming from the multilateral system, at least in the short term. Even though little progress has been made up to now to solidify these non-monetary components, there has been considerable success in gathering voluntary contributions, which has enabled funds to be passed along to smallholder farmers in developing countries. These efforts have provided the foundation through which developing countries, and previously marginalized actors, have managed to consume the public goods that the multilateral system has helped to make available while, at the same time, creating valuable national public goods.

It is important to understand that the ITPGRFA has made a valiant, but fragile, attempt to provide a pluralistic framework that encompasses genetic resources in all of their dimensions. It has attempted to encompass *in situ* and *ex situ* material; conservation and use; the formal seed sector and the informal seed sector; farmers' rights and breeders' rights and so on. The ITPGRFA is not just a juxtaposition of these different dimensions, but also provides an operational and functional structure, the multilateral system, around which these other aspects can be integrated. Looking at the multilateral system in this way – taking into consideration its complexity and comprehensiveness (that is, looking beyond its 'accession' side, which is very often promoted by the gene bank managers) is important when dealing with its variety of partners around the world.

Widening the range of objectives regarding the management and use of PGRFA

In looking at the multilateral system in this way, what are the consequences for the management and use of PGRFA, particularly for the research sector? To respond to this question, it is necessary to look more deeply at how the research on PGRFA, and the innovation that is arising from it, is reaching its potential beneficiaries. Research on PGRFA combines elements with a high economy of scale with those with a low economy of scale:

> Economies of scale and scope are realized in the production of improved agricultural technologies when knowledge, methods, processes, and information are mobilized in ways that individual countries are not able to do because of a lack of funds or a lack of scientists or institutional infrastructure.
>
> *(World Bank, 2003)*

Since it requires a substantial fixed investment, the *ex situ* conservation of germplasm has the potential to generate substantial economies of scale (Gardner and Lesser, 2003). It is indeed one of the *raison d'être* of the ITPGRFA, and the Global Crop Diversity Trust has built its legitimacy and actions precisely on the realization of these economies of scale.

When it comes to plant breeding research, things become less clear. Strong cases have been made, in many reviews and evaluations, about the global public good nature of plant breeding research (Byerlee and Dubin, 2010; CGIAR Interim Science Council, 2005; Dalrymple, 2006a; Sagasti and Timmer, 2008). For instance, Dana Dalrymple (2006b) has noted that wheat and rice breeding research enjoys significant payoff because they 'tend to be raised in similar agroecological zones (often irrigated) over wide areas of the world. And it has been facilitated where the centres breed for broad adaptability.' However, it is fair to say that because breeding research is usually limited to a crop or a sector, its outputs (in terms of improved lines at least) are by definition excludable from other sectors or crops. In other words, the mono crop-based approach of breeding programmes and/or mono-trait selection is sometimes a hindrance to the broad application of the innovation that results from them.

As the intellectual property rights regime in most countries expands its reach, the GPG nature of public breeding is being questioned since it enables the private sector to invest in PGRFA research and could lead to some forms of restriction on access to material and information. In addition, the specialization of the nature of knowledge and technology that has been mobilized and generated increases the ability of farmers to tailor applied research products to more specific crops, locations and traits. This new environment offers new opportunities to respond to the main criticisms initiated by the Green Revolution in developing crops that are more adapted to specific conditions. However, it also requires that more attention be paid to the institutional aspects of adequate adaptive research and the development capacity of recipient organizations and countries in order to minimize the occurrence of second-generation problems such as restrictive policies and institutions, the lack of credit for processing products and the lack of a market for produce (Tesfaye, 2009). In this context, generating outputs that have a large market, broad applicability and spillovers across several regions requires accrued efforts. In particular, it necessitates more than ever many highly coordinated research networks involving different partners with a complement of many different disciplines (Hall et al., 2001; Hocdé et al., 2006).

The concept of innovation systems (as opposed to a linear system that leads to productivity enhancement based on maximum genetic progress) helps to cover a broad range of objectives regarding the management and use of PGRFA.[7] This concept acknowledges the continuum of activity from conservation to use and the interfaces between the *ex situ* and on-farm management of genetic resources and their improvement. It also accounts for the dynamic interplay between several forms of knowledge necessary to build experience that is applicable across various locations: knowledge of local implementation; knowledge based on global research and knowledge on 'best practice' that draws on both local and global sources (see van Kerkhoff and Lebel, 2006). A key feature of the innovation systems is that, although they all have common goals and a similar network of actors/organizations, these actors all have different values, norms and practices, which vary in their spatial positions and in their access to resources, knowledge and power (Ashby, 2009; Tesfaye, 2009).

Investing in such a variety of networking activities is sometimes felt to be cumbersome by a lot of scientists since the costs involved in preparing the actors, many of whom do not have much in common, to communicate, exchange information and exchange resources for innovation are sometimes very high, and the returns are often uncertain (Horton et al., 2009). It is true that the high transaction costs associated with these activities could lead to a fairly modest benefit-cost ratio if they are only measured against the final output (product or variety). However, as Bashah Tesfaye (2009, 6) explains, 'the improved capacity of actors and the entire system that is achieved though the process of learning and use of this experience for the future should not be overlooked.'

Developing the IARCs into brokers between the different stakeholders and approaches (horizontal articulation)

If we agree that the ITPGRFA is offering a unique framework that enables several actors/institutions to work together towards common goals, we can then argue that the IARCs probably need to pay at least as much attention to fostering complementary activities in partnership with other organizations than they spend in expanding their comparative advantages in terms of breeding. As Jim Ryan (2005, 16) points out, 'complementary advantage implies that the particular role[s] of the CGIAR centres be such that they maximize the synergies and multiplicative effects from interactions with other actors/partners/suppliers in the R & D continuum, consistent with their mandates and comparative advantages.' It is worth noting that these complementary activities very often exist within the IARCs themselves, even though they probably are under-exploited (CGIAR Interim Science Council, 2002, 2008). In this regard, it might be useful to build a map of research partnerships and innovation networks within the CGIAR system. For instance, linking knowledge generation and knowledge use might require building more linkages between the System-Wide Program on Collective Action and Property Rights (CAPRi) and the Participatory Research and Gender Analysis Program, the Institutional Learning and Change Program (ILAC) or the Natural Resources Management Research Program.

CAPRi or ILAC have developed useful concepts, tools and case studies on how to explicitly integrate knowledge and learning dimensions in order to best design action-oriented institutional arrangements.

For instance, the concept of innovation systems referred in the previous section has been described and analysed under the ILAC programme (see Lundy et al., 2005; Hall et al., 2005; Watts et al., 2007; Tesfaye, 2009; Watts and Horton, 2008). These authors offer some solutions on how innovation systems can be initiated and facilitated in order to reorient the production of research outputs designed to be adapted and modified by end users instead of strengthening the pipeline approach (which tends to consider any intervention from users on finalized products as being harmful).

These tools and concepts are meant to better design necessary boundary spanning actions to level the playing field. These actions could take several forms: creating internal and external networks, issuing identification, creating common understanding by combining different forms of knowledge, influencing and educating internal and external stakeholders, identifying champions who would align themselves with the purpose of new initiatives and so on. In summary, these activities will help to generate hybrid, co-created knowledge and to deal with the often significant (and largely hidden) asymmetries of power felt by the stakeholders (Horton et al., 2009). They help the IARCs to act as catalysts, stimulating additional investments by others to capture the various comparative advantages of the partners and stakeholders on agreed priorities (Ryan, 2005).

Similarly, it is crucial in this context that the IARCs also play a facilitative and enabling function to help make the activities of the partners and stakeholders easier to manage (Ryan, 2005). In this regard, it is important that the IARCs use their convening power, professional skills and unique network of relationships with governments, the private sector and civil society to provide platforms for international dialogue and global action. Eventually, it is also important to try to balance the needs of stakeholders in the IARCs' decision-making processes and activities and engage with the stakeholders at the grassroots level of technical projects and daily work in order to make them feel part of the system. All of these activities are crucial in order to avoid the perception that facilitated access simply means that the IARCs have better access to a broader range of diversity to produce outputs (seeds and technologies) and continue their 'business as usual'.

The non-commercial benefit-sharing activities described in the ITPGRFA are good examples of this approach in the context of the implementation of the Treaty.

Capacity building, the exchange of information and access and transfer of technology are very often catchwords in international texts. We have already mentioned in the previous section (see Box 16.2) how important the non-commercial benefit-sharing component is for maintaining the overall political coherence of the Treaty. These components are taking on their full meaning in the context of implementing the multilateral system – they are the means and the vehicle through which greater benefit could be generated, a broader range of stakeholders could be encouraged and a wider range of concerns could be dealt with. It is argued that these provisions (Articles 13.2 (a), (b) and (c)) have not yet received adequate attention. So far, the Governing Body has done very little to address and advance the implementation of this very important part of the multilateral system. Unlike the monetary benefits, how the non-monetary benefits can, or must, be shared has not been clearly established by the Treaty, and all types of user need to define and experiment with the possible ways of sharing. The centres are in the best position to give these fuzzy concepts the opportunity to take shape and influence the policy process. Non-monetary benefits mechanisms need to be established in the day-to-day work of the IARCs' scientists and properly documented.

Multilateral system and GPG: towards a meta-integrated approach

Box 16.3

Fact

By addressing simultaneously several global challenges, the ITPGRFA helps to raise the profile of PGRFA in public policies.

Collective action problem

Addressing several global challenges through a mechanism that has been established only for a sub-set of species within a single specific sector of PGRFA.

Solution

In dealing with projects involving PGRFA use and exchange, the IARCs should try to refer to a broader research paradigm that emphasizes potential spillover effects on several global challenges. Advocacy for a more integrated approach encompassing all PGRFA could have more impact on the provision of these GPGs.

The ITPGRFA at a crossroads of several global challenges

The main political dimension of the multilateral system relates to access and benefit-sharing discussions. The ITPGRFA has been developed in a politically charged environment where the main concern has been to respond to benefit-sharing issues expressed mainly by developing countries. Now that the multilateral system is up and running, one could think that these problems are now behind us. However, the Treaty is not operating in a vacuum, and the main concerns expressed by countries in relation to the implementation of the multilateral system pertain to access and benefit-sharing issues, in particular: (1) a lack of confidence in the exchange of germplasm, which is still present despite the existence of dispute settlement rules (third party beneficiary); (2) the perception that the multilateral system does not benefit the country because of a lack of capacity to utilize the resources coming from it; and (3) high expectations related to the monetary benefits.

However, the functioning of the multilateral system of access and benefit sharing is more than an end in itself. It is also a means by which to achieve more efficiently broader global objectives related to the use and management of PGRFA such as conservation, evaluation, genetic enhancement, plant breeding innovations, the promotion of seed diversity and the diversification of crops. Indeed, the ITPGRFA addresses simultaneously several global challenges:

- genetic erosion and biodiversity loss,
- food security,
- the adaptation of crops to climate change,
- the rural poverty of small-hold farmers and
- access to knowledge and innovation.

These challenges, however, are poorly acknowledged and addressed in the implementation of the multilateral system. The Treaty, of course, is only one small step in addressing these challenges. In addition to it, there is a real need to establish more clearly, both conceptually and practically, the links between the Treaty and the other policy components related to these global challenges.

From a sectoral perspective to global challenges

The multilateral system has been established only for a sub-set of species within a single specific sector of PGRFA. Under these conditions, it is not always easy to go beyond the rhetorical reference to global challenges such as food security or climate change adaptation. Evidence still has to be established to demonstrate the impact of the multilateral system on these global challenges. It is crucial to better articulate the link between PGRFA use and food security in relation to the functioning of the multilateral system. Existing breeding research capacity in each country is critical to ensure this link (Trommetter, 2009). For countries with existing research capacities, the ITPGRFA's facilitated access mechanism allows countries to mobilize a larger pool of diversity, together with a larger range of technologies, to address a wider array of constraints, which thereby enable production to rise. Consequently, the Treaty is a powerful tool in helping countries with existing breeding research capacities enrich the gene pool of crops essential for national food security and agricultural productivity (Eng-Siang, 2008).

For countries with limited or non-existent capacities of breeding research, the Treaty, ideally, should be able to create the conditions for developing cooperative research on local varieties and under-utilized crops based on the most advanced technologies. The involvement of farmers in the different phases of research (as foreseen in Articles 6.2 and 9), the development of policy and legal tools to enhance the value of local food production (as foreseen in Articles 6.2(a) and (b)), the mobilization of additional funding under the Treaty's funding strategy to finance plant breeding in favour of local plant varieties, and the realization of non-commercial benefit sharing (especially Article 13.2(b) on the access to, and transfer of, technology) are examples of the combination that is needed to build a link between the use and exchange of PGRFA and food security in poor developing countries.

Taking the most advantage of the benefits provided by the ITPGRFA requires having these elements well settled at the national level under the broader food security strategy. It also requires that the national decision makers deploy differentiated (but overall coherent) strategies according to the type of Annex 1 crop: high-economic value crops, crops with less commercial value but that are essential for self-subsistence and the livelihood of a marginalized population and crops that have a functional value for nutrition. To follow this approach would require not only that parties focus on creating a legal space for the Treaty in national access and benefit-sharing legislation (on which capacity-building strategies are still too often focusing) but also in food security legislation and/or strategies.

Making the IARCs brokers between several GPG beyond the PGRFA community

The multilateral system is often presented as the first and only operational access and benefit-sharing mechanism, which in addition, while being in harmony with the Convention on Biological Diversity, is based on a multilateral approach.[8] However, by doing so, the Treaty constantly holds the risk of being only associated with the specific needs of a single sector (the plants sector) and, within this sector, with only a selected number of crops (Annex 1) and, within this sub-set of crops, with a specific category of actors (mainly gene bank managers, as previously discussed) or crops. In this context, therefore, the question becomes how to give the Treaty the high profile it deserves as an innovative international agreement, while simultaneously addressing several global challenges, including the most immediate one: food security.

It is absolutely crucial that the IARCs acknowledge the fact that the ITPGRFA (and, beyond it, the use of PGRFA in general) is actually at the crossroads of several global challenges. In order to deal with these issues directly, specific efforts must be made in the exchange of genetic resources to ensure a clear and, above all, a functional link with one or more of these challenges. This requires going beyond the direct application of the knowledge/technology envisaged in each project and looking more in-depth at their potential impacts, both *ex ante* and *ex post*, on those global challenges. The relevance of any project to these socio-economic and sustainable development global challenges is very often mentioned. However, such attention is often only on a general level – the concrete mechanisms by which each project is actually able to impact on these challenges is seldom made explicit and, even less often, properly initiated. Doing so would probably require designing projects in a different way than scientists are used to doing – in particular, by incorporating social science dimensions in earlier phases of the project. It would also require sometimes establishing new strategic alliances with partners beyond the individual crop or sector within each project.

Conclusion

The ITPGRFA goes beyond the functioning of the multilateral system in its restricted meaning (that is, the pooling and facilitated use of PGRFA). It could be seen as an attempt to provide an enabling framework for people with different sensitivities to work together. Starting at its core with the exchange of material mainly from *ex situ* gene banks, the multilateral system aggregates several interests that help to address simultaneously several global challenges: genetic erosion and biodiversity loss, food security, crop adaptation to climate change, the rural poverty of small-hold farmers, a bottom-up approach to development policy in agriculture, access to knowledge and innovation.

In this sense, it is a very pluralistic framework, which encompasses different sensitivities previously presented as being irreconcilable. These linkages are made possible thanks to the benefit-sharing mechanisms, both monetary (through the benefit-sharing fund projects) and, just as importantly, through non-monetary mechanisms. If properly managed, these non-monetary benefit-sharing mechanisms could increase the cooperative capacities of different communities by building a series of linkages between initiatives in conserving, using and valorizing PGRFA.

Beyond the strategic motives of implementing the multilateral system (access to resources and the improved efficiency of use), which are the natural motives of scientists, one has also to think about the political motives such as improved accountability to stakeholders, greater leverage and political legitimacy. It is also essential to pay due attention to the need to foster and accelerate, through networking and boundary-spanning activities, behavioural changes that might benefit the whole genetic resource community in the longer term and help to raise awareness of the ITPGRFA as a model for other sub-sectors of genetic resources for food and agriculture.

Notes

1 International Treaty on Plant Genetic Resources for Food and Agriculture, 29 June 2004, www.planttreaty.org/texts_en.htm (last accessed 11 May 2011).
2 Standard Material Transfer Agreement, 16 June 2006, ftp://ftp.fao.org/ag/agp/planttreaty/agreements/smta/SMTAe.pdf (last accessed 11 May 2011).
3 Monitoring and controlling the flow and use of plant genetic resources for food and agriculture is a task that has proven to be burdensome and costly (if not technically impossible) for most countries.
4 Scientists may feel that these policy processes are distracting them from their main objectives and work goals. However, the division of labour on which such perception is built is less and less operative to deal with global common concerns (Brahy and Louafi, 2007).

5 At its first session, for example, the Governing Body adopted the Standard Material Transfer Agreement, and it has recently, during the session held in June 2009, considered and adopted the third party beneficiary procedures.
6 This Article 15 status has allowed some parties to overcome some of the problems linked to the governance structure of the IARCs, which have not always made them accountable to the international community. In a sense, this governance problem was partially resolved first by the 1994 in trust agreement with the Food and Agriculture Organization and then by the ITPGRFA by providing a international legal umbrella to the IARCs' collections.
7 The World Bank (2007) defines innovation system as a network of organizations focused on bringing new processes and new forms of organization into social and economic use, together with the institutions and policies that affect their behaviour and performance (Tesfaye, 2008; see also Hall et al., 2005).
8 Convention on Biological Diversity, 31 ILM 818 (1992).

References

Anderson, J.R. (1998) 'Selected Policy Issues in International Agricultural Research: On Striving for International Public Goods in an Era of Donor Fatigue', *World Development* 26 (6): 1149–62.

Ashby, J. (2009) 'Fostering Farmer First Methodological Innovation: Organizational Learning and Change in International Agricultural Research', in I. Scoones and J. Thompson (eds), *Farmer First Revisited: Innovation for Agricultural Research and Development*, Practical Action Publishing, Bourton on Dunsmore, UK.

Brahy, N., and S. Louafi (2007) *The Role of the Research Sector in ABS Governance*, Idées pour le débat, no. 9, Iddri, Paris online: www.iddri.org/Publications/Collections/Idees-pour-le-debat/Id_0709_Brahy&Louafi_ABS.pdf (last accessed 11 May 2011).

Byerlee, D., and H.J. Dubin (2010) 'Crop Improvement in the CGIAR as a Global Success Story of Open Access and International Collaboration', *International Journal of the Commons* 4(1): 452–80.

Consultative Group on International Agricultural Research (CGIAR), Interim Science Council (2002) *Lessons Learned in the Implementation of System-Wide Programmes*, CGIAR Interim Science Council Secretariat, FAO, Rome.

——(2005) *CGIAR Research Priorities 2005–2015*, Draft, Science Council Secretariat, FAO, Rome.

——(2008) *The Role of System-Wide Initiatives in Implementing the CGIAR's Research Agenda: An Assessment of Current System-Wide and Eco-Regional Programs (SWEPs)*, CGIAR Science Council Secretariat, Rome.

Dalrymple, D.G. (2006a) 'Setting the Agenda for Science and Technology in the Public Sector: The Case of International Agricultural Research', *Science and Public Policy* 33(4): 277–90.

——(2006b) 'Social Science Knowledge as Public Good for Agriculture', in M.M. Cernea and A.H. Kassam (eds), *Researching the Culture in Agri-Culture: Social Research for International Development*, CABI Publishing, London.

——(2008) 'International Agricultural Research as a Global Public Good: Concepts, Experience, and Policy Issues', *Journal of International Development* 9(4): 459–82.

Dedeurwaerdere, T. (2009) 'The Role of Law, Institutions and Governance Processes in Facilitating Access to Genomics Research', in G. Van Overwalle (ed.), *Gene Patents and Clearing Models: From Concepts to Cases*, Cambridge University Press, Cambridge.

Dryzek, J., D. Downes, C. Hunold and D. Schlosberg (2005) 'Green Political Strategy and the State: Combining Political Theory and Comparative History', in J. Barry and R. Eckersley (eds), *The State and the Global Ecological Crisis*, MIT Press, Cambridge, MA.

Eng-Siang, L. (2008) *Implementing the Treaty, Developing Country Concerns*, Sawtee Policy Brief no. 16, 85(3): 692–97.

Gardner, B. (2003) *Global Public Goods from the CGIAR, Impact Assessment*, Thematic Working Paper, World Bank (Operations Evaluation Department), Washington, DC.

Gardner, B., and W. Lesser (2003) 'International Agricultural Research as a Global Public Good', *American Journal of Agricultural Economics* 85: 692–97.

Grindle, M., and J. Thomas (1990) 'After the Decision: Implementing Policy Reforms in Developing Countries', *World Development* 18(8): 1163–81.

Hall, A., G. Bockett, S. Taylor and M. Sivamohan (2001) 'Why Research Partnerships Really Matter: Innovation Theory, Institutional Arrangements and Implications for Developing New Technology for the Poor', *World Development* 25(5): 783–97.

Hall, A., L.K. Mytelka and B. Oyelaran-Oyeyinka (2005) *The Innovation Systems Concept: Implications for Agricultural Research Policy and Practice*, ILAC Policy Brief no. 2, International Plant Genetic Resources Institute, Rome.

Hocdé, H., B. Triomphe, G. Faure and M. Duleire (2006) *From Participation to Partnership, a Different Way for Researchers to Accompany Innovations Processes: Challenges and Difficulties*, Paper presented at the Innovation Africa Symposium, Kampala, Uganda.

Horton, D., G. Prain and G. Thiele (2009) *Perspectives on Partnership: A Literature Review*, Social Sciences Working Paper no. 2009-3, International Potato Center, Lima, Peru.

Kanbur, R., T. Sandler, with K.M. Morrison (1999) *The Future of Development Assistance: Common Pools and International Public Goods*. Overseas Development Council, Policy Essay no. 25, Washington, DC.

Kaniaru, D. (2002) *The Role of Institutions and Networks in Environmental Enforcement*, 6th INECE Conference Proceedings, vol. 2, online: www.inece.org/conf/proceedings2/10-Role%20of%20Institutions.pdf (last accessed 11 May 2011).

Koontz, T.M., and C.W. Thomas (2006) 'What Do We Know and Need to Know about the Environmental Outcomes of Collaborative Management?', *Public Administration Review* 66: 111–21.

Louafi, S. (2011) *Co-management and Network Governance: The Role of the International Agronomic Research Centres in the Emergence and the Implementation of the International Treaty on Plant Genetic Resources for Food and Agriculture*, Paper presented at the XIème congrès de l'Association Française de Sciences Politiques, Strasbourg, September 2011.

Lundy, M., V. Gottret and J. Ashby (2005) *Learning Alliances: An approach for Building Multistakeholder Innovation Systems*, Institutional Learning and Change (ILAC) Brief no. 8, ILAC Initiative, Rome.

McNie, E.C. (2007) 'Reconciling the Supply of Scientific Information with User Demands: An Analysis of the Problem and Review of the Literature', *Environmental Science and Policy* 10: 17–39.

Olson, Mancur (1971; originally published in 1965) *The Logic of Collective Action: Public Goods and the Theory of Groups*, Harvard University Press, Cambridge, MA.

Ostrom, E. (1990) *Governing the Commons: The Evolution of Institutions for Collective Action*, Cambridge University Press, Cambridge.

Pinstrup-Andersen, P., and T. Mengistu (2008) 'Some Implications of Globalization for International Agricultural Research', in J. von Braun and E. Diaz-Bonilla (eds), *Globalization of Food and Agriculture and the Poor*, Oxford University Press, New Delhi.

Pistorius, R. (1997) *Scientists, Plants and Politics: A History of the Plant Genetic Resources Movement*, International Plant Genetic Research Institute, Rome.

Raustiala, K. (2002) 'The Architecture of International Cooperation: Transgovernmental Networks and the Future of International Law', *Virginia Journal of International Law* 43(1): 1–92.

Ryan, J. (2005) *International Public Goods and the CGIAR Niche in the R for D Continuum: Operationalising Concepts*, Revised draft of a discussion paper prepared for a brainstorming session between some members of the Science Council of the CGIAR and the Centre Deputy Directors' Committee of the Future Harvest Centres of the CGIAR at the CGIAR Annual General Meeting in Morocco, 3 December.

Sagasti, F., and V. Timmer (2008) *A Review of the CGIAR as a Provider of International Public Goods*, System-wide review of CGIAR, Bioversity International, Rome.

Sandler, T. (2002) 'Demand and Institutions for Regional Public Goods', in A. Estevadeordal, B. Frantz and T.R. Nguyen (eds), *Regional Public Goods: From Theory to Practice*, Inter-American Development Bank and Asian Development Bank, Washington, DC.

System-wide Genetic Resources Programme (SGRP) (2010) *Workshop on National Programmes and CGIAR Centres' Cooperation to Implement the Multilateral System of Access and Benefit Sharing of the International Treaty on Plant Genetic Resources for Food and Agriculture*, 15–16 February, Bioversity International, Rome.

Tesfaye, B. (2008) *How to Get Innovation System Work in Agriculture and Rural Development? Reflection on Methodological Issues*, Paper presented at APPRI 2008 International Workshop, Ouagadougou, Burkina Faso, 21–24 October.

——(2009) *Innovation Systems in Agriculture and Rural Development*, Technical Center for Agricultural and Rural Cooperation, online: http://knowledge.cta.int/en/Dossiers/Demanding-Innovation/Innovation-systems/Articles/Innovation-Systems-in-Agriculture-and-Rural-Development (last accessed 11 May 2011).

Trommetter, M. (2009) *Intellectual Property Right, International Treaty on Plant Genetic Resources for Food and Agriculture and the Stakes for Food and Nutrition Security*, Submission to the Special Rapporteur on Right to Food, online: www.fundp.ac.be/droit/crid/propriete/M.TROMMETTER_food%20and%20nutrition%20security (last accessed 11 May 2011).

van Kerkhoff, L., and L. Label (2006) 'Linking Knowledge and Action for Sustainable Development', *Annual Review of Environmental Resources* 31: 445–77.

Watts, J., et al. (2007) *Institutional Learning and Change: An Introduction*, ILAC Working Paper 3, Bioversity International, Rome.

Watts, J., and D. Horton (2008) *Institutional Learning and Change: An Initiative to Promote Greater Impact through Agricultural Research for Poverty Alleviation*, ILAC Working Paper 5, Bioversity International, Rome.

World Bank (2003) *The CGIAR at 31: An Independent Meta-Evaluation of the Consultative Group on International Agricultural Research*, vol. 1, Overview report, World Bank (Operations Evaluation Department), Washington, DC.

——(2007) *Enhancing Agricultural Innovation: How to Go beyond the Strengthening of Research Systems*, Agriculture and Rural Development Department, World Bank, Washington, DC.

17

INTERNATIONAL AND REGIONAL COOPERATION IN THE IMPLEMENTATION OF THE INTERNATIONAL TREATY ON PLANT GENETIC RESOURCES FOR FOOD AND AGRICULTURE

Gerald Moore

Introduction

International cooperation is at the very heart of the International Treaty on Plant Genetic Resources for Food and Agriculture (ITPGRFA).[1] Indeed, it underlies the whole multilateral system of access and benefit sharing (multilateral system), which is itself a major component of the Treaty. In the preamble to the ITPGRFA, the contracting parties recognize that plant genetic resources for food and agriculture (PGRFA) are a common concern of all countries, in that all countries depend very largely on PGRFA that have originated elsewhere.[2] The preamble also recognizes that PGRFA are essential to food security. In fact, the Treaty set up the multilateral system in order to facilitate the continued exchange of these essential resources without incurring unsustainable transaction costs. The multilateral system is essentially a cooperative system that allows all contracting parties to continue to have access to the most important PGRFA for their agricultural development on the basis of common multilaterally agreed terms, thus avoiding the need to negotiate each exchange on a bilateral basis.

Access to PGRFA and benefit sharing is only one of the areas in which international cooperation is needed. Article 5.1(e), for example, requires the contracting parties to cooperate to promote the development of an efficient and sustainable system of *ex situ* conservation. More generally, Article 7 calls for international cooperation in the conservation and sustainable use of PGRFA, including cooperation in capacity building and enhancing international activities to promote conservation, evaluation, documentation, genetic enhancement, plant breeding and seed multiplication. International cooperation should also be directed at sharing, providing access to and exchanging PGRFA as well as appropriate information and technology. Article 7 further requires that international cooperation be directed to maintaining and strengthening the institutional arrangements dealing with supporting components such as the Global Plan of Action (GPA) for the Conservation and Sustainable Use of PGRFA and the implementation of the funding strategy. Article 16 calls for cooperation to develop international PGRFA networks,[3] and Article 17 calls for cooperation to develop and strengthen a global information system.

This chapter will review developments in international cooperation in the implementation of the ITPGRFA at both the global and regional levels in the various areas indicated earlier and explore ways in which this cooperation may be further developed in the future. In general, significant advances are being made in the areas of *ex situ* conservation, the development of capacity building, on-farm conservation, sustainable utilization and the development of a global information system. More needs to be done in the areas of *in situ* conservation, the improvement of plant-breeding capacities and seed distribution.

Cooperation in the development of an efficient and sustainable system of *ex situ* conservation

One of the areas that is basic to the multilateral system and in which most obvious gains can be achieved through international cooperation is *ex situ* conservation. Once the principle of free availability of PGRFA is accepted, or at least the free availability of the most important PGRFA for food security and on which countries are most interdependent, then the field is open for the rationalization of *ex situ* conservation at the global level and the consequent improvement in the quality and availability of conserved PGRFA as well as the releasing of funds and effort that can be fruitfully redirected towards other activities.

Ex situ conservation is an area in which there has been an increasing amount of cooperation at both the global and regional levels. Historically, the International Agricultural Research Centres of the Consultative Group on International Agricultural Research (CGIAR) have been both a focus and a product of international cooperation with respect to *ex situ* conservation as well as plant breeding and other elements of the sustainable utilization of PGRFA.[4] Responding during the 1970s and 1980s to the threat of genetic erosion, the CGIAR centres increased their efforts to collect and conserve endangered PGRFA. Today, the 11 gene banks operated by the CGIAR contain almost 700,000 accessions of some of the most important PGRFA for food security.[5] Although the CGIAR collections contain only about 12 per cent of the accessions held worldwide, they are particularly important because they conserve a comparatively high proportion of landraces and wild relatives.[6] The gene banks run by the CGIAR centres complement the *ex situ* collections held by some 1,300 gene banks at both the national and regional levels. The total number of accessions held by gene banks has been estimated at 6 million, but the number of unique accessions (that is, those not duplicated in other collections) is much lower.

The collections operated by the CGIAR centres were placed in trust for the benefit of the international community under agreements signed between the centres and the Food and Agriculture Organization (FAO) in 1994, and their importance has been expressly recognized in the Treaty.[7] In accordance with Article 15 of the ITPGRFA, all of the CGIAR centres holding *ex situ* collections have signed agreements with the

Governing Body of the Treaty, placing those collections within the purview of the Treaty. Samples of Annex 1 crops are made available under the terms and conditions of the multilateral system using the Standard Material Transfer Agreement (SMTA).[8] Following a decision of the Governing Body of the ITPGRFA at its second session in November 2007, non-Annex 1 material collected before the entry into force of the Treaty is also to be made available under the SMTA.[9] Non-Annex 1 material collected after the entry into force of the Treaty is to be made available on terms and conditions consistent with those agreed with the country of origin of the material or other country that acquired it in accordance with the Convention on Biological Diversity or other applicable law.[10]

While the ITPGRFA is clear in calling for international cooperation to promote the development of an efficient and sustainable system of *ex situ* conservation, little guidance is given as to what this system should look like. More guidance is given in the GPA, which was adopted by the Leipzig Technical Conference in 1996. The Treaty itself recognizes the continuing importance of the GPA.[11]

The GPA provides that full use should be made of appropriate existing facilities, including national, regional and international centres, that conserved materials should be, as appropriate, replicated and stored in long-term facilities meeting international standards, and that unintended and unnecessary duplications between collections within the networks should be reduced to promote cost efficiency and effectiveness in global conservation efforts.[12] Agreements should be drawn up to safeguard diversity in *ex situ* collections to allow those countries so desiring to place collections voluntarily in secure facilities outside their boundaries.[13]

While much of the funding for international gene banks, including those run by the CGIAR centres, has traditionally been provided by bilateral donors, one of the leading players now emerging in international cooperation under the ITPGRFA with respect to *ex situ* conservation is the Global Crop Diversity Trust (GCDT). The GCDT was established under an international agreement that came into force in 2004 and concluded a relationship agreement with the Governing Body of the Treaty at its first session in 2006. The objectives of the GCDT, as set out in Article 1 of its constitution, mirror the provisions of both the Treaty and the GPA. Its broad objective is to ensure the long-term conservation and availability of PGRFA with a view to achieving global food security and sustainable agriculture. To this end, the GCDT is required to endeavour to safeguard collections of unique and valuable PGRFA held *ex situ*, with priority being given to PGRFA included in Annex I to the Treaty or referred to in Article 15.1(b) of the Treaty. It is also required to promote an efficient goal-oriented, economically efficient and sustainable global system of *ex situ* conservation in accordance with the Treaty and the GPA, to promote the regeneration, characterization, documentation and evaluation of PGRFA and the exchange of related information, to promote the availability of PGRFA, and to promote national and regional capacity building, including the training of key personnel.

The GCDT operates within the framework of the ITPGRFA and under the overall policy guidance of the Governing Body. Under this relationship agreement, the Governing Body recognizes the GCDT as an essential element of the Treaty's funding strategy in relation to the *ex situ* conservation and availability of PGRFA. The trust, which has an endowment fund now amounting to some US $108 million, is already providing some US $2.3 million a year in long-term funding for *ex situ* collections of unique germplasm of global significance as well as substantial project funding for regeneration, characterization and evaluation.[14] While the programme of the GCDT has hitherto focused mainly on *ex situ* conservation and the regeneration and evaluation of threatened PGRFA, it is now planning to extend its activities to encompass the whole span of conservation and sustainable utilization with respect to crop wild relatives (see the discussion later in this chapter).

The long-term funding activities of the GCDT have been concentrated on ensuring the conservation and availability of globally significant *ex situ* collections of unique PGRFA. These include the collections held by the CGIAR centres, as well as other important collections such as those held by the Secretariat of the Pacific Community (the Centre for Pacific Crops and Trees (CePaCT)). The GCDT is also funding the Global Seed Vault as a security back-up deposit of last resort. The Global Seed Vault is located in Svalbard in the Arctic archipelago of Spitsbergen, and the seeds are stored in permafrost conditions in a vault constructed by the government of Norway in a mountain. The vault already contains over 600,000 accessions. The seeds will be conserved at minus 18 degrees Celsius.

International cooperation at the regional level also presents a number of opportunities for improving *ex situ* conservation and achieving economies in operational costs. The kind of cooperation varies from the provision of security back-up facilities, through cooperation to increase the quality and efficiency of *ex situ* conservation operations, to the provision of regional gene bank facilities serving all aspects of *ex situ* conservation and distribution for the region as a whole.

The Plant Genetic Resources Centre (SPGRC), which was set up in 1988 by the member states of the South African Development Community (SADC), is an example of the first two types of cooperation. The main responsibilities of the SPGRC are to keep the SADC base collection, which involves the maintenance of long-term storage facilities, and to ensure the efficient coordination of plant genetic resources (PGR) work within the region. The SADC base collection provides back-up storage facilities at the regional level for the active collections maintained by the individual countries.

The creation of a more rational regional system of *ex situ* conservation is the objective of the scheme now being set up in the west Asia and northern Africa region. The regional strategy for the conservation of PGRFA developed in 2006 calls for the strengthening and renewal of coordination both within and among countries in the region (Zehni, 2006). As recommended in the strategy, the countries of the region are intending to establish a new PGRFA regional network within the framework of the Association of Agricultural Research Institutes in the Near East and North Africa. It is envisaged that membership in the new network will include non-governmental organizations, private sector and farmers' associations as well as public institutions. Membership in the new network will be subject to formal endorsement by the highest competent authority in the country seeking membership, possibly through the signing of a letter/memorandum of understanding (MOU) that would spell out the benefits and would commit members to defined obligations. The network would be country driven and would be primarily concerned with the conservation and sustainable use of PGRFA in line with the Treaty. The objectives will include fostering conservation and sustainable use of PGRFA in the region, promoting the exchange of PGRFA, scientific and technical experience and information, and strengthening national and regional research capabilities. An important basis of a rationally developed regional collaboration will be the sharing of expertise and experience as well as responsibilities for conservation activities. The strategy makes it clear that it is not aimed necessarily at encouraging all countries to have full conservation capacity for all of the crops of importance in the region. In this context, it will be important to encourage regional collaboration in order to understand the capacities, strengths and comparative advantages of each country in possibly taking on responsibility at the regional and international level.

A practical step towards the realization of a rational system of cooperation in the conservation and sustainable use of PGRFA for the region has recently been taken by the 45 countries of the European region.[15] The new system is called the European Genebank Integrated System (AEGIS), and it will establish a European collection, consisting of selected accessions designated by the individual countries. Material designated as part of the European collection will continue to be conserved in the individual gene banks concerned. But it will be maintained in accordance with agreed quality standards and will be made freely available, both within Europe and outside, in accordance with the terms and conditions set out in the ITPGRFA.[16] In so doing, the countries plan to share responsibilities relating to the conservation and sustainable use of PGRFA and, thus, to develop a more efficient regional system in Europe. The new system will allow individual gene banks to rely on the work being undertaken by other gene banks, without having to duplicate it themselves.[17] The AEGIS initiative is the brain child of the European Cooperative Programme for Plant Genetic Resources (ECPGR) (ECPGR, 2009).

The legal mechanism for establishing AEGIS is an MOU entered into by eligible countries of the region and the European Community, setting out their commitments as full members of AEGIS and the main lines of AEGIS. To be eligible for membership, the countries listed must be members of the ECPGR and either a contracting party to the ITPGRFA or otherwise willing to make PGRFA under their jurisdiction available under the conditions of the Treaty. The MOU came into force in August 2009 upon the signature of ten countries, and it now has 26 member countries. The MOU is supplemented by a series of associate member agreements for the individual gene banks that wish to become part of AEGIS. The associate member agreements are entered into by the gene banks, along with the ECPGR national coordinator for the country concerned, who undertakes to work with the gene banks in implementing AEGIS and who is ultimately responsible for overseeing the gene banks' compliance with its obligations under the agreement. So far, some 19 individual gene banks have accepted to become part of AEGIS.

All European accessions must be made freely available for research, breeding and training for food and agriculture. Accessions of crops listed in Annex 1 to the

ITPGRFA must be made available under the SMTA. Accessions of non-Annex 1 crops are to be made available under the SMTA with explanatory notes, in much the same way as the Governing Body of the Treaty has authorized for the CGIAR centres. European accessions accessed for non-food and agricultural purposes are to be made available according to the terms and conditions agreed upon on a case-by-case basis between the recipient and the associate member concerned. In all cases, public domain accession-level information will be made available as well as non-confidential characterization and evaluation data.

The main benefits expected to be gained from AEGIS are:

- improved collaboration among European countries and a stronger unified Europe;
- increased cost-efficiency of conservation activities;
- reduced redundancy in European collections;
- the improvement of quality standards across Europe;
- more effective regeneration;
- facilitated access to all of the germplasm included in AEGIS;
- improved security of germplasm through normal commitments and safety duplication;
- improved linkages between *ex situ* and *in situ* conservation as well as linkages with users; and
- the improved sharing of knowledge and information.

The countries of West and Central Africa, under the auspices of the *Conseil Ouest et Centre Africain pour la Recherche et la Developpement Agricole* (West and Central African Council for Agricultural Research and Development – CORAF/WECARD), are also looking at ways to create a regional system of *ex situ* conservation for priority PGRFA in West and Central Africa. The system would be based on the establishment of a number of nodal centres of excellence (NCEs) in the region for the gene banking of priority species and collections in the region. As described in the Ouadadougo Declaration, the NCEs would be located in selected countries/organizations but would hold their collections on behalf of their partners in the region.[18] To get started, the NCEs will build on existing functional national or international facilities and networks of partners. Initially, four NCEs were envisaged, focusing on the following plant species and agro-ecological zones:

- cereals and related plant species in the Sahelian zone;
- root and tuber crop species in the coastal West African zone;
- coconut collections and other industrial crop species in the coastal West African zone; and
- banana and plantain collections in the Central African zone.

The NCEs would not be dedicated solely to conservation but would also have research and training responsibilities. Research at the NCEs would add value to the work being carried out in the national gene banks and would focus on the development of regional public goods. Responsibilities for multiplication, evaluation, characterization and other research would be spread out among a range of partners in West and Central Africa and beyond, including countries that are not hosting NCEs as well as in the NCEs themselves. The countries of West and Central Africa as well as CORAF/WECARD itself would be called on to facilitate the movement of experts to work in the NCEs. The regional economic community organizations, such as the Economic Community of West African States and the Economic Community of Central African States, are to be involved in the oversight of the NCEs. Over time, it is expected that the NCEs will increase in size and capacity and gain sufficient financial support for the regional economic organizations to be able to run their own facilities on a sustainable basis.

To be eligible to host a NCE, a country must be a party to the ITPGRFA and a member in good standing of CORAF/WECARD, be willing to host the NCE and boast a national institution with a good track record in PGRFA conservation and management as well as in financial management. As a minimum, the NCE host countries are expected to ensure the current levels of funding for their national PGRFA institutions. If necessary, the selection of hosting countries will be decided on a competitive bidding process. The Ougadougo Declaration also includes provisions designed to secure adequate financial resources for the regional system, including long-term commitments from countries hosting NCEs.

An even tighter system of regional cooperation is already in place in the Pacific region where the Pacific Island countries face problems related to their remoteness and the scarcity of technical and administrative resources as well as the costly nature of long-term storage of vegetatively propagated crops such as taro, yams, sweet potatoes and bananas, which are among the most important crops in the region. The region relies heavily on the regional *in vitro* gene bank at CePaCT, which is run by the Secretariat of the Pacific Community and is closely linked with the Pacific Agricultural Plant Genetic Resources Network. CePaCT has a unique collection of taro from the Pacific and southeast Asia region and works closely with the taro breeding programmes in Papua New Guinea and Samoa. In 2009, the Secretariat of the Pacific Community (SPC) entered into an agreement with the Governing Body of the ITPGRFA, placing the regional collection of taro and yams (over one thousand accessions) held by CePaCT within the purview of the Treaty under Article 15.5.

The potential benefits of international and regional cooperation are well demonstrated in the recent initiative to introduce blight-resistant taro breeding material from the Pacific region to West Africa. Samples of resistant taro (cocoyams) are now being transferred from the SPC's CePaCT collection to the International Institute of Tropical Agriculture's gene bank in Nigeria for evaluation. If the results are satisfactory, they will eventually be multiplied in various breeding programmes and distributed to farmers in West Africa.

Cooperation in capacity building, on-farm conservation, sustainable utilization and the implementation of the funding strategy

The funding strategy adopted by the Governing Body at its first session in 2006 is the main mechanism for promoting cooperation among contracting parties in the implementation of the ITPGRFA. Article 18.3 of the Treaty provides that the Governing Body is to establish funding targets for the mobilization of funding for priority activities, plans and programmes, in particular in developing countries and countries with economies in transition. The Governing Body established the first such target at its third session in June 2009, setting the figure at US $116 million for the five-year period between June 2009 and December 2014.

One element of the funding strategy, along with the GCDT, is the benefit-sharing fund established by the Governing Body under Article 19.3(f) of the ITPGRFA. The Governing Body, at its second session in 2007, agreed on the initial priorities for the use of funds from the benefit-sharing fund, reflecting priority activities set out in the GPA. These were (1) information exchange, technology transfer and capacity building; (2) managing and conserving plant genetic resources on farm; and (3) the sustainable use of plant genetic resources. The first 11 projects addressing those priorities, costing a total of US $500,000, were announced at the third session of the Governing Body in June 2009.

Since then, significant new funds have been deposited in the benefit-sharing fund, and the focus and scope of projects to be funded have been refined to ensure greater impact. The theme chosen for the 2010 call for proposals is assisting farmers in adapting to climate change through the conservation and sustainable use of PGRFA. The call for proposals, which is for a total investment of more than US $10 million for the 2010–11 biennium, concentrates on the first priority area of the benefit-sharing fund (information exchange, technology transfer and capacity building) and on three inter-related activities to address the other two priority areas (on-farm conservation and management of plant genetic resources, participatory plant breeding and the distribution of appropriate seed and planting materials).

The second initiative aimed at developing cooperation in capacity building and sustainable utilization is the Global Partnership Initiative for Plant Breeding Capacity (GIPB), a multi-party initiative of knowledge institutions around the world that have a track record in supporting agricultural research and development, working in partnership with country programmes committed to developing stronger and more effective plant-breeding capacity. The establishment of the GIPB represents an attempt to improve the balance between supporting the conservation of PGRFA and support for its effective use. Its mission is to enhance the capacity of developing countries to improve crops for food security and sustainable development through better plant breeding and delivery systems. Its objectives are to provide support for policy development on plant breeding and associated scientific capacity and to help allocate resources to strengthen and sustain developing countries' capacity to use PGRFA. The GIPB also provides education and training in plant breeding and related scientific capacities, facilitates access to technologies in the form of tools, methodologies, know-how and facilities for finding genetic solutions to crop constraints and facilitates the exchange, from public and private breeding programmes, of plant genetic resources that can enhance the genetic and adaptability base of improved cultivars in developing countries. It further promotes the sharing of information focused on plant-breeding capacity building to deliver newly available knowledge to national policy makers and breeders in developing country programmes.

A third initiative that seeks to draw together many elements of conservation and sustainable utilization is the new project now being launched by the GCDT on adapting agriculture to climate change – collecting, protecting and preparing crop wild relatives. The project, which is being supported by the government of Norway to the tune of some US $50 million over a ten-year period, will be aimed at collecting and conserving crop wild relatives that may contain traits essential to adapting agriculture to the challenges of climate change and evaluating those traits and making them available to modern agriculture through a process of pre-breeding.

The fourth initiative in the area of capacity building is the joint programme of assistance to developing countries in the implementation of the ITPGRFA. The programme is run jointly by the FAO and Bioversity International, and it provides assistance to countries on request in the implementation of the Treaty, with particular reference to the multilateral system. So far, the joint programme has organized a number of workshops on the implementation of the Treaty and its multilateral system in individual countries and regions.

One of the main outcomes of these workshops has been the development of common regional approaches towards the implementation of the ITPGRFA and the harmonization of national legislation. A second is the development of regional capacities and mechanisms to help countries in the region to interface with the Treaty and its multilateral system.

Common approaches to the implementation of the Treaty's multilateral system

Many of the regions in developing parts of the world are placing great importance on the promotion of harmonized legislation and national procedures for the implementation of the ITPGRFA. One of the most advanced examples is the West Africa and northern Africa region, where the regional strategy singles out the harmonization of national legislation and procedures in the region as being of significant importance. A workshop on the Treaty organized in 2007 by the Arab Organization for Agricultural Development (AOAD) reviewed the status of national implementation of the Treaty in the various states of the region and encouraged Arab countries to issue and implement national legislation on PGRFA in harmony with related international treaties, taking as a basis model legislation developed by the AOAD. National legislation for the implementation of the Treaty, including the multilateral system, has been developed for Syria and was the focus of a workshop in Morocco in July 2008. The workshop also called for the organization of training courses to build up capacities and exchange experiences and information on PGRFA. A committee of five countries has been established to coordinate the initiative. A further workshop was held in Cairo at the end of March to consider, *inter alia*, guidelines for the development of PGRFA legislation in the region. The workshop recommended that the draft guidelines prepared by the AOAD be revised and used as a model for future assistance to countries in the region. A revised version of the guidelines including draft model legislation was prepared in September 2009 and has now been circulated to member countries of the AOAD.

The call for a harmonized approach to the implementation of the ITPGRFA has also been reiterated in the southern Africa region served by the SADC Plant Genetic Resources Centre (SPGRC). At a recent workshop organized under the joint programme, workshop participants discussed options for a SPGRC strategy to facilitate the regional implementation of the Treaty in the SADC region.[19] The main elements for such a strategy included capacity building, including policy and legal capacity as well as technical capacity for plant breeding and seed production, the regional harmonization of policies and legislation and the exchange of information, awareness raising and reporting under the multilateral system.

With respect to capacity building, the workshop suggested that national workshops be organized on the domestication of the ITPGRFA and that legal experts be recruited to advise a regional project management team. A survey of regional technical capacity should also be undertaken. In regard to the regional harmonization of policies and legislation, the workshop recognized that the efficient use of limited institutional resources (for example, legal expertise) and the adoption of common conservation and sustainable use measures would be the main benefits resulting from a regional approach. The establishment of a SPGRC group of legal experts, whose mandate would be to consider the legal space for a regional policy within the existing SADC institutional mechanisms and the sensitization of the SADC Council of Ministers, is one of the serious options the workshop is considering. The SPGRC could initiate and lead the process of regional harmonization.

A similar approach is being taken by the countries of the Pacific region. A workshop on the implementation of the ITPGRFA was held in Fiji in September 2009. The workshop recommended that approaches to the implementation of the Treaty and its multilateral system by contracting parties be harmonized in the region and that provisions for a draft model to incorporate the Treaty into the national legislation as well as other policies be drawn up by the SPC. A study has been completed that reviews the experiences of the contracting parties in the region in implementing the Treaty, with a view to ascertaining whether or not there are gaps in the policy and legislative regimes and the need for reform. Draft model provisions, concentrating on the implementation of the Treaty's multilateral system within the more general context of model access and benefit-sharing legislation, have now been prepared.

Development of regional cooperative mechanisms

At the second session of the Governing Body of the ITPGRFA in October/November 2007, the SPC observer explained the difficulties that the Pacific Island countries faced in implementing the Treaty, given the scarcity of their administrative resources. The SPC informed the Governing Body that Pacific Island countries were considering placing samples of all PGRFA of Annex 1 crops that are presently under their management and control as well as in the public domain into the SPC gene bank (CePaCT) and were considering the possibility of asking CePaCT, on their behalf, to perform the administrative responsibilities required for making that material freely available under the SMTA in accordance with the Treaty.[20] As a start in this direction, the Governing Body approved the request of the SPC to authorize the Secretariat to enter into an agreement with the SPC, under Article 15.5 of the Treaty, placing its collection of PGRFA within the purview of the Treaty. The agreement was accordingly signed on 1 June 2009.

The workshop on the implementation of the ITPGRFA, which was held in Fiji in September 2009, recommended that countries consider taking full advantage of the services of SPC as an agent for the Pacific Island countries in implementing the multilateral system on their behalf and, when requested, in granting access to materials to countries outside the region. Proposals formalizing such a scheme have now been drawn up and were endorsed by the Pacific heads of agriculture and forestry services (HOAFS) in September 2010. Meanwhile, the SPC is also developing recommendations for the HOAFS regarding the possible use of the SMTA for the exchange of all Annex 1 material by countries within the region, whether they are contracting parties to the Treaty or not, as well as reviewing the multilateral system currently in use in the region for the exchange of non-Annex 1 material and developing appropriate recommendations for the HOAFS at the next session.

In other regions, countries are also considering strengthening the interface between the regional cooperation activities and the ITPGRFA and its multilateral system, particularly with respect to reporting functions. In this regard, the workshop organized under the Joint Programme for the SADC region in September 2008 noted that the SPGRC data and information system could assist in reporting what is in the system but that the SPGRC capacity in this regard should be strengthened. Since reporting on the use of SMTAs is the responsibility of the contracting party, a review of the SPGRC mandate would be required to be able to report movement on behalf of the members. Additional measures that may be undertaken to enhance information sharing are improving the information technology infrastructure, linking the SPGRC website to the Treaty's website and establishing rosters of experts and specialist working groups within the SPGRC.

Increasing the efficiency of the ITPGRFA and its multilateral system is also one of the objectives of the AEGIS system in the European region. AEGIS is being developed and will operate within the framework of the ITPGRFA in a manner consistent with the Treaty's objectives.[21] Indeed, AEGIS is the response of the European region to the Treaty's call for cooperation to establish an efficient and sustainable system of *ex situ* conservation. Wherever possible, AEGIS will use the Treaty's mechanisms, procedures and instruments, including the SMTA, and will thereby contribute to its effective implementation. The parties to AEGIS are also considering the ways in which the AEGIS system can cooperate with respect to communication between the Treaty and the parties on the operation of the multilateral system.

Development of a global information system

A third main area of global cooperation, which complements the first two areas, and, indeed, is also a component of all of the initiatives in these areas, is that of information sharing. The ITPGRFA in its Article 17 calls for cooperation to develop and strengthen a global information system to facilitate the exchange of information on scientific, technical and environmental matters relating to PGRFA. Information on the location and availability of PGRFA is only one dimension of the global information system, but it is an important one. PGRFA can only be used in plant breeding if there is adequate information on what PGRFA is available and where it is located. There are a number of information systems on PGRFA already in existence. The main examples are the System-Wide Information Network for Genetic Resources (SINGER), which links the gene banks of the CGIAR centres, Genetic Resources Information Network (GRIN), which is the US information network now being expanded with the support of the GCDT, Bioversity International and the Treaty's Secretariat to become an information management system available to countries throughout the world, and the European Genetic Resources Search Catalogue, which is the European information network.

What is lacking is the development of common data standards and a global information portal to link these existing systems together. Such a system is now being provided through a cooperative project (GENESYS) between Bioversity International, the GCDT and the Treaty's Secretariat. The project will deliver data standards for agro-biodiversity and data exchange protocols; the development of GRIN-Global and an on-line global germplasm information portal that will serve as a one-stop entry point on the characteristics and collecting sites of accessions held in gene banks worldwide (Arnaud et al., 2010). Currently, the GENESYS portal provides access to accession-level information on over two million accessions of the world's most important food crops, with a focus on the following 22 crops: bananas, barley, beans, breadfruit, cassava, chickpeas, coconuts, cowpeas, faba beans, finger millet, grass peas, lentils, maize, pearl millet, pigeon peas, potatoes, rice, sorghum, sweet potatoes, taro, wheat and yams. While the initial work on the establishment of the GENESYS portal is encouraging, there is still a good way to go before the portal is fully functional.

Areas in which further international cooperation is needed

The ITPGRFA is still young, and more in-depth cooperation at both the global and the regional levels is needed to ensure that its full benefits can be brought to fruition. Mention has already been made of the crucial importance of continuing the work on the development of a global information system. Critical to the success of the Treaty will also be continued cooperation in establishing, and where necessary strengthening, regional and crop-based networks.

Areas in which further global and regional cooperation will be essential includes *in situ* and on-farm conservation and plant breeding. The GCDT is tangible evidence of what can be achievable in the area of *ex situ* conservation. However, some long-term sustainable mechanism must also be found for *in situ* and on-farm conservation. The multilateral system provides a mechanism for facilitating access to important PGRFA and reducing the transaction costs of seeking access on a bilaterally negotiated basis. However, more needs to be done to ensure that tangible benefits reach the farmers in developing countries. This means developing new cooperative systems to ensure that adequate financial resources are available to the Treaty and to the farmers themselves. New cooperative schemes that would allow for the generation of these financial resources need to be sought, both from the food industry and from the seed industry.

Equally important, more resources need to be sought in order to ensure that developing countries have the access to breeding capacity that will enable them to take full advantage of the potential provided by facilitated access to PGRFA. What is the point of establishing a system of facilitated access to the most important PGRFA for food security, if developing countries lack access to breeding capacity? Indeed, even improved breeding capacity will not be sufficient if seed distribution systems are not in place that will allow improved seeds to reach the farmers themselves in the most appropriate form.

To a certain extent, the multilateral system established by the ITPGRFA is already working. Over 1.15 million samples of PGRFA, mainly improved material, have been distributed by the CGIAR Centres alone over the first three years of operation (ITPGRFA, 2011). Most of this material (84 per cent) has gone to developing countries. Large amounts of PGRFA have also been distributed by both developed and developing countries. However, many countries are still looking for help in setting up their own implementation of the multilateral system as well as, indeed, other important elements of the Treaty. Further international cooperation at both the global and regional levels will be required to ensure that this help is available where requested.

Conclusions

The ITPGRFA places great emphasis on the importance of international cooperation in achieving its objectives. One of the main areas for cooperation laid down in the Treaty is providing access to, sharing and exchanging PGRFA and appropriate information and technology, as covered by the multilateral system according to Part IV of the Treaty. Some seven years after the entry into force of the Treaty, the multilateral system is still in the early stages of implementation. One element of this system, the distribution of PGRFA, is already functioning quite effectively.

Perhaps the most obvious target area for cooperation has been *ex situ* conservation, where significant efficiencies are to be gained through international cooperation. Action has already been taken to set up mechanisms for ensuring long-term secure funding of the most important collections of unique PGRFA essential for food security. Developments at the regional level demonstrate an understanding of the need to share responsibilities and resources and to take advantage of the opportunities for rationalization of the system of *ex situ* conservation provided by the Treaty's multilateral system.

Cooperation is now focusing more on the area of capacity building, on-farm conservation and sustainable utilization as well as information sharing. One encouraging development is the development of new initiatives, such as the Crop Wild Relatives Project, which span the whole spectrum of the ITPGRFA's activities, from conservation to sustainable utilization and capacity building, to meet the final objectives of improving sustainable agriculture and achieving food security in the face of the challenges of climate change. Already there are examples of the benefits that can be achieved through international cooperation in the framework of the Treaty, such as the cooperative initiative between the Pacific and West African regions to mitigate the effects of taro leaf blight.

Further work is needed to establish an effective global information system, capacity building in plant breeding and seed distribution systems and *in situ* and on-farm conservation. Cooperation in strengthening regional and crop networks is also essential as well as assistance in the establishment of national systems for the implementation of the Treaty.

Notes

1 International Treaty on Plant Genetic Resources for Food and Agriculture, 29 June 2004, www.planttreaty.org/texts_en.htm (last accessed 15 May 2011) [ITPGRFA].
2 ITPGRFA, *supra* note 1, preamble, para. 3: 'Cognizant that plant genetic resources for food and agriculture are a common concern of all countries, in that all countries depend very largely on plant genetic resources for food and agriculture that originated elsewhere.'
3 An Inter-Regional Workshop on Crop Genetic Resources Networks, convened by the European Cooperative Programme on Plant Genetic Resources, the European Community-funded project Diverseeds and Bioversity International and held at the Bioversity headquarters in December 2007, described some of the benefits provided by regional networks in the light of new developments, including the entry into force and implementation of the ITPGRFA.
4 The breeding programmes of the International Agricultural Research Centres (IARCs) of the Consultative Group on International Agricultural Research are a major source of improved PGRFA on which many developing countries are heavily dependent. Indeed, the amount of improved material distributed by the IARCs dwarfs the amount of unimproved material distributed by a factor of three.
5 Bioversity International, the International Center for Tropical Agriculture, the International Centre for Maize and Wheat Improvement, the International Potato Center, the International Centre for Agricultural Research in the Dry Areas, the World Agroforestry Centre, the International Crop Research Institute for the Semi-Arid Tropics, the International Institute of Tropical Agriculture, the International Livestock Research Institute, International Rice Research Institute and the West Africa Rice Development Association.
6 On average, land races and wild relatives make up 16 per cent of national collections. The figure for collections held by IARCs is 73 per cent (Moore and Tymowski, 2005, 119).
7 ITPGRFA, *supra* note 1, Article 15.
8 Standard Material Transfer Agreement, 16 June 2006, ftp://ftp.fao.org/ag/agp/planttreaty/agreements/smta/SMTAe.pdf (last accessed 15 May 2011) [SMTA].
9 The Governing Body approved a series of footnotes clarifying that the use of terms such as 'multilateral system' or 'Annex 1' in the SMTA should not be interpreted as precluding the use of the SMTA for non-Annex material.
10 Convention on Biological Diversity, 31 ILM 818 (1992) [CBD].
11 ITPGRFA, *supra* note 1, Article 14. The Constitution of the Global Crop Diversity Trust requires the trust to promote 'an efficient goal-oriented, economically efficient and sustainable global system of ex situ conservation' in accordance with both the ITPGRFA and the Global Plan of Action.
12 GPA, *supra* note 11 at para. 82.

13 *Ibid.* at para. 84.
14 A total of US $4.2 million to date.
15 Albania, *Armenia, Austria,* Azerbaijan, Belarus, *Belgium,* Bosnia and Herzegovina, *Bulgaria, Croatia, Cyprus, Czech Republic, Denmark, Estonia, Finland, France,* Georgia, *Germany, Greece, Hungary, Iceland, Ireland,* Israel, *Italy, Latvia, Lithuania, Luxembourg,* Macedonia, Malta, Moldova, Montenegro, *Netherlands, Norway, Poland, Portugal, Romania,* Russian Federation, Serbia, Slovakia, *Slovenia, Spain, Sweden, Switzerland, Turkey, United Kingdom* and Ukraine (countries that are in italics are currently contracting parties to the ITPGRFA).
16 Thirty-one of the 45 countries in the region plus the European Community are contracting parties to the ITPGRFA, *supra* note 1.
17 At the present time, plant genetic resources for food and agriculture (PGRFA) in Europe are conserved in some 500 institutions scattered over more than 40 European countries.
18 Ouagadougou Declaration on Regional Cooperation for Effective and Efficient ex situ Conservation of PGRFA in West and Central Africa, adopted in September 2006. It was adopted by a regional conference organized by the West and Central African Council for Agricultural Research and Development with the support of Bioversity International, the Genetic Resource Policy Initiative and the Global Crop Diversity Trust.
19 The South African Development Community's Plant Genetic Resources Centre's Treaty Implementation Workshop, Lusaka, Zambia, 18 September 2008.
20 Regional cooperation in the conservation of PGRFA is particularly relevant in the Pacific region, given the importance of vegetatively propagated crops, which are expensive to conserve.
21 The objectives of the ITPGRFA, as set out in Article 1, are the conservation and sustainable use of PGRFA and the fair and equitable sharing of the benefits arising out of their use, in harmony with the CBD, *supra* note 10, for sustainable agriculture and food security.

References

Arnaud, E., Dias, S., Mackay, M., Cyr, P.D., Gardner, C.A., Bretting, P.K., Kinard, G.R., Guarino, L. (2010) 'A global portal enabling worldwide access to information on conservation and use of biodiversity for food and agriculture' in L. Maurer and K. Tochtermann (eds), *Information and Communication Technologies for Biodiversity Conservation and Agriculture*, Shaker Verlag, Aachen, Germany.

European Cooperative Programme for Plant Genetic Resources (ECPGR) (2009) *A Strategic Framework for the Implementation of a European Genebank Integrated System (AEGIS)*, Discussion Paper, Bioversity International, Rome.

International Treaty on Plant Genetic Resources for Food and Agriculture (ITPGRFA) (2011) *Experience of the International Agricultural Research Centres of the CGIAR with the implementation of the agreements with the Governing Body, with particular reference to the use of the standard material transfer agreement for Annex 1 and non-Annex 1 crops*, report submitted to the Fourth Session of the Governing Body of the Treaty, Doc. IT/GB-4/11/Inf.05.

Moore, Gerald, and Witold Tymowski (2005) *Explanatory Guide to the International Treaty on Plant Genetic Resources for Food and Agriculture*, IUCN, Gland, Switzerland and Cambridge, UK.

Zehni, M.S. (2006) *Regional Strategy for the Conservation of Plant Genetic Resources in West Asia and North Africa (WANA)*, IPGRI, the Global Crop Diversity Trust, ICARDA and AARINENA, Rome, Aleppo and Amman.

18

THE EVOLVING GLOBAL SYSTEM OF CONSERVATION AND USE OF PLANT GENETIC RESOURCES FOR FOOD AND AGRICULTURE

What is it and where does the Treaty fit in?

Toby Hodgkin, Nicole Demers and Emile Frison

Introduction

Collaboration between organizations and institutions in different countries has always been a characteristic of work on the conservation and use of plant genetic resources for food and agriculture (PGRFA). This collaboration has ranged from informal efforts between individual researchers, conservation workers and plant breeders in different countries to more formal attempts to establish collaborative arrangements and develop activities and institutions able to provide a regional or global oversight or support activities that require the participation of different countries. The importance of transnational collaboration on PGRFA to the development and maintenance of national agricultural production and food security has been clearly demonstrated (Palacios, 1998), and the arguments for ensuring that mechanisms are needed that facilitate collaboration in regard to conservation and use are well known (Cooper et al., 1994). Collaboration between institutes and organizations around the world increases the effectiveness of conservation since the work can be shared between gene banks and conservation agencies in different countries. Such cooperation enables countries to focus conservation efforts on what they consider to be their own most important priorities with respect to the conserved species, populations and accessions, knowing that other countries complement their efforts. Collaboration can also help improve the security of conservation through planned duplication of collections and cooperation (most often through regional or crop networks) on the various activities associated with conservation (for example, identification of species, populations and varieties for conservation, collection of materials, storage, regeneration and distribution). International collaboration further contributes to the availability and accessibility of information and of conserved materials and facilitates the use of such materials in research, breeding or, more directly, by other users. Increasingly, varieties of major crops involve the use of parental material from several different countries – for example, the pedigree of the Veery wheats distributed by the International Centre for Maize and Wheat Improvement includes varieties or lines from over 15 countries (Fowler and Hodgkin, 2004).

Over the last 40 years, international cooperation on the conservation and use of PGRFA has developed significantly and taken a range of forms, which have included the establishment of institutional mechanisms and structures, sharing information, identifying priority activities and creating frameworks that can further strengthen cooperation (Halewood and Nnadozie, 2008). The various mechanisms that have been developed to support collaboration have included ad hoc project-based groupings, crop and regional networks, institutional collaborative arrangements such as the Consultative Group on International Agricultural Research's (CGIAR) System-wide Genetic Resources Programme (SGRP) and international programmes, bodies and treaties.[1]

The Food and Agriculture Organization (FAO) has been the focal point for many of these international initiatives and, since 1983, has provided the administrative and technical support for the Commission on Genetic Resources for Food and Agriculture (CGRFA), an intergovernmental forum that has overseen the creation and development of a 'Global System for the Conservation and Sustainable Use of Plant Genetic Resources' (FAO, 2010). The *Second Report on the State of the World's Plant Genetic Resources (SOWPGR-2)* states that '[t]his system, managed and coordinated by FAO, aims to ensure the safe conservation and promote the availability and sustainable use of PGR' (*ibid.*). The elements of the global system as described in the *SOWPGR-2* report include the Global Plan of Action (GPA) for the Conservation and Sustainable Use of PGRFA, the periodic state of the world reports (FAO, 1998, 2010), the international network of *ex situ* collections, the International Treaty on Plant Genetic Resources for Food and Agriculture (ITPGRFA), and the CGRFA itself.[2] The report notes as important developments of the global system in the last ten years the establishment of the Global Partnership Initiative for Plant Breeding Capacity Building (GIPB) and the establishment of a formal memorandum of cooperation between the FAO and the Convention on Biological Diversity (CBD).[3] These elements of the global system as described by the FAO, together with other elements that make a contribution to international collaboration on the conservation and use of PGRFA, are further described later in this chapter and in Table 18.1.

Although the FAO's global system includes many of the international efforts on conservation and use of PGRFA that have resulted from various decisions by countries to cooperate, it does not include all of the different organizations, operations and activities that are supporting this international cooperation. A more comprehensive view of the global system would include these additional elements. This would also provide the necessary basis for an analysis of international collaboration using a system analysis perspective that would focus on the effectiveness of international collaborative elements in terms of their contribution to a set of generally agreed system functions. Using this approach, the FAO's global system can be seen as a part of a larger system that includes all of the different elements that contribute to the different international objectives of PGRFA's conservation and use. This more comprehensive systems approach would allow for the development of an overall framework for a global system in which the identification of the desired outcomes of plant genetic resources conservation and use helps determine system functions and operations. This approach, in turn, would allow for an analysis of the roles and effectiveness of all of the different elements in terms of their contribution to system function and overall objectives. In this chapter, we present a framework that allows the different components of this wider system to be identified and consider some of the current features of its operation with respect to the ITPGRFA.

The nature of a conservation and use system

Broadly speaking, a system can be described as any set of interconnected elements that carry out a particular function or set of functions. For example, the international postal system or air traffic control system constitute examples of global systems with established mechanisms of cooperation, practices, institutions and administrative arrangements aimed at achieving the specific outcomes of, respectively, the international movement of letters or the safe and efficient regulation of air travel and the benefits that flow from these outcomes.

A system may involve deliberate design with respect either to specific elements or to the whole, but usually it involves a mixture of designed elements and those that have developed in a more ad hoc way.[4] A central aspect of understanding a system is an emphasis on system operations and a concern with the extent to which the overall system objectives are achieved. The focus on the elements themselves (for example, gene banks, protected areas, accessions, descriptors, sample numbers and databases) is replaced by a concern with the contribution that these elements make to the overall operation of the system and to achieving the identified objectives. Viewed in this way, the different elements of the system can be identified and described, not just as a set of individual components but, rather, out of a concern for the ways in which they function and are connected, for the pathways that connect them, for the effectiveness of these pathways and, ultimately, for the contribution that they make to the overall outcomes or benefits that are desired from the system.

One feature of this systems-based approach is its concern with the effectiveness of the linkages and connections within the system. Another feature is the recognition that different users and participants in the system will differ in their evaluation of it and reflect their own specific interests in particular outcomes and components. For example, in the case of a global system of conservation and use of PGRFA, plant biology research workers, plant breeders, conservationists, farmers and rural communities have different needs and interests and may view the functionality of the system from quite different perspectives. They may also come to different conclusions about its effectiveness in meeting their expectations or their needs.

The objectives of the conservation of PGRFA are to ensure that the necessary diversity of crops, their wild relatives and other useful plants is available to meet present and future societal needs and is being used to this end. The inclusion of a specific reference to the use of plant genetic resources reflects the view that conservation is not an end in itself and that the use of plant genetic resources to improve production is both desirable and necessary. It also reflects a view that conservation activities should be carried out in a way that supports the use of genetic resources. Achieving these objectives involves different activities carried out in different places by various actors and organizations. As argued earlier, the activities and organizations can be thought of as being linked together, to a greater or lesser extent, and, thus, as constituting a functioning system.

It is worth noting that this systems approach could be applied at various scales, including the functioning of individual gene banks and national plant genetic resources systems. However, in this chapter, the approach is used to explore the multinational dimensions of the system and the contribution of identifiably international components in achieving agreed global objectives, outcomes and benefits. Figure 18.1 presents a conceptual framework for the global plant genetic resources system.[5] It identifies the main benefits that are desired from the system, the outcomes needed to achieve these benefits and the kinds of activity that are commonly undertaken to achieve the outcomes. Two classes of benefit can be identified from PGRFA conservation and use: social or human benefits and environmental benefits. Social or direct human benefits arising from the conservation and use of PGRFA include improved production and productivity from agriculture or from harvesting useful wild plants for medicines, fuel and other purposes. These benefits provide improved incomes and livelihoods for rural communities and can support the maintenance of traditional cultures. They can also provide improved food security, food safety and nutrition to both rural and urban communities (Frison et al., 2011).

Environmental benefits that result from the conservation and use of PGRFA include contributions to ecosystem service provision – for example, control of pests and diseases, pollination, maintenance of soil health and control of erosion (Frison et al., 2011; Hajjar et al., 2008). Other environmental benefits that have been identified include resilience and adaptability. For example, diversity in production systems has been shown to contribute to resilience by buffering agro-ecosystems that experience shocks such as hurricanes (PAR, 2010), and rural communities meet the challenge of climate change in part through the enhanced use of genetic resources that provide the required adaptability.

It is expected that in increasing total food production (by 70 per cent by 2050) (FAO, 2006), improving agricultural sustainability and meeting the challenges of climate change, the demand for PGRFA will increase, and the movement of genetic resources (particularly its international movement) will grow in importance (Burke et al., 2009). Three outcomes from the conservation system are necessary to ensure that it is able to contribute to the benefits that have been identified. The first outcome is maintenance, and it involves the conservation of the diversity that is present in PGRFA in ways that ensure completeness of coverage for the available diversity, security for the materials conserved and durability for the conservation effort over time. Simply conserving the material is not in itself sufficient, it must be accessible for those who have a use for it (Fowler and Hodgkin, 2004). The second outcome is the availability of material. It involves ensuring that the material conserved is healthy and viable and can be distributed to all users as required. Ideally, users should also be helped to identify what material may be most relevant to their needs. Thus, accessibility and availability are active concepts that should involve activities designed to enhance use. The third outcome is the use of PGRFA. The use of PGRFA includes the identification of materials with desired traits through characterization and evaluation, the use of PGRFA in research, pre-breeding and breeding programmes and the direct use of accessions in production systems by farmers, communities and other individuals. The extent of use is, in part, reflected by the presence of PGRFA in cultivars as well as the amount of diversity in production systems. Ideally, PGRFA flows through the system to provide this diversity and the varieties needed to secure the desired environmental and social benefits. The connections that exist between the different components that contribute to the identified outcomes are therefore crucially important to system function.

While it is not identified as an outcome in this scheme, a comprehensive knowledge of the materials within the system is an important prerequisite for its effective operation. This knowledge includes understanding the extent and distribution of diversity, the determination of the effectiveness of conservation procedures, the development of information systems that support availability and use and an improved understanding of the genetics of useful traits and the ways they can be optimally used in new varieties. It also involves understanding the social, economic and cultural aspects of the maintenance and use of PGRFA. Such understanding is necessary for all of the outcomes, and, thus, a strong knowledge generation component is an essential part of the system.

Achieving the identified outcomes involves a range of activities that are usually carried out by a number of different organizations and institutions that need, therefore, to be more or less linked to each other. As Figure 18.1 indicates, these activities

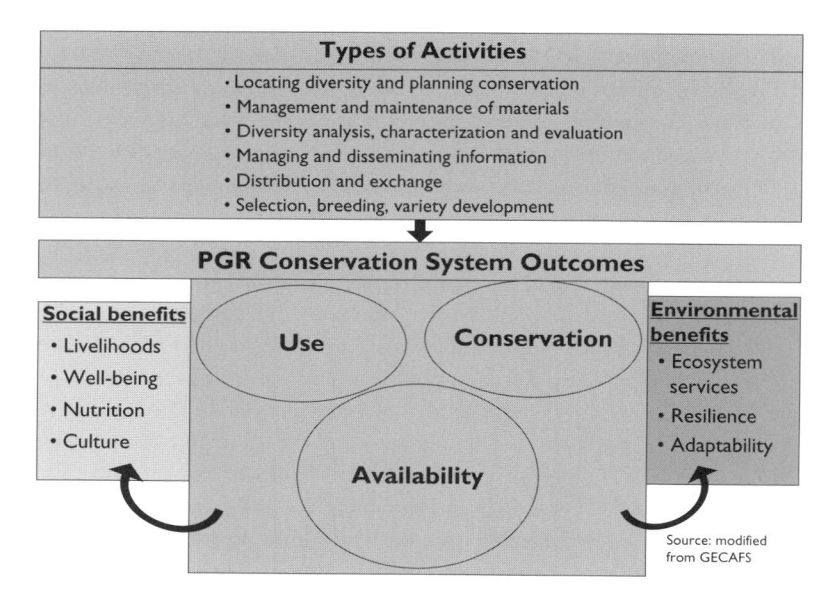

FIGURE 18.1 System-level description of conservation of PGRFA in terms of benefits and outcomes.

include locating diversity that needs to be conserved; undertaking *in situ* conservation activities or collecting and conducting the *ex situ* conservation, characterization and evaluation of PGRFA accessions and populations; the development and maintenance of information systems; the multiplication and distribution of materials; and the various activities associated with breeding, variety release and seed distribution. Each of these actions, of course, can themselves be broken down into a number of different activities so that *ex situ* conservation may involve seed storage, cryopreservation, tissue culture or the use of field gene banks. Each also involves specific organizations and institutions including gene banks, protected areas, universities, research institutes and the breeding programmes of the public and private sectors. The extent to which these different organizations are linked to each other will affect the overall functioning of the system.

The operational elements of the global system

Most of the efforts to conserve and use PGRFA are undertaken as part of the work of organizations and institutions that constitute national programmes.[6] National gene banks are responsible for the conservation of about 90 per cent of the *ex situ* conserved materials held around the world. *In situ* conservation of crop wild relatives and on-farm conservation efforts are undertaken as part of the work of national conservation programmes or directly by farmers and rural communities. The relevant organizations of national programmes are responsible for the bulk of within-country distribution of genetic resources, although international centres may play an important role in transnational distribution. The research and breeding work associated with utilization is carried out by universities, research centres and private breeding companies. From a conservation perspective, the major exceptions to the national nature of most of the activities are the conservation, distribution and use activities of the International Agricultural Research Centres (IARCs) of the CGIAR and other international or regional centres (for example, the Asian Vegetable Research and Development Center (AVRDC) and the Centro Agronómico Tropical de Investigación y Enseñanza (CATIE)), which also undertake a variety of crop improvement activities for a selected set of crops. International plant-breeding companies also operate within an international framework, although their contribution to global conservation and the global availability of the resources they conserve is small (FAO, 2010).

Table 18.1 lists the different elements of the global system for the conservation and sustainable use of PGRFA, as identified by the FAO (2010), as well as other components that contribute to the global system in terms of its function and the achievement of outcomes that are regional or fully international in scope. Most of these elements are described in the FAO's two reports on the state of the world's PGRFA (FAO, 1998, 2010, chapters 6 and 7). These different elements or components have often developed in a largely ad hoc manner and are still evolving in regard to both their nature and function. In describing the global system, the FAO has identified the different elements as belonging to one of the following groups: (1) agreements, (2) cooperation, (3) information and (4) action. In this typology, the CGRFA is seen as the hub

TABLE 18.1 Multinational elements of the current global system

Role or function in the system	Element*	Aims and notes
	International agreements	
Agreement	Food and Agriculture Organization's (FAO) Commission on Genetic Resources for Food and Agriculture (CGRFA) www.fao.org/nr/cgrfa/cgrfa-home/en/ (last accessed 4 April 2011)	The Commission strives to halt the loss of PGRFA and to ensure world food security and sustainable development by promoting their conservation, sustainable use, including exchange and the fair and equitable sharing of the benefits arising from their use. It covers animal, fish, forest and microbial genetic resources as well as crosscutting issues and ecosystem perspectives. It has developed a multi-year programme of work to guide its work.
	International Treaty on Plant Genetic Resources for Food and Agriculture (ITPGRFA) www.planttreaty.org/ (last accessed 4 April 2011)	Its objectives are the conservation and sustainable use of PGRFA and the fair and equitable sharing of the benefits arising out of their use. The Treaty aims at recognizing the enormous contribution of farmers to the diversity of crops that feed the world; establishing a global system to provide farmers, plant breeders and scientists with access to plant genetic materials and ensuring that recipients share benefits they derive from the use of these genetic materials with the countries where they have originated.
	International Code of Conduct for Plant Germplasm Collecting and Transfer www.fao.org/nr/cgrfa/cgrfa-global/cgrfa-codes/en/ (last accessed 5 April 2011)	It aims to promote the rational collection and sustainable use of genetic resources, prevent genetic erosion and protect the interests of both donors and collectors of germplasm. It sets out minimum responsibilities of collectors, sponsors, curators and users of collected germplasm in the collection and transfer of plant germplasm. It was adopted by the FAO conference in 1993 and was negotiated through the CGRFA, which also has the responsibility of overseeing its implementation and review.

TABLE 18.1 (continued)

Role or function in the system	Element*	Aims and notes
	Convention on Biological Diversity's (CBD) Programme of Work on Agricultural Biodiversity www.cbd.int/agro/pow.shtml (last accessed 4 April 2011)	These programmes aim to promote the positive effects and mitigate the negative impacts of agricultural practices on biodiversity in agro-ecosystems and their interface with other ecosystems; the conservation and sustainable use of PGRFA and the fair and equitable sharing of benefits arising out of the utilization of these genetic resources. The CBD Programme of Work on Agricultural Biodiversity was last reviewed in 2008. Decision no. X/34 at the tenth Conference of the Parties (COP-10) drew attention to the importance of work on crop wild relatives and agreed on collaboration with the CGRFA, the ITPGRFA and the FAO on identified activities. At COP-10, the parties agreed to adopt the Nagoya Protocol on Access and Benefit Sharing.
	Regional agreements in Asia, Africa, South America and Europe	
Cooperation	Regional networks	This division includes about 18 regional and sub-regional networks identified in the second *Report on the State of the World's Plant Genetic Resources*.
	Crop networks	The objectives are usually to support all work on a particular crop. They often have a strong emphasis on genetics and breeding.
	Thematic networks	These include, for example, Crops for the Future, which is concerned with underutilized species; the International Union for the Conservation of Nature's (IUCN) specialist group on crop wild relatives; and Botanic Gardens Conservation International.

TABLE 18.1 (continued)

Role or function in the system	Element*	Aims and notes
	International fora and associations with interests in PGRFA	Such fora include, for example, Diversitas, the IUCN and the Global Forum on Agricultural Research's Intergovernmental Science-Policy Platform on Biodiversity and Ecosystem Services.
	Regional fora and associations with interests in PGRFA	These exist for each region although with somewhat differing structures and organizations and concerns for PGRFA. They include the Forum for Agricultural Research in Africa (FARA), the Asia-Pacific Association of Agricultural Research Institutions (APAARI), the Forum for the Americas on Agricultural Research and Technology Development (FORAGRO), the Central Asia and the Caucasus Association of Agricultural Research Institutions (CACAARI) and the Association of Agricultural Research Institutions in the Near East and North Africa (AARINENA).
Information	**Global Information and Early Warning System on PGRFA** www.fao.org/giews/english/index.htm (last accessed 5 April 2011)	Its mandate is to keep the world's food supply/demand situation under continuous review, issue reports on the world's food situation and provide early warnings of impending food crises in individual countries. For countries facing a serious food emergency, the FAO, the Global Information and Early Warning System and the World Food Programme also carry out joint crop and food security assessment missions.
	State of the World's Plant Genetic Resources for Food and Agriculture www.fao.org/nr/cgrfa/cgrfa-global/cgrfa-globass/en/ (last accessed 4 April 2011)	It assesses the state of plant genetic diversity and capacities at the local and global levels for *in situ* and *ex situ* management, conservation and utilization of PGRFA.

TABLE 18.1 (continued)

Role or function in the system	Element*	Aims and notes
	GENESYS www.genesys-pgr.org/ (last accessed 4 April 2011)	GENESYS is currently being developed to improve global information exchange of PGRFA in the attempt to secure and enhance biodiversity throughout the world. It aims to give breeders and researchers a single access point to information on approximately one-third of the world's gene bank accessions.
Resource mobilization	**International Fund and Financial Mechanism for Plant Genetic Resources** www.globalplanofaction.org/servlet/CDS Servlet?status=ND1zb3VyY2Vzlj̇Y9ZW4mMzM9KiYzNz1rb3M~ (last accessed 4 April 2011)	The objective of this funding strategy is to enhance the availability, transparency, efficiency and effectiveness of the provision of financial resources to implement activities under the ITPGRFA. The aims of the funding strategy are, among others, the development of ways and means by which adequate resources are available for the implementation of the ITPGRFA, in accordance with Article 18 of the Treaty.
	Global Crop Diversity Trust www.croptrust.org/main/ (last accessed 4 April 2011)	The trust is raising funds from individual, foundation, corporate and government donors for an endowment fund that will support the conservation of key crop collections in perpetuity.
	Global Environment Facility (GEF) www.thegef.org/gef/home (last accessed 4 April 2011)	An independent financial organization, it provides grants to developing countries and countries with economies in transition for projects related to biodiversity, climate change, international waters, land degradation, the ozone layer and persistent organic pollutants. Although it mostly supports country projects, the GEF has an agreed global strategy and strategic objective on mainstreaming conservation, which is relevant to the conservation and use of PGRFA. The UN Environment Programme's GEF has provided over US $100 million in support of multinational projects over the last ten years.

TABLE 18.1 (continued)

Role or function in the system	Element*	Aims and notes
Concerted action	**International Network of *Ex Situ* Collections (including the Consultative Group on International Agricultural Research (CGIAR), in-trust collections plus the Centro Agronómico Tropical de Investigación y Enseñanza and the International Coconut Genetic Resources Network)** www.fao.org/nr/cgrfa/cgrfa-about/cgrfa-history/cgrfa-internnet/en/ (last accessed 5 April 2011)	In 2006, in accordance with Article 15 of the ITPGRFA, the International Agricultural Research Centres placed their *ex situ* gene bank collections under the ITPGRFA. The Article 15 agreements replace the former agreements concluded between the centres and the FAO in 1994.
	Network of *in situ* conservation areas	Two relevant networks that already exist are the Globally Important Agricultural Heritage Systems and the Man and Biosphere Programme.
	Svalbard Global Seed Vault www.regjeringen.no/en/dep/lmd/campain/svalbard-global-seed-vault.html?id=462220 (last accessed 4 April 2011)	It is designed to store duplicates of seeds from seed collections around the globe. Many of these collections are in developing countries. If seeds are lost – for example, as a result of natural disasters, war or simply a lack of resources – the seed collections may be re-established using seeds from Svalbard.
	Global Partnership Initiative for Plant Breeding Capacity Building (GIPB) http://km.fao.org/gipb/ (last accessed 4 April 2011)	Its mission is to enhance the capacity of developing countries to improve crops for food security and sustainable development through better plant breeding and delivery systems. The longer-term vision of success of this initiative is the improvement in crop performance and food security based on the establishment of enhanced sustainable national plant breeding capacity.

TABLE 18.1 (continued)

Role or function in the system	Element*	Aims and notes
	Global Plan of Action (GPA) for the Conservation and Sustainable Use of Plant Genetic Resources for Food and Agriculture www.globalplanofaction.org/index_en.jsp (last accessed 4 April 2011)	It might also be classed as an agreement but is placed here because of its emphasis on the actions that need to be undertaken to support global conservation objectives. The GPA's main objectives are to: ensure the conservation of PGRFA as the basis of food security; promote sustainable use of PGRFA to foster development and reduce hunger and poverty; promote the fair and equitable sharing of the benefits arising from the use of PGRFA; assist countries and institutions to identify priorities for action; strengthen existing programmes; and enhance institutional capacity.
	National collections placed under the multilateral system	Included here to emphasize the point that national collections are as much part of the global system as international collections and, once placed under the multilateral system, become an effective global resource.
Other support (often in more than one area)	International non-governmental organizations	These organizations include the IUCN, the Botanic Gardens Conservation International as well as civil society organizations with a commitment to specific objectives in regard to PGRFA conservation, such as ETC Group (Action Group on Erosion, Technology and Concentration http://www.etcgroup.org), GRAIN, Practical Action and so on.
	International research efforts	Include the research and breeding activities of the CGIAR and other international and regional centres.

Notes:
There are a number of international agreements that affect the use of PGRFA through their effect on the release, availability and distribution of crop varieties and seeds. These include the International Convention for the Protection of New Varieties of Plants, the Convention on the Prior Informed Consent Procedure for Certain Hazardous Chemicals and Pesticides in International Trade, world trade regulations and a range of seed certification schemes that operate internationally and regionally. Although not usually regarded as part of the global system of PGRFA conservation and use, their effect on PGRFA use and on the amount of diversity likely to be found in production systems may be significant.

* In bold if included in the FAO's description of the global system.

of the global system, but it would presumably be better classed under agreements. In Table 18.1, we classify these elements slightly differently: (1) agreements, (2) cooperation arrangements, (3) information-providing facilities, (4) resource mobilization provisions and (5) concerted actions. These are the main international mechanisms used in the global system to achieve the agreed objectives. While a number of organizations make contributions to different mechanisms (as in the case of the Global Crop Diversity Trust (GCDT) providing funding and implementing a global project on regeneration), most have a major focus in one area.

The components of the global system listed in Table 18.1 complement and support the work of national programmes (and of farmers and rural communities), assisting with, and facilitating, the multinational or international aspects of their work. In this sense, the global system and its various different elements can be perceived as adding to the efforts of national programmes, supporting specific areas of work necessary to achieving the outcomes both nationally and internationally. Many of the elements of the global system seek to strengthen and support national programmes and other country-based activities (for example, the GPA and the International Fund and Financial Mechanism for Plant Genetic Resources), while others, such as the FAO's *SOWPGR* or the Global Information and Early Warning System, reflect or inform on the state of development and activity of national programmes.

The operations and functions of the global system

Certain features of conservation, availability and use seem particularly important from a global point of view and thus likely to be of particular concern to the international elements of the system (see Table 18.2). With respect to conservation, these features would include the overall representativeness, or completeness,[7] of conservation for

TABLE 18.2 Some aspects of PGRFA conservation and use where international elements can play a significant role beyond that of the national programmes

Desired outcome	Global system's contribution or function
Conservation	• representativeness
	• completeness
	• security
	• efficiency of resource use
	• sustainability
Availability	• proportion of conserved material available
	• extent of availability to all users
	• completeness of information systems
	• accessibility of information to users and exchange of information
	• responsiveness to global or regional threats
Utilization	• capacity for pre-breeding and breeding
	• collaboration on crop improvement and use programmes
	• detection and reduction of vulnerability

any crop's gene pool in regard to total diversity. Supporting the security of the collections through appropriate duplication is another area of international concern since safety duplication arrangements often involve different countries. Other aspects that have been identified include the efficiency of conservation and the optimum use of available resources. For the availability outcome, the components of the global system would be concerned with ensuring universal availability of the conserved materials throughout the world, helping to ensure that all materials are indeed available and supporting the transnational movement of materials. Information systems that enable users to locate materials and find out about their properties are also considered essential to availability. For utilization, the components of the global system are particularly concerned with supporting collaboration on, for example, pre-breeding or breeding efforts and with actions that support the maintenance of sufficient diversity in production systems so as to reduce the threat of genetic vulnerability (FAO, 1998).

While a full analysis or description of the operation of all of these different components of the global system, the ways in which they are connected to support the system's overall objectives, and their ultimate effectiveness, are beyond the scope of this chapter, some observations on the elements in relation to the global functioning of the system may be useful.

Conservation

The elements that support conservation are the most developed aspects of the global system and have seen continuing evolution over the past ten years. Examples include the entry into force of the ITPGRFA, the establishment of the GCDT, the formalization of the international status of the international collections, the release of funds for conservation work using globally developed perspectives and the preparation of the *SOWPGR-2*. During 2011, the CGRFA has approved a revised GPA, which has also been adopted by the 143rd FAO Council.

The elements are probably now sufficiently developed that they are beginning to have an effect on the functioning of the global system although the effect is probably still fairly patchy. Completeness and comprehensiveness of collections may have improved since the assessment made by Fowler and Hodgkin (2004), but the data are not yet available to assess the situation. The quality of a number of the collections may well have improved as a result of a project led by the GCDT on regeneration, which has also increased the numbers of accessions in the multilateral system on access and benefit sharing and, thus, improved the availability of materials. The security of collections has also improved as a result of the establishment of the Svalbard Global Seed Vault and work on the duplication of international and other collections over the past decade.

One objective of the global system has been identified as improving the efficiency of conservation (FAO, 2010). A particular concern has been with what the *SOWPGR-2* has called unplanned or redundant duplication. It notes that analyses based on the country of origin suggest that only about 25–30 per cent of the total number of accessions worldwide are distinct but notes that there are large differences according to species (FAO, 2010). The most advanced system of regional collaboration to support conservation and address this aspect of efficiency is in Europe where the European Cooperative Programme for Plant Genetic Resources is developing A European Gene Bank Integrated System (AEGIS) for PGRFA, aimed at conserving the genetically unique and important accessions for Europe and making them available for breeding and research. In addition to its work on comprehensiveness and efficiency of conservation efforts with respect to European materials, the initiative also addresses aspects of quality. The *ex situ* conservation of germplasm is carried out according to common, agreed quality standards, independently of where the germplasm is physically located, in such a way as to facilitate close linkages with *in situ* conservation.[8]

The international gene banks of the CGIAR, together with those of the AVRDC and CATIE, make perhaps the most substantial direct international contribution to the global system's conservation objectives. Currently, the IARCs conserve some 700,000 accessions of 3,446 species of the world's most important crops, forages, agroforestry species and crop wild relatives. In 2006, agreements were signed to bring all of the *ex situ* collections held by the IARCs under the framework of the ITPGRFA. These agreements are described in the *SOWPGR-2* as superseding the International Network of Ex Situ Collections. The Svalbard Global Seed Vault, which currently holds some 747,000 accessions, acts as a global repository of last resort.[9]

Most of the global system elements that support conservation (for example, the GPA, the International Fund and the GEF) have a strong country-based focus in regard to the work that they encourage or financially support. However, the GCDT has a particular concern with global aspects and explicitly seeks to ensure the comprehensiveness and security of the global conservation effort. Its funding is targeted at accessions that contribute to global comprehensiveness for crops listed under Annex 1 of the ITPGRFA (those that are part of the multilateral system).[10] While *ex situ* conservation involves significant fixed investment in gene banks, it also offers potential economies of scale, especially where global collaboration can be used to reduce the extent of what appears from a global perspective to be unnecessary duplication (many gene banks holding the same accession). However, this view may be rather simplified since it takes little account of how countries view their holdings, of the relatively small costs of holding additional accessions in already established gene banks and of the benefits that can result.[11]

There is very little global collaboration on *in situ* conservation, and global system function is clearly underdeveloped in this area with the exception of the Globally Important Agricultural Heritage Systems (GIAHS) initiative, which, at present, includes only a few specific agricultural landscapes around the world (Koohafkan, 2009). The GIAHS is largely concerned with the conservation of production systems and not explicitly with PGRFA per se. More recently, there have been explicit calls for the development of a global network of reserves that would support the conservation of crop wild relatives. The UN Educational, Scientific and Cultural Organization's Man and the Biosphere Programme, which is explicitly concerned with combining use and conservation, might be an example of how such a network could be developed on a more formal basis that could ensure the effective conservation of key landscapes and areas.

In terms of function, the global system seems underdeveloped in two other areas. The first concerns the ways in which different approaches to conservation can complement each other to achieve overall comprehensiveness. In what ways can *in situ* complement *ex situ* activities around the world and be used to maximize the conservation benefit and how can different actors and countries best collaborate? Another functionality of the global system that would benefit from further development is that of responsiveness – both to changing situations and circumstances. Areas for development could include emergency responses to specific threats of genetic erosion or the development of more coherent strategies that support response. An important development in the latter area has been the recent call for conservation proposals that respond to climate change by the International Fund under the ITPGRFA.

Availability

The contribution of the global system to the transnational availability of materials lies in the creation of agreed procedures and practices that support the access to, and movement of, PGRFA. Ensuring the availability of plant genetic resources to all bona fide users (together with appropriate benefit-sharing arrangements) was a central concern in the negotiations that resulted in the ITPGRFA (FAO, 2002). Its multilateral system provides the agreed framework of procedures and practices for parties, international organizations and legal individuals to make material available. It could be argued that the multilateral system constitutes the central achievement of the developing global system to date. While the Treaty and its multilateral system potentially ensure universal availability of the PGRFA of certain gene pools and supports availability to all users of such material, there remain a large number of crops and useful species outside the multilateral system by virtue of their non-inclusion in Annex 1 of the Treaty. In this respect, therefore, the global system remains imperfect in regard to its contribution to availability (Chapter 7 by Daniele Manzella in this volume includes a more detailed analysis of what materials are automatically included in the multilateral system and what is not in the system until it is proactively or voluntarily placed in it by the parties, international organizations or legal individuals).

Many countries that have ratified the ITPGRFA have still to put in place the necessary policies and mechanisms that will make PGRFA fully available. Improving this aspect of functionality may involve both capacity building in different countries and technical support – that is, a focus on the different operational aspects of availability. Complementing the development of the multilateral system are a number of other practical aspects of enhancing availability where the global system can play a significant role, such as the different protocols developed for the safe movement of crops and aimed at avoiding the spread of pests and pathogens while supporting germplasm exchange (FAO/IPGRI, 1989–2002). These were largely developed during the 1980s and 1990s and may require some revision to take account of changes such as the increasing concern over the spread of alien and invasive species and the expectation that climate change will require significant increases in transnational movement of PGRFA. Other aspects of availability concern the capacity of gene banks and other suppliers of PGRFA to maintain sufficient quantities of viable materials that are correctly identified so as to respond to users' requests.

The availability of information about PGRFA is central to the availability of the plant genetic resources themselves. The development of the Internet has enabled gene banks (both national and international) to ensure global availability of information about their holdings. A major initiative is underway to create a global information system.[12] GENESYS, which is supported by the ITPGRFA, the GCDT and the CGIAR, already provides access to more than 2.4 million accession-level records compiled from the System-Wide Information Network for Genetic Resources, the EURISCO web catalogue of European gene bank accessions, and the Germplasm Resources Information Network of the US Department of Agriculture. Additional information, including characterization and evaluation data, is continually being added. GENESYS not only provides users of genetic resources with direct access, but it also provides the Governing Body of the ITPGRFA with information for monitoring Treaty implementation and the GCDT with the basis for important funding decisions.

Use

While there are identifiable global system elements that contribute to conservation (particularly *ex situ*) and availability, the systems that support use are much less developed and are more disaggregated and less clear. Although the GPA and the *SOWPGR-2* both deal explicitly with aspects of use, the activities identified are generally seen as taking place at the country level. Of course, the IARCs do undertake use activities within a global framework. The materials that they develop using PGRFA are made available throughout the world, and, in this sense, they directly undertake activities that contribute to utilization as part of their role in the global system.

Most of the global collaboration on use is probably essentially crop based and is undertaken as part of the work of crop networks. There are a large number of these networks, and some of the major ones are described in the *SOWPGR-2* (FAO, 2010). Although these activities may be the most appropriate focus for the global system in support of use, they are largely disconnected from other elements of the global system, and, thus, there may be benefits in finding other ways to strengthen the linkages between these networks and other elements. There may also be a need for further initiatives, and the recent establishment of the GIPB suggests that this is the case. The GIPB is currently developing support for capacity development in pre-breeding, which is often a necessary first step in the use of plant genetic resources.

Article 6 of the ITPGRFA specifically deals with the sustainable use of PGRFA and lists a range of activities that are considered to constitute aspects of sustainable use. These include the maintenance of diverse farming systems, the promotion of appropriate plant-breeding efforts, the expansion of the use of locally adapted crops and underutilized species and the broadening of the genetic base of crops. Thus, the

Treaty explicitly recognizes the importance of use and could provide a framework for strengthening the links between conservation and use and expanding the global system in ways that take more account of utilization. The non-commercial benefit-sharing aspects of the multilateral system with their emphasis on technology transfer and capacity development might be useful in this respect.

Knowledge

The mechanisms for sharing knowledge about PGRFA, the extent and distribution of diversity, the state of conservation and the extent of use are relatively well developed at the global level, even if the available information is often incomplete with respect to specific aspects such as the extent and distribution of the diversity of many crops or the scale of genetic erosion. The *SOWPGR-2*, like its forerunner the *SOWPGR-1*, was based on country reports that were analysed and synthesized to provide a picture of the current status of conservation and the use of all PGRFA on which information was available. The Global Information and Early Warning System maintained by the FAO and updated on a continuing basis also provides a readily accessible perspective on PGRFA work around the world as do (to a limited extent) the country reports periodically submitted to the CBD, which are available through its clearing-house mechanism. Adding to this information is a continuing flow of review articles on different aspects of conservation and use, which provide up-to-date analysis of specific aspects (for examples, see Brehm et al., 2010; Fowler and Hodgkin, 2004; Jarvis et al., 2011) as well as information and reports concerning relevant activities to the Governing Body of the ITPGRFA, the CGRFA and the CBD (for example, SGRP, 2009).

The role of the ITPGRFA and its multilateral system

The entry into force of the ITPGRFA has a potentially important impact on the global system and its operations. The Treaty places considerable emphasis on international aspects of conservation and use and contains a number of provisions on international collaboration. Not only is the creation of the multilateral system (Articles 10–13) a centrepiece of the Treaty, but so are a number of the additional elements, which explicitly include reference to international collaboration. These include, *inter alia*, cooperation on *ex situ* conservation (Article 5); cooperation to strengthen the capabilities of developing countries and economies in transition and enhance international activities to promote conservation, evaluation, documentation, genetic enhancement and other activities (Article 7); provision of technical assistance (Article 8); international collaboration to promote implementation of the GPA (Article 14); recognition of the importance of international collections held by the CGIAR and other international institutes (Article 15); encouragement of participation in international networks (Article 16); and cooperation to develop and strengthen a global information system (Article 17).

Given the emphasis placed in the ITPGRFA on international collaboration, one might expect that, as implementation activities develop both nationally and internationally,

Treaty-related activities would make an increasingly important contribution to overall system functionality and to the different outcomes. Thus, for conservation, there is explicit reference to the importance of collaboration to improve efficiency and sustainability of conservation, to support evaluation and documentation and to recognize the role of international collections. The multilateral system has the potential to make an important contribution to availability as does the development of the global information system. The value of collaboration to support utilization is also recognized in Article 7, which outlines the support for networks, the implementation of the GPA and cooperation on capacity building and technology transfer. The development of the International Fund provides the Treaty's Governing Body with the opportunity to provide support to its parties in ways that strengthen the operation of the global system and help achieve its outcomes. It is currently planning to disburse significant funds to country activities 'with a thematic focus to help ensure sustainable food security by assisting farmers to adapt to climate change through a targeted set of high impact activities on the conservation and sustainable use of plant genetic resources for food and agriculture'.[13] It would be interesting to assess the extent to which the funded proposals contribute to global system outcomes and functionality.

However, there are a number of questions that remain to be answered in relation to the role of the ITPGRFA in the global system, particularly in regard to the multilateral system and the relation of the Treaty to other international components of the system. The multilateral system might become one of the most significant elements of the developing global system. It includes a specific set of practices that are meant to govern the international exchange of PGRFA placed under its auspices and provides a genuinely international framework for exchange and use of a subset of global PGRFA. However, the multilateral system only applies to those crops listed in Annex 1 of the Treaty (some 35 crops and 80 forage species), and there is concern that this limitation will create a 'two-tier' system for PGRFA and divide those crops that are in the list from those that are excluded. Will international collaborative activities and even the funds disbursed either from the International Fund or from other international bodies such as the GCDT create a situation in which non-Annex 1 crops become increasingly neglected and ignored by the different elements of the global system? This possibility needs to be carefully assessed, as does the actual functioning of the multilateral system itself in case there are unforeseen problems with respect to its operation.[14]

The operation of the ITPGRFA in relation to other elements of the global system is also not entirely clear. The FAO's global system places the CGRFA at the centre, although one might argue that the Treaty is likely to play a more significant role in regard to many aspects of PGRFA conservation and use. However, the GPA (which is supposed to be a guiding framework for the Treaty's implementation) was developed and has been revised through the CGRFA, which also oversees the updating of the first and second reports on the state of the world's PGRFA. The CBD explicitly recognizes the contributions of both the CGRFA and the ITPGRFA in implementing its Programme of Work on Agricultural Biodiversity. While global system function may well benefit from the commitment of these different bodies, it might also lead to confusion about roles and responsibilities or, worse, to gaps, duplication and conflict in the discharge of international responsibilities. The recent adoption at COP-10 to the CBD of the Nagoya Protocol on Access to Genetic Resources and the Fair and Equitable Sharing of Benefits Arising from Their Utilization is perhaps an important development in this regard as it explicitly mentions the Treaty in its preamble.[15]

Notwithstanding the attention given to international elements of cooperation, the Treaty depends ultimately for its effectiveness on national implementation of its different provisions. There is concern that this area is not receiving enough attention and that countries are not finding it easy to develop the different policies and mechanisms needed to implement the Treaty in full. This is perhaps particularly true for the provisions on *in situ* conservation, sustainable use and farmers' rights. While it is too early to assess how the ITPGRFA will contribute to the functioning of the global system, it is perhaps worth noting some possibilities and issues. One possibility is that the Treaty will become an increasingly dominant organizing focus in the global system, stimulating, facilitating and overseeing collaboration between parties and supporting international activities. It might even go further, developing agreed practices and standards for the global system, providing a degree of structural formality as well as a set of standards for participation. Such developments do not necessarily always help a system to function. Any system can also benefit from flexibility and the ability to grow, develop and evolve. At the other extreme, the emphasis on Annex 1 crops might lead to a restriction of Treaty-related operations to a particular subset of PGRFA. In functional terms, it would then become one component of the global system that is contributing to all of the outcomes but only for a restricted list of crops.

Of course, there are many possibilities that lie between these two extremes, and where the ITPGRFA will end up depends not only on the future decisions of the parties in regard to the Treaty's operations but also on the actions and decisions of the other international elements as well as the activities of the parties themselves as they implement the Treaty. Monitoring and evaluating the implementation of the Treaty in terms of its effect on function, on the way in which the different mechanisms develop and on the impact it has on other elements will be an essential part of building a more effective global system. At the moment, while the ITPGRFA can be said to recognize the importance of various different elements of the global system, it is not clear which mechanisms will be the ones to be ultimately developed by the Treaty's Governing Body to strengthen the various connections and cooperative activities and what will be the effect of the actions already undertaken – for example, through its International Fund.

Limitations of the global system

The most developed parts of the global system are concerned with *ex situ* conservation and with information about *ex situ* conserved materials. The multilateral system of the ITPGRFA provides a framework for global availability, while aspects of use are dealt with largely on a crop-by-crop basis and reflect the global significance of the different crops (or at least the interests of institutions in a number of different countries and their need to cooperate). There are also established components providing information and knowledge about PGRFA. In contrast, the financial arrangements that support the system are underdeveloped. The existing instruments, which include the International Fund, the GCDT and the GEF's support for agricultural biodiversity, are limited in scope and do not constitute a coordinated strategic approach to providing the resources needed for the system as a whole. Such an approach would need to be open and all encompassing so as to avoid some of the problems noted earlier that might restrict, rather than enhance, the development of the system.

The connections between the different components are, arguably, much less developed than the components themselves. This may reflect the relatively underdeveloped nature of the system. The emphasis has been on the individual elements and on the components within them. The approach has often been to deal with each part separately. For example, the former GPA was concerned with all of the outcomes but deals with them separately under 20 different activities with little regard for the importance of the linkages and connections between them. Thus, it deals with *in situ* and *ex situ* conservation as two separate activities rather than as an integrated effort to achieve the effective conservation of a crop gene pool. This distinction tends to hamper the development of an effective systems approach that seeks to strengthen the connections between the different parts of the system and sees functionality in terms of the effectiveness of the system as a whole. The further development of the global system may involve identifying which international activities might support not only a more comprehensive approach to specific outcomes but also a more effective linkage between them, particularly with respect to the links between conservation and use.

Table 18.1 identifies a number of more informal components of the system such as international and regional fora and non-governmental organizations (NGOs). These play important roles in setting the agenda and in contributing to the activities that are actually being undertaken. For example, the Botanic Gardens Conservation International has supported significant conservation efforts, the International Foundation for Science has made important contributions in terms of use and NGOs such as Practical Action have supported on farm maintenance and the use of diversity by rural communities. There are also international farmers' organizations such as Via Campesina and NGOs such as ETC, which have played an active role in the debate about PGRFA conservation and use and the way in which the system should function. In developing a systems approach, the roles and contribution of these participants need to be more fully appreciated and their contribution recognized.

One major weakness of the global system has been its failure to make progress on the conservation, availability and use of minor or neglected and underutilized crops. There is strong evidence that these crops are being under-conserved and under-used, and there is also evidence that they are likely to become increasingly important contributors to agricultural sustainability, dietary diversity and adaptation to climate change. The vast majority of these crops and their wild relatives fall outside the remit of the international gene banks, and the limited resources available to countries for conservation work mean that they are often not the target of work by national programmes. Perhaps one of the functions of an effective global system would be to use its different components and instruments to address these kinds of problem.

Conclusions

A 'systems' conceptual framework provides a way of assessing the roles of the different components of the FAO's global system and the other organizations, agreements and activities that contribute to the effectiveness of the international dimensions of PGRFA conservation and use and overall system functionality. The different components are largely focused on aspects of conservation, while aspects of use are much less well supported internationally. These differences may reflect the relatively well-developed crop-based mechanisms for international collaboration and the relatively long history of support by public agencies for the collection and *ex situ* conservation of PGRFA in support of research and plant breeding. However, it does suggest a possible disconnect in the global system that needs to be further considered. The concern of the ITPGRFA with sustainable use (Article 6) may be one entry point for correcting this imbalance.

A 'systems' conceptual framework changes the focus from a concern with establishments and numbers (for example, the number of gene banks and their attributes and the number of species and accessions stored under particular conditions), which has characterized much of the discussions of global conservation (see, for example, FAO, 2010), to a focus on outcomes and the activities and relationships between entities that can help achieve these outcomes. It also draws attention to the importance of links between these outcomes. It therefore creates a useful tool for further analysing the global system.

One important aspect of the global system's function that deserves further consideration is the relation between the national elements and the international components. National elements are still responsible for 90 per cent of the *ex situ* conservation and all *in situ* conservation as well as much of the effort on sustainable use. National gene banks may also meet most of a country's utilization needs. In terms of overall system function, the effectiveness of national elements is likely to remain critical and, thus, the role that can be played by international agreements or operations in strengthening national capabilities to deliver on all three outcomes will remain central to the success of the system.

The continuing analysis and discussion of the roles of national and international elements of the system will help identify the properties of the global system in which international components are particularly important. The systems framework proposed in this chapter can help to further evaluate the effectiveness of these components. This is particularly the case with respect to the completeness of conservation efforts and availability. However, there may be other characteristics that could be identified as being of particular importance at a global level, such as overall genetic vulnerability for a crop or a region.

The ITPGRFA contains important and useful provisions that can provide links between the activities undertaken by national programmes and the functioning of the global system. Its provisions reflect both the importance of implementing conservation and use activities by parties and their international responsibilities. It also describes a number of the components of the global system and situates them in relation to the work of the parties. At this early stage, it is not entirely clear how (or if) this process will work in practice, but it should prove to be an important area of future work. However, even at this early stage, it is possible to hypothesize that the effect of the multilateral system on availability is likely to be particularly significant and that ways will need to be developed to monitor these achievements closely.

Notes

1 System-wide Genetic Resources Programme, www.sgrp.cgiar.org (last accessed 5 April 2011).
2 International Treaty on Plant Genetic Resources for Food and Agriculture, 29 June 2004, www.planttreaty.org/texts_en.htm (last accessed 15 February 2011) [ITPGRFA].
3 Global Initiative on Plant Breeding, http://km.fao.org/gipb (last accessed 4 April 2011). Convention on Biological Diversity, 31 ILM 818 (1992).
4 In the case of plant genetic resources, parts of the Food and Agriculture Organization's (FAO) global system could be regarded as 'designed', while other elements have developed more haphazardly. As discussed later in this chapter, this is probably an oversimplification.
5 The development of the conceptual framework owes much to the thinking of the Global Environmental Change and Food Systems and the framework they have developed for the global food system (Ericksen, 2008).
6 National programmes include the various organizations, operations and arrangements that contribute to the conservation of plant genetic resources for food and agriculture (PGRFA) in countries. Typically, they can include gene banks, *in situ* reserves, information systems and management arrangements that ensure the oversight of national PGRFA conservation activities.
7 It may be useful to distinguish representativeness and completeness. Representativeness would be concerned with the extent to which conservation actions adequately cover the range of diversity – by including the representatives of different ecotypes or morphological types and by covering the known geographical, ecological and botanical range of a crop, species or gene pool. The degree of completeness would be concerned with the proportion of known genetic diversity that has been conserved.
8 A European Gene Bank Integrated System (AEGIS), *Objectives*, http://aegis.cgiar.org/about_aegis/objectives.html (last accessed 19 January 2011).
9 Svalbard Global Seed Vault, www.nordgen.org/sgsv/ (last accessed 19 June 2012).
10 Global Crop Diversity Trust (GCDT), www.croptrust.org/main/lprioritycrops.php (last accessed 15 February 2011).
11 A single holding of an accession will inevitably tend to become homozygous over successive regenerations, which will lead to a loss of accession diversity in that holding. Random drift might lead to the fixation of different alleles in different gene banks, thus helping to maintain total diversity of the particular accession.
12 GENESYS, www.genesys-pgr.org (last accessed 15 February 2011).
13 ITPGRFA, *supra* note 2.
14 To date, GCDT funding, for example, has been more or less entirely provided for the support of crops listed in Annex 1.
15 Nagoya Protocol on Access to Genetic Resources and the Fair and Equitable Sharing of Benefits Arising from Their Utilization, 29 October 2010, www.cbd.int/abs/text/ (last accessed 15 February 2011).

References

Brehm, J.M., N. Maxted, M.A. Martins-Loução and B.V. Ford-Lloyd (2010) 'New Approaches for Establishing Conservation Priorities for Socio-Economically Important Plant Species', *Biodiversity and Conservation* 19(9): 2715–40.

Burke, M.B., D.B. Lobell and L. Guarino (2009) 'Shifts in African Crop Climates by 2050, and the Implications for Crop Improvement and Genetic Resources Conservation', *Global Environmental Change* 19(3): 317–25.
Cooper, D., J. Engels and E. Frison (1994) *A Multilateral System for Plant Genetic Resources: Imperatives, Achievements, and Challenges'*, Issues in Genetic Resources, no. 2, International Plant Genetic Resources Institute, Rome, online: http://www.bioversityinternational.org/fileadmin/bioversity/publications/pdfs/464_A_multi-lateral_system_for_plant_genetic_ resources.pdf (last accessed 21 June 2012)
Ericksen, P. (2008) *Global Environmental Change and Food Systems*, Global Change Newsletter, no. 71, International Geosphere-Biosphere Programme, Stockholm, Sweden.
Food and Agriculture Organization (FAO) (1998) *State of the World's Plant Genetic Resources for Food and Agriculture*, Report no. 1, FAO, Rome.
——(2002) *The International Treaty on Plant Genetic Resources for Food and Agriculture: A Global Treaty for Food Security and Sustainable Agriculture*, FAO, Rome.
——(2006) *World Agriculture: Towards 2030/2050: Prospects for Food, Nutrition, Agriculture and Major Commodity Groups*, Interim Report, Global Perspective Studies Unit, FAO, Rome.
——(2010) *Second Report on the State of the World's Plant Genetic Resources for Food and Agriculture*, Commission on Genetic Resources for Food and Agriculture (CGRFA), FAO, Rome.
FAO/International Plant Genetic Resources Institute (IPGRI) (1989–2002) *Technical Guidelines for the Safe Movement of Germplasm*, Series published from 1989 to 2002, FAO, Rome.
Fowler, C., and T. Hodgkin (2004) 'Plant Genetic Resources for Food and Agriculture: Assessing Global Availability', *Annual Review of Environmental Resources* 29: 143–79.
Frison, E.A., J. Cherfas and T. Hodgkin (2011) 'Agricultural Biodiversity Is Essential for a Sustainable Improvement in Food and Nutrition Security', *Sustainability* 3: 238–53.
Hajjar, R., D.I. Jarvis and B. Gemmill (2008) 'The Utility of Crop Genetic Diversity in Maintaining Ecosystem Services', *Agriculture, Ecosystems and Environment* 123: 261–70.
Halewood, M., and K. Nnadozie (2008) 'Giving Priority to the Commons: The International Treaty on Plant Genetic Resources for Food and Agriculture', in G. Tansey and T. Rajotte (eds), *The Future Control of Food: A Guide to International Negotiations and Rules on Intellectual Property, Biodiversity and Food Security*, Quaker International Affairs Programme, International Development Research Centre, Earthscan, London.
Jarvis, D.I., T. Hodgkin, B.R. Sthapit, C. Fadda and I. López-Noriega (2011) 'A Heuristic Framework for Identifying Multiple Ways of Supporting the Conservation and Use of Traditional Crop Varieties within the Agricultural Production System', *Critical Reviews in Plant Science*, 30: 125–176.
Koohafkan, P. (2009) 'Conservation and Adaptive Management of Globally Important Agricultural Heritage Systems', *Resources Science* 31(1): 4–9.
Palacios, X.F. (1998) *Contribution to the Estimation of Countries' Interdependence in the Area of Plant Genetic Resources*, Report no. 7, Review 1, CGRFA, FAO, Rome.
Platform for Agrobiodiversity Research (PAR) (2010) *The Use of Agrobiodiversity by Indigenous and Traditional Agricultural Communities in: Adapting to Climate Change*, Synthesis paper, PAR, Rome.
System-wide Genetic Resources Programme (SGRP) (2009) *Submissions by International Organizations: Report from the International Agricultural Research Centres of the Consultative Group on International Agricultural Research on their Policies, Programmes and Activities and Potential Contributions to the MYPOW Implementation*, Report presented at the twelfth regular session of the CGRFA, Rome, Italy, Doc. CGRFA-12/09/Inf.6 Add.1 (19–23 October), FAO, Rome.

19

INSTITUTIONALIZING GLOBAL GENETIC RESOURCE COMMONS FOR FOOD AND AGRICULTURE

Tom Dedeurwaerdere

New challenges for the use and exchange of plant genetic resources for food and agriculture

The purpose of this chapter is to provide an overview of the development of new theories and case study material in the field of the evolving global commons in plant genetic resources for food and agriculture (PGRFA). The exploration of various emerging global genetic resource commons in fields such as animal genetic resources and microbial genetic resources is still in its infancy, compared to the global crop commons (Byerlee, 2010) or to the locally managed natural resource commons more generally (Ostrom, 1990). Nevertheless, the commonalities between these various forms of commons have caught the attention of a growing number of scholars and managers of common pool resources. Important common features, such as the role of social norms in complying with common rules and the presence of clearly defined boundaries, deserve to be analysed in a systematic manner with the aim of supporting the development of appropriate institutional and legal frameworks.

The first major instance of a formal commons on a global scale in the field of genetic resources for food and agriculture was the pool of PGRFA established and governed by various International Agricultural Research Centres (IARCs) under the leadership of the Collaborative Group on International Agriculture Research (CGIAR). Over the last 40 years, the IARCs have played a leading role in promoting open access to biological resources through the organization of a network of specialized *ex situ* conservation facilities throughout the world. However, legal uncertainty over the status of the collections in the early 1990s, and the growing recognition of the specific nature of PGRFA, called for the development of a specific legal instrument to regulate the conservation and use of these resources. Lengthy negotiations eventually led to the adoption of the International Treaty on Plant Genetic Resources for Food and Agriculture (ITPGRFA) on 3 November 2001, which provided further support for the collections' open access policy.[1] This open access policy is clearly reaffirmed in the 2003 CGIAR's policy guidelines:

> The germplasm [that is, the seeds or the parts of a plant that enable reproduction] designated by the Centres is held in trust for the world community in accordance with the agreements signed with the FAO ... Based on the conviction that their research will continue to be supported by public funds, the Centres regard the results of their work as international public goods. Hence full disclosure of research results and products in the public domain is the preferred strategy for preventing misappropriation by others.
>
> (SGRP, 2003)

The recent history of the global crop commons, however, increasingly shows the limits of an approach that is solely focused on plant genetic resources. New challenges have led public and private research entities, farmers and industry to consider pooling genetic resources together beyond only PGRFA. Examples of these challenges are the management of global infectious diseases affecting food crops, animals and sometimes humans, the development of crop and non-crop biofuels and the opportunities arising from biotechnology for increasing the nutritional value and safety of food products. The creation of such common pools has also benefited from the combination of technological progress in life sciences and the information sciences (Parry, 2004). On the one hand, the development of innovative methods for the identification, long-term conservation (for example, freezing and freeze-drying) and shipping of genetic resources has enhanced interest and international cooperation in global life science research. On the other hand, the information technology revolution has dramatically expanded the possibilities of distributed coordination as well as diminished the search costs for locating genetic resources held in collections throughout the world.

The positive impact of these technological advances on the development of the global genetic resource commons, however, has been attenuated by a set of counterbalancing factors, which could jeopardize the whole enterprise (Reichman et al., 2011). The important commercial value of a small subset of genetic resources, especially in the field of pharmaceutical product development, has put pressure on the sharing ethos that is at the basis of the exchange of resources within the commons. In particular, communalism and norms against secrecy have been eroded by delays in publication and restrictions on the sharing of research materials and tools, which, in turn, have often been caused by concerns about intellectual property rights (Rai, 1999). Another hurdle is the heterogeneity of the legal frameworks, which raises the costs of designing appropriate institutional rules that can operate on a global scale. A major obstacle in this respect consists of divergent national access and benefit-sharing legislation across countries as well as a lack of international coordination in the implementation of these legal provisions in a way that is consistent with the needs of public science in developing and industrialized countries (Jinnah and Jungcurt, 2009; Roa-Rodríguez and van Dooren, 2008).

In recognition of these obstacles, science policy makers and genetic resource managers have increasingly focused on devising new methods for organizing and integrating vast and diverse collections of resources, with a view to better securing the research needs of the various user communities. The need for institutionalizing the common pools of genetic resources used in food and agriculture was already acknowledged during the negotiation of the Convention on Biological Diversity (CBD) in 1992.[2] At this time, it became clear, and was accepted, that a purely bilateral system between single providers and users of given genetic resources, such as what is used in the case of medicinal plants, would not be possible for the food and agriculture sector due to the fact that the plant-breeding process calls for a broad range of genetic resources from many different providers to be input into any one product (Burhenne-Guilmin, 2008).

The choice of a commons-based governance of genetic resources, however, should be justified in terms of its costs and benefits compared to other institutional modalities for governing genetic resources (such as private intellectual property rights or exclusive use license contracts) as well as its success in addressing the specific needs in each of the sub-sectors. Much is to be learned in this respect from a systematic comparative assessment of over two decades of experimentation with genetic resource commons, in terms of their commonalities and differences, the specific patterns of use and exchange in the various sub-sectors, the features of the regulatory environment dealing with specific sets of resources (microbial versus animal, for example) and the importance of the pre-existing social norms. Further, where there is evidence of similar challenges and policy needs in the various sub-sectors (especially in regard to intellectual property rights and access to genetic resources for pre-competitive research purposes), these should be systematically explored to support the introduction of common strategies for the access and sustainable use of genetic resources.

This chapter shows that, when the research and innovation processes are based on screening or breeding from common pools with multiple inputs from various sources, commons-based innovation offers an interesting institutional option, as an alternative to both private proprietary and state-based solutions. Indeed, commons-based innovation in genetic resources allows for both the barriers of case-by-case contracting over every single entity in a system of exclusive property rights (Dedeurwaerdere, 2005) as well as the rigidity of centralized governmental and intergovernmental organizations (Halewood, 2010) to be overcome, without compromising downstream commercial applications (Lessig, 2008; Reichman et al., 2011). Therefore, it seems relevant to explore whether, and if so how, the principles of commons-based production can also be applied to the specific case of global and regional genetic resource governance.

The chapter is structured in the following way. The first section briefly presents the recent history of global genetic resource commons, followed by an examination of the commonalities between the various theoretical models, ranging from the global genetic resource commons to the digital commons, with a view to enhancing our understanding of how commons-based design principles can offer an alternative to the private and state-based control of genetic resources. The next section applies these insights to analyse a focused set of case studies on the institutional design of global exchanges with microbial, animal and plant genetic resources. On the basis of this analysis, the final section summarizes some key features of the genetic resource commons that can guide efforts in developing appropriate legal and institutional frameworks.

Emerging models of global genetic resource commons

Global collections of genetic resources emerged in response to the needs of research infrastructure faced with the challenges of food security, global health issues and the biodiversity crisis more generally. Similarly, the genomics revolution and the broader impact of the globalization of research in the life sciences enhanced the interest and cooperation in the collection of genetic resources. As a result, vast amounts of human, animal, plant and microbial genetic material were collected throughout the world from various regions and habitats and exchanged in collaborative research networks. For instance, in the 1980s, Africa faced the destruction of a major crop, cassava (also known as manioc), by a scale insect, the mealybug (Hammond and Neuenschwander, 1990). Through research on the natural enemies of this bug, which took place in one of the IARCs in Colombia (the International Center for Tropical Agriculture (CIAT)), tiny wasps were identified as an effective predator. They were imported into Africa and successfully used in a major biological control programme that took place in collaboration with the International Institute for Tropical Agriculture. As a result, millions of dollars of food crops were saved. More recently, the CIAT, with the support of the Nippon Foundation, coordinated a similar biological control programme in Thailand and other countries in Southeast Asia. Other well-known examples of the worldwide sharing of biological resources involve microbial materials. For instance, soybean production throughout the world has been improved through the use of nitrogen-fixating bacteria, known as the root nodule bacteria. Some well-characterized and high-performing isolates of these bacteria, derived from the worldwide exchange of materials, are currently used in public and private research, for training and education, and commercially produced in large quantities in various countries around the world (Dedeurwaerdere et al., 2009).

The increase in the exchange of genetic materials in relatively open global networks, however, has also raised a set of new collective action problems. One of the main problems is the increase in practices that potentially create new threats to food and agriculture as well as to human health (Doyle et al., 2005) and quality management (Stern, 2004). The regional and international shipment of agriculture products means that pathogens do not necessarily require natural dispersion. Further, intensive agricultural practices bring with them a set of undesirable consequences, which are likely to increase these threats, such as resistance due to the overuse of antibiotics and pesticides or new variants of zoonotic animal diseases. The importance of these and other problems has lead to initiatives to promote the further institutionalization of exchange networks in globally distributed common pools with common quality standards, clear rules for entry into the pool and coordinated management, such as the creation of the Global Biological Resources Centres Network in the microbial field (Smith, 2007).

In this context, digital infrastructures create a new set of mechanisms for restructuring the collaborative enterprise. More specifically, the use of computational methodologies within the life sciences makes it possible to build large knowledge repositories and to develop data-mining tools for integrating the huge accumulation of data in these repositories into a virtual collection (Dawyndt et al., 2006). Further, digital networks make it possible to directly improve the global exchange of materials by implementing machine-readable material transfer agreements directly into the digital catalogues of the collections (Nguyen, 2007). Finally, by systematically documenting the source and history of the materials deposited in these genetic resource collections, and releasing this information online, the digital information infrastructures also become a tool for making the reciprocity of exchanges clearly visible (Fowler et al., 2001).

At present, most genetic resource collections are taking advantage of the proliferation of these new mechanisms, by networking the existing infrastructure of the physical collections into global digital data and information infrastructures. As they stand, however, the measures that have been taken so far do not go far enough in implementing the kind of infrastructure that would be needed to realize the full potential of a digitally networked genetic resource commons. Moreover, existing initiatives remain fragmented, incomplete and limited in scope, with the risk that they may succumb to adverse legal, economic and political pressures over time.

For the design of a worldwide microbial commons, however, a more systematic approach, based on an analysis of the structure of the exchange practices and the terms and conditions of exchanges between the collections, and between the collections and the provider countries, is needed. In particular, in order to improve the current state of affairs, a better understanding is needed of the costs and benefits of the alternative institutional frameworks, which would serve to harmonize the conditions of exchange and put the emerging worldwide microbial commons onto a solid legal and institutional basis. The objective of this chapter is therefore to conduct a comparative assessment of the use and exchange practices in the various sub-sectors of the genetic resource commons (microbial, animal and plant) and to identify commonalities, differences and user community needs under different social, institutional and technical conditions.

Theoretical models for designing a genetic resource commons on a global scale

The design of a global genetic resource commons should take into account the specific characteristics of the genetic resources. Genetic resources are complex goods, with both a biological (the biological entity) and an informational component (the genetic information and information on the biochemical pathways). As biological entities, most genetic resources are widely dispersed, whether originally in nature (Beattie et al., 2005) or as a result of human domestication (Braudel, 1992). As a result, it is often costly (or simply difficult) to exclude users from accessing these resources *in situ*.

In many cases, however, biological entities are accessed not for direct exploitation of the entity itself but, rather, for access to the informational components (Dedeurwaerdere, 2005; Goeschl and Swanson, 2002b). For example, large quantities of biological entities are collected in order to screen the biological functions and properties that they exhibit against certain targets. Once a new property or function has been discovered, the search for genetic similarity can identify the genetic sequences that are involved in the expression of these properties. This search may, in turn, lead to further research on these genes or properties without having to access the specific organism that led to the discovery of the new informational inputs. Nevertheless, accessing specific entities such as microbial isolates with unique properties for bioremediation, enzyme production or to produce food additives becomes important at the end of the research and innovation chain when biological entities are used in commercial applications. Therefore, any regime for regulating access to these resources should take into account both the broad informational features of the pool of resources as well as the potential commercial uses of specific biological entities.

In general, genetic resources act as informational inputs in the process of research and innovation, both as stock (in the form of accumulated traits of known usefulness in natural environments) and as generators of new flows of information (the discovery of new useful features) (Swanson and Goeschl, 1998). The current literature on natural resource commons, however, only partially takes these features of global genetic resource networks into account. To take the specific informational features of the networked genetic resources into account, it might be more fruitful to look at the commonalities that they have with the institutional solutions and models that have been developed in the digitally networked information commons. Digital information commons have been proven to offer a set of robust and successful models for the production of informational goods and services (Benkler, 2006; Boyle, 2008; Hess and Ostrom, 2007; Lessig, 2001). In their case, a 'hybrid' regime has been developed that addresses both the commercial and non-commercial uses of multiple contributions to the development of knowledge goods (Benkler, 2006, 122–27; Lessig, 2008). Moreover, there is already considerable experience with these global commons, and systematic research on generic design principles has been conducted. This research can provide elements for a systematic comparative analysis with the genetic resource commons. This section focuses on two key common design principles of successful commons that came out of this research, which are the role of non-market motivations and the modular character of the organizational architecture.

The main institutional feature that is common to all successful digital information commons is the design of complex incentive schemes that are driven more by social and intrinsic motivations than by monetary rewards (Benkler, 2006). Mixed motivations are common in a heterogeneous set of initiatives such as open-source software communities, global genetic sequence databases and distributed peer-to-peer computational infrastructures. Since it is difficult to put a precise monetary value on the creative inputs of a vast and distributed network of contributors, it has proven to be more effective to rely on non-market motivations for organizing the networks (Deek and McHugh, 2008). Moreover, extensive empirical research has shown that, when social motivations are involved, such as increasing recognition in a collaborative group or increasing the satisfaction of intrinsic motivation with respect to furthering general interest objectives, monetary rewards can decrease the willingness to contribute to the global pool (Frey and Jegen, 2001). Further, there are hidden costs involved in moving from a social to a monetary reward. These costs are related to a clear delineation of the tasks to be paid for (Deci, 1976) as well as a monetary evaluation of the value of each and every contribution to these tasks (Benkler, 2006).

The exchange of genetic resources in the global commons is clearly a case where social and intrinsic motivations will play an important role. Indeed, the attribution of a monetary value to each entity is especially hard, or simply impossible, when the genetic resource that is used as an input for collaborative research in a global exchange network has to be assessed. Many innovations result from the combination and comparison of information gained from a wide variety of genetic resources from different sources, which all play a certain varying role in the progress of the research. Furthermore, the value of the resources only becomes apparent late in the research and innovation process, while its theoretical monetary value at the beginning of this process is likely to be extremely low (Simpson et al., 1996). Finally, in some cases, the initial value of the resource is increased by the presence of informational components that are difficult to quantify, such as associated know-how and traditional knowledge, but which can make a major contribution to research into environmental, food or health-related properties (Blakeney, 2001).

The second feature, which plays a role in the success of the commons-based production of knowledge in the digital commons, has been the adoption of modular technical and organizational architectures. Modular architectures have allowed efforts and contributions from many human beings, which are diverse in their quality, quantity, focus, timing and geographical location, to be pooled in an effective manner (Benkler, 2006, 100). Modularity presupposes the presence of a set of independently produced components that can be integrated into a whole. The fine-grained character of the modules determines the number of potential contributors to the network. If there is a large set of relatively small contributors, each of whom only has to invest a moderate amount of additional effort and time in the network, the potential benefits of taking part in a global exchange network is likely to be high. However, if even the smallest contributors are relatively large and if they each require a large investment of additional time and effort to take part in the collaborative network, the potential reciprocity benefits of being part of the network, and the cost-effectiveness of doing so, will diminish, and the universe of potential willing contributors will probably decrease.

Modularity was clearly also present in the major successful collaborative projects in the field of genetic resource commons, such as the collaborative sequencing of the worm genome by a network of teams distributed around the world in the early days of the genomic revolution (Sulston and Ferry, 2003) or the networks of crop improvement established by the various members of the CGIAR (Byerlee, 2010).

The importance of non-market motivation is a necessary condition for the emergence of effective commons-based production, but it is clearly not sufficient. It is the combination of the potential of non-market production of collective goods and the effectiveness of an organizational form that allows widely dispersed contributions to be integrated and that makes effective commons-based innovation possible on a global scale. Research on these general design principles shows that, under conditions of appropriate quality control and through an initial investment in the creation of social networks (Benkler, 2006), commons-based production and management of informational goods can be a desirable and effective institutional modality, which can co-exist with a market or state-based production of knowledge goods. This is especially true in the early stages of research on the innovation and product-development chain, when access to multiple inputs is required.

Results of case studies of institutional choices

Materials and methods

Except for the analysis of some specific initiatives (Beck, 2010; Hope, 2008; Halewood and Nnadozie, 2008), few case studies have addressed the institutional characteristics of commons-based production with genetic resources. This section presents a comparative analysis of three such case studies, in the fields of microbial genetic resources, plant genetic resources and animal genetic resources. These cases have been selected because of their broad similarity in regard to the characteristics of the innovation process and their common concern with agricultural biodiversity.

The methodology of this comparative assessment builds upon the well-established literature on the natural resource commons, by acknowledging that the biophysical characteristics of the resource, the governance arrangements and the characteristics of the users all impact upon the management of the commons (Agrawal, 2001; Ostrom, 2005). Therefore, the discussion of commonalities and differences between the three cases will address the variations along these three dimensions (see also Table 19.1). However, at the outset, it is necessary to point to some broad similarities that justify the treatment of these cases under the common category of the genetic resource commons. First, with respect to the biophysical characteristics of the resource, a substantial part of the innovation process in all three cases is based on access to multiple inputs in order to produce a single output, whether this process is for screening for interesting entities in a pool of resources (in all three fields) or for breeding new entities by combining multiple inputs from the pool (in the animal and plant fields). From a governance perspective, institutional modalities have been developed in all three fields, in order to gain facilitated access to these multiple research inputs. In particular, commons-based innovation is one of the emerging institutional modalities of these sectors, although it has to cope increasingly with the pressures from the global intellectual property regime, as described earlier; the recourse to restrictive contractual practices; the adverse impact of national access and benefit-sharing legislation on scientific research; the competition from vertical integration; and the centralization of the inputs within global companies, among others (Reichman et al., 2011). Finally, in these three cases, the user communities are driven by a mix of market and non-market motives, even if in some sectors the commercial pressures have led to an erosion of some of the non-market components.

TABLE 19.1 Biophysical, institutional and community characteristics of the genetic resource commons

		Microbial genetic resources	Plant genetic resources	Animal genetic resources
Biophysical properties		High genetic variation within a species and high mutation rates	Well-defined varieties and a high degree of genetic stability	Relatively well-defined breeds
Characteristics of the user groups	Type of users	Mix of public and private	Mostly public sector	Mostly private sector
	Main non-market motivations	Contribution to scientific research ethos and conservation of biodiversity	Food security and contribution to training and research	Animal health and conservation of genetic variety
Features of the governance arrangements	Governing bodies	Global and regional federations	Secretariat of international treaty	National and local breeding organisations
	Organization of distributed collaboration	Shared resources among a distributed network of microbial collections	Shared resources among plant breeders and centralized collections for key species	Shared resources among farmers of the same breed
	Forms of exchange	Mix of formal and informal	Mainly formal	Mix of formal and informal
	Main challenges to the commons-based exchange	Restrictive license policies in high commercial value collections	Broad patents on plant genetic resources	Vertical integration of breeding operations in transnational companies

Source: Capgènes, www.capgenes.com.

The analysis of the case studies is based on the original surveys and semi-structured interviews by a research group at Cambridge University conducted in 2005 and 2009 with members of the World Federation for Culture Collections (WFCC) (Stromberg et al., 2007; Stromberg et al., 2006) and at Université catholique de Louvain (Dedeurwaerdere et al., 2009) as well as on expert interviews with officials at the Policy Research and Support Unit at Bioversity International and the Commission on Genetic Resources for Food and Agriculture at the Food and Agriculture Organization (FAO). The findings of these surveys and interviews were combined with information taken from the notes of internal meetings and official reports and compared to the results of previous studies.

For each of the three cases, the analysis will focus on three elements: (1) analysis of the institutional arrangements governing the exchange practices in that field; (2) synthesis of the data on commons-based production obtained through in-depth case studies of major collections; and (3) evaluation of the contribution of non-market motivations and modular organization to effective commons-based innovation practices.

Microbial genetic resources

The study and commercial exploitation of microbial genetic resources requires the systematic authentication of pure micro-organisms in *ex situ* collections and the preservation and exchange of certified biomaterials for cumulative follow-on research (WFCC, 2010). As a result, many countries are actively involved in collecting and exchanging micro-organisms on a global scale.

Well-known examples of the worldwide sharing of biological resources involve microbial materials in the field of food and agriculture. For instance, soybean production throughout the world has been improved through the use of a nitrogen-fixating bacteria, known as the root nodule bacteria. Through the exchange of some well-characterized and high-performing isolates of this bacterium, soybeans can now be used in public and private research, for training and education, and commercially produced in large quantities in various countries (Genetic Resources Policy Committee, 2007). Another example is related to the management of the threats from pathogenic micro-organisms for agriculture and food production systems such as fungi-causing root rot and stem rust diseases or mycotoxin-producing fungi, which are harmful for animal and human health. Some of these fungal pathogens can be transported by the wind, while others move with the international shipment of agricultural products. Through international collecting efforts, diagnostic and identification tools have been developed that can be used in the early detection of these pathogens (Smith et al., 2008) as well as in the detection of contamination in agriculture and food commodities (Doyle et al., 2005). Another case that is more centred in one region is the use of microbial ferments in food processing, such as yoghurt. For decades, these microbial strains have been exchanged and managed through farmers' markets as a common heritage of the local farmers (see Figure 19.1). However, commercial interest from Japanese companies who are selling this local yoghurt variety in Japan under their own brand name has placed pressure on these local collections to limit such exchange in order to ward off

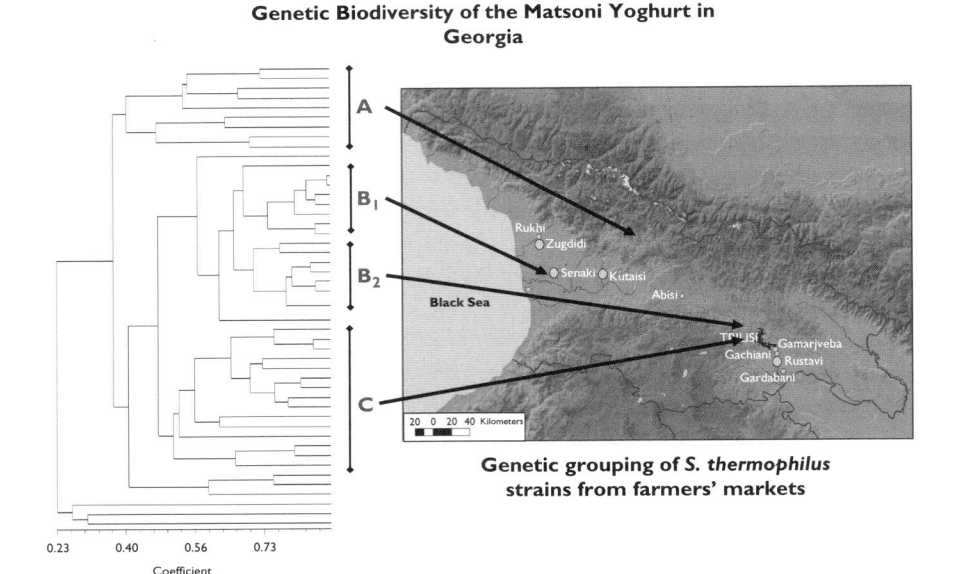

FIGURE 19.1 Genetic biodiversity of the Matsoni Yoghurt in Georgia.

misappropriation. Thus, these new restrictions might work against the open exchange practices that have produced the local diversity in the first place.[3]

Currently, more than half a million microbial samples, which have been collected from various countries, are distributed throughout the world every year by the public *ex situ* collections that are members of the WFCC, mostly for the marginal costs of distribution (Dedeurwaerdere et al., 2009). Each of these collections contains a very substantial set of unique materials (an average of 40 per cent of the strains in the WFCC's culture collections that are referenced on StrainInfo are unique).[4] Intense collaboration and exchange among these culture collections is a necessary consequence of this situation. It is difficult to estimate how many *ex situ* materials are exchanged between the research collections outside the WFCC collections on an informal basis, but it is fair to say that the volume of materials exchanged between these collections is probably even greater (Dedeurwaerdere et al., 2009).

Nevertheless, the exchange of biological materials within a global commons, which prevailed during the early days of the modern life sciences, now seems to be reversed. More and more biological materials are enclosed behind national and privatized fences or only accessible under very restrictive license conditions. For instance, recent research on avian bird influenza has been hampered by countries such as Indonesia that refuse to provide access to samples of the H5N1 virus.[5] In this case, the Indonesian government feared that foreign companies would acquire the rights to any vaccine that might eventually be developed, without proper guarantees of low-cost access to this vaccine for developing nations. After an agreement in 2007 to start negotiations to define the terms of references for fair and equitable sharing in any possible profits, Indonesia resumed sharing the H5N1 avian influenza virus samples.

In order to get a better picture of the institutional arrangements within the microbial commons, a set of original surveys and interviews were conducted in 2005 and 2009. In 2005, P. Stromberg, U. Pascual and T. Dedeurwaerdere (2006) surveyed the 499 public collections that were members of the WFCC (119 completed survey forms). In 2009, T. Dedeurwaerdere, M. Iglesias, S. Weiland and M. Halewood (2009) undertook a quantitative assessment of the entire accession database of a geographically representative set of nine major collections (totalling more than 15,000 single accessions), conducted semi-structured interviews with administrators of these collections, organized a short complementary e-mail survey on access and benefit-sharing measures with 238 WFCC collections (43 completed questionnaires), and completed 16 in-depth phone interviews with scientists from both public and laboratory culture collections.

The most advanced institutional arrangement for formal commons-based innovation and research is the viral license adopted in February 2009 by the 61 members of the European Culture Collection Organization, which permits distribution to other culture collections and collaborating scientists under the condition that recipients use the same license when further distributing the microbial strains. The open-access license thus travels with the micro-organisms, which prevents misappropriation of the resource by other players in the network (in a similar way to open-source licensing of computer software). Many developing countries' culture collections, such as the All-Russian Collection of Micro-Organisms in the Russian Federation and BIOTEC in Thailand, have also adopted formal licenses that permit non-exclusive use and further distribution by other culture collections for non-commercial uses. Nevertheless, in spite of these examples of coordinated and standardized formal arrangements for facilitated access among culture collections, which are used both in developing and industrialized countries, many culture collections still use ad hoc formal arrangements or even have recourse to more restrictive license conditions, as discussed later in this chapter.

At the same time, many resources are still distributed on an informal basis, especially within the United States where, with some major exceptions such as the Agricultural Research Collection, there is only a loosely organized network of public culture collections. Informal exchanges occur without written contracts, which serve to permit all uses of the material as well as the further distribution of the material to collaborating scientists or other third parties. In spite of the obvious advantages, in terms of costs and rapid dissemination, the informal system has some major disadvantages (Reichman et al., 2011). These include the absence of a clear tracking procedure for microbial resources, which is present in the formal arrangements, and the possibility that scientists will limit the distribution of the resources to a small group of close collaborators, on the basis of informal case-by-case arrangements, instead of providing a set of standard conditions that make the materials easily available to all possible users. Moreover, in the absence of formal agreements, it is unclear how cross-border exchange can be organized in compliance with the access and benefit-sharing provisions of the CBD.

The microbial commons is a clear case where non-market motivations and modular organization play a key role in contributing to the effectiveness of distributed collaboration in the management and conservation of microbial material, both in the formal and informal exchange regimes. Strong norms of reciprocity and a general shared conception that the collections hold the materials in trust for humankind as a whole underlie the exchange practices (Dedeurwaerdere et al., 2009). This conception is especially strong in taxonomic research, where a copy of every type of strain – the official reference strain used in the definition of the species – is present in, on average, seven different collections worldwide (based on the WFCC collections that are referenced on StrainInfo). The community also invests in strengthening social norms – for example, by regularly revising the common guidelines for operating the culture collections drawn up by the WFCC (WFCC, 2010). However, there have been some breaches in these social norms, which were communicated in the interviews. These include the competitive pressures for rapid publication (which has led to an increase in secrecy), a lack of compliance with the norm of depositing microbial strains once the research is published, and concerns about biopiracy when collecting microbial strains from developing countries. On the positive side, however, the vast majority of deposits are still made without attaching any specific conditions.

The modular organization, based on collaboration and specialization in the worldwide network of culture collections, is mainly a response to the high cost of conserving *ex situ* microbial genetic resources and to the fact that vast amounts of resources are still being collected from *in situ* sources and are being added to the existing pool. This distributed collaborative infrastructure has recently been digitally empowered, primarily by the establishment of the StrainInfo bio-portal. Initially conceived as a one-stop open-access portal for digitally linking and integrating the information content of the databases in different collections, StrainInfo has now also developed a common standard for interoperability – the microbial commons language and XML standard – which will allow automated knowledge generation based on the decentralized efforts of all of the individual data contributors (Verslyppe et al., 2010). At present, 62 collections (holding more than 300,000 strains) have joined the open-data portal, while 13 of these have moved towards using the new common XML standard and other collections have already shown an interest.[6]

The commercial pressures on life science research have, however, also led to an alternative system for exercising property rights, based on exclusive license contracts. One prominent example is the American Type Culture Collection, which distributes its holdings under a contract, thereby prohibiting the further distribution of its microorganisms by recipients until a new license has been negotiated with the collection, even if the recipient is a public service collection with high quality management requirements for the handling and distribution of microbial material. Such exclusive licensing has had a negative impact on the effectiveness of microbial research and innovation. Since there are high mutation rates for micro-organisms, cumulative scientific research is only possible when it is based on access to, and distribution of, identical micro-organisms within a network of collaborating scientists. This is especially true within taxonomy. Tracking the further distribution of identical copies is part of the overall aim of quality management and is a basic scientific procedure (OECD, 2007; WFCC, 2010). The rise of exclusive contracting poses a real threat to public microbial research.

Finally, exclusive ownership based on patents is also an important institutional modality within the field of microbial genetic resources. The impact of patents on the access to genetic resources, however, is much less than it is in other areas, such as plant genetic resources. Indeed, microbial species are characterized by high internal genetic variation between organisms in the same species complex as well as by high mutation rates upon reproduction (Staley, 2002). The patenting of one microorganism within a species complex provides exclusive rights to that one microbe, which is selected for its balanced expression of a cluster of interesting properties. However, many other organisms within the species complex, which are not covered by the patent, may have a similar set of properties, and they can often be accessed for further research and innovation through the culture collection system.

In sum, the field of microbial genetic resources is characterized by a strong and lively commons-based innovation sector, which has recently been empowered by new digital means for distributed collaborative research. Nevertheless, a large part of this sector is still informal, which poses serious threats to its long-term sustainability. Moreover, commercial pressures have also led to the adoption of exclusive-use contracts by a small number of collections, which present a major breach in the traditional sharing norms of the global microbial community. On the other hand, some collections are moving towards the adoption of formal non-exclusive licenses in order to safeguard the benefits of the pre-existing informal arrangements for the exchange of the bulk of resources, which still have unknown scientific and/or commercial potential but which are the building blocks for future scientific research and innovation.

PGRFA

The impact of the intellectual property regime on access to genetic resources is much greater in the plant genetic resources field than in the microbial or animal genetic resources field (Chen and Liao, 2004; Tvedt et al., 2007). Plants have well-defined varieties and much greater genetic stability on reproduction than either microorganisms or animals, which means that exclusive rights can be extended to direct offspring, and the results of all cross-breeding from these offspring will have sufficient genetic similarity (for protected varieties) or will contain a specific gene (for patented genes). On the one hand, the ease of transferring traits between crops makes it very hard to protect the proprietary information contained in the improved varieties or to stimulate private investment in the absence of intellectual property rights (Swanson and Goeschl, 2005). On the other hand, intellectual property rights favour the innovators who are already situated on the innovation frontier (Goeschl and Swanson, 2002a), under-represent the needs of poor countries (Benkler, 2006) and do not provide appropriate incentives for collaborative investment in the long-term informational values associated with the resource (Dedeurwaerdere et al., 2007).

These features of plant genetic resources, in conjunction with the commoditizing pressures of the last quarter of the twentieth century, have led to a conflict between the plant breeding industry and developing countries (Reichman et al., 2011). The former has succeeded in obtaining tough, patent-like *sui generis* protection of new plant varieties, without either a research exemption or the farmers' right to reuse their own seeds. Industry also continues to breed off the stocks that have been drawn – with or without permission – from developing countries. These latter countries, in turn, have asserted sovereign rights over their genetic resources through the CBD, which has jeopardized the viability of the heretofore internationally public collections. Only by a major effort undertaken under the auspices of the FAO has it become possible to rescue the public collections held by the CGIAR and entrust them to the oversight of an intergovernmental organization operating under the 2002 ITPGRFA. This treaty organization, which is affiliated with the FAO, has stabilized the CGIAR repositories and avoided the risk of chaos that threatened their survival. Its biggest success lay in the de facto creation of a legally formalized global commons for basic plant materials of primary concern for global food security and in stimulating ever-growing contributions to this basic resource. However, the implementation of the FAO treaty is still underway, and – as discussed in other chapters of this volume – many of its core features such as its Standard Material Transfer Agreement are still under scrutiny.[7]

Extensive research has been conducted on the institutional arrangements adopted within the global crop commons (Halewood and Nnadozie, 2008; Helfer, 2005). At the time of writing, the crop commons, formalized through the ITPGRFA, pools over 1.2 million accessions conserved in the collections and gene banks of contracting parties all over the world. The majority come from the 11 international collections of the CGIAR, some from other international collections, while more and more national public collections are officially joining the multilateral system of exchange as the Treaty is implemented.[8] The plant genetic resources that are within this pool are all exchanged with the viral license of the Treaty for research, breeding and education purposes. When commercial applications are developed, the Treaty offers two options: (1) commercialization with a non-exclusive-use license that permits further use for non-commercial research, breeding and education purposes or (2) commercialization with an exclusive-use license and the payment of a fixed royalty to a multilateral fund.

Non-market values and modular organization also play an important role in making the crop commons a sustainable institutional form. Strong bonds and trust among scientists from many countries involved in crop improvement programmes underlie the exchange practices and promote the sharing of information as well as the integration of regional efforts (Byerlee, 2010). The shared commitment to the common goal of increasing food production and reducing global poverty is a key driver of the whole system (*ibid*.). From the early days of the crop improvement programme, the community also invested in strengthening these social norms. The core activity has involved six months of field-based training for young scientists. In the field of wheat improvement alone, over 1,360 individuals from 90 countries have participated in these training courses, and 2,000 more have visited the International

Centre for Maize and Wheat Improvement (CIMMYT) in Mexico. In addition, participating countries are allowed to give their own names to the varieties they release, and this process produces a sense of ownership and ensures that the international seed banks are seen as honest brokers with respect to germplasm and information sharing (*ibid.*). Finally, the CGIAR has developed policy guidelines that broadly reflect these values, both before and after the entry into force of the ITPGRFA on 29 June 2004 (CGIAR, 2009; SGRP, 2003).

Experimental breeding is a clear example of a worldwide modular and distributed organization of research and innovation. One well-documented case is the international nursery network, which was organized by the CIMMYT. Every year, the CIMMYT dispatches improved germplasm to a global network of wheat research cooperators who evaluate wheat germplasm in experimental trials targeted at specific agro-ecological environments. From 1994 to 2000, the CIMMYT distributed 1.2 million samples to over 100 countries, corresponding to an average of 500 to 2,000 globally distributed field trials per year (Byerlee, 2010). Data from the field trials are returned to the CIMMYT for analysis, and the results are returned to the network of collaborating scientists. In this way, the crop commons builds an iterative collaborative platform that collects environmental and local feedback in a way that is similar to the system used by free software projects to collect bug reports (Benkler, 2006, 344).

In sum, the analysis of these institutional characteristics shows many similarities between the microbial commons and the global crop commons. In both fields, institutional arrangements have established a globally networked commons that is open to new users and contributors to the system under a standardized non-exclusive contract. Based on the quantitative data, the scope of the crop commons seems to be more limited than that of the microbial commons, which covers far more individual collections and has a larger number of holdings. Nevertheless, within the crop commons, all of the material is exchanged under a formal viral license because of the major threat of exclusion from key research resources in the form of patents. In the microbial commons, a mix of formal and informal contracts is used, depending on the circumstances and the commercial pressure on the collections.

Farm animal genetic resources

Three major institutional arrangements are in use for the management and exchange of genetic resources in animal breeding. The first was developed in the hybrid-breeding sector. Hybrid breeding is based on crosses of very different parent or grandparent lines. Since innovators do not disclose the parent and grandparent lines that are used to produce the hybrid, unauthorized reproduction of animals can be effectively prevented by keeping the information on the parent lines and grandparent lines that are crossed secret. As a result, in areas where hybrid breeding is a well-developed technological option – mainly in poultry and pigs – an exclusive-access regime has developed within a centralized and large-scale breeding industry (Pilling, 2009).

Most livestock breeding is, however, based on experimental breeding within a pool of animals that are managed in an open commons. On the one hand, when animals are exchanged between livestock keepers, the assumption normally is that the owners of the breeding animals (or other genetic material) acquired through such exchanges can use the genetic resources involved for further breeding as they wish (Pilling, 2009). On the other hand, sustainable breeding requires a high level of coordination between the breeders in the pool, both for information exchange on the parent lines and for developing common health and sanitary standards, and the breeding objectives have to be adjusted to local consumption patterns and available feed resources. Therefore, the majority of experimental breeding programmes that are run by farmer-owned cooperatives and breeder organizations operate in the context of national breeding programmes or farmer-driven societies with a regional scope (*ibid.*). Such programmes are often developed by one country or one region alone, even if there is often an important level of cross-breeding with imported animals to improve the genetic quality of the pool. As such, the institutional arrangement for commons-based management is not a globally interconnected pool as in the case of plants and microbials but, rather, is better characterized as a global network of exchange among limited (national or regional) commons.

Commons-based experimental breeding is, however, coming under increasing pressure from international companies that are taking over farmer-owned cooperative schemes, especially cattle-breeding schemes (Mäki-Tanila et al., 2008). The growth of transnational commercial breeding operations has led to the development of a third institutional regime, based on the operations of centralized commercial-breeding companies, with high expectations for quick profits and a unilateral focus on productive traits. Centralizing breeding operations may raise new challenges, such as the reported decline in the reproduction and health traits of the Holstein breed (one of the most widely used dairy cows), possibly due to a long-standing emphasis on production yield (*ibid.*, 35). Other challenges posed by the global commercial-breeding companies are the introduction of new business practices, such as patents. However, these patents mostly concern certain genetic mutations that cause genetic defects, while patents on productive traits at present only have a minor impact because of the multi-locus nature of most economically important traits (*ibid.*, 24).

Non-market values play an important role in the organization of the traditional commons-based production sector in animal breeding, which is described earlier, although these values have to be combined with the productivity constraints of the private farms that are breeding the animals (CGRFA, 2009, 7 and 20; Mäki-Tanila et al., 2008, 21). Animal breeding is part of national food security, and cooperative breeding programmes are set up to promote collective goals such as animal health and the conservation of genetic variety within populations and breeds (which is essential to meet future challenges in the development of livestock). These non-market motives have to be shared by most of the members of the collective pool to be effective, which explains why most schemes are developing common guidelines for quality management and sustainable breeding. They are also actively promoting these guidelines among individual farmers, through information campaigns and quality-assurance contracts provided by the breeders' cooperatives. Finally, in many countries, legal rules have been adopted to strengthen the general interest objectives of the breeders' organizations (FAO, 2007). For instance, under current regulations, existing breeding organizations cannot claim property rights on the basis of which they could breed the animal in question exclusively. Moreover, any new breeding organization has to be approved by the state, comply with a set of quality-management standards and undertake conservation-breeding programmes.

The modular organization of the experimental breeding programme has been developed as a solution to deal with the specific problems of animal breeding, such as the need to limit in-breeding and to maintain a sufficiently diverse breeding base for disease management. The goat improvement programme developed in France by Capgènes aptly illustrates this modular organization.[9] In this programme, a yearly selection of the 1,000 best-performing animals is made from a pool of 170,000 goats on 800 farms. From this 1,000, 40 male goats are selected after a lengthy process of quality checking and off-breeding. These males then serve as the starting point for the following year's artificial insemination programme for breed improvement. This collective breeding process is illustrated in Figure 19.2.

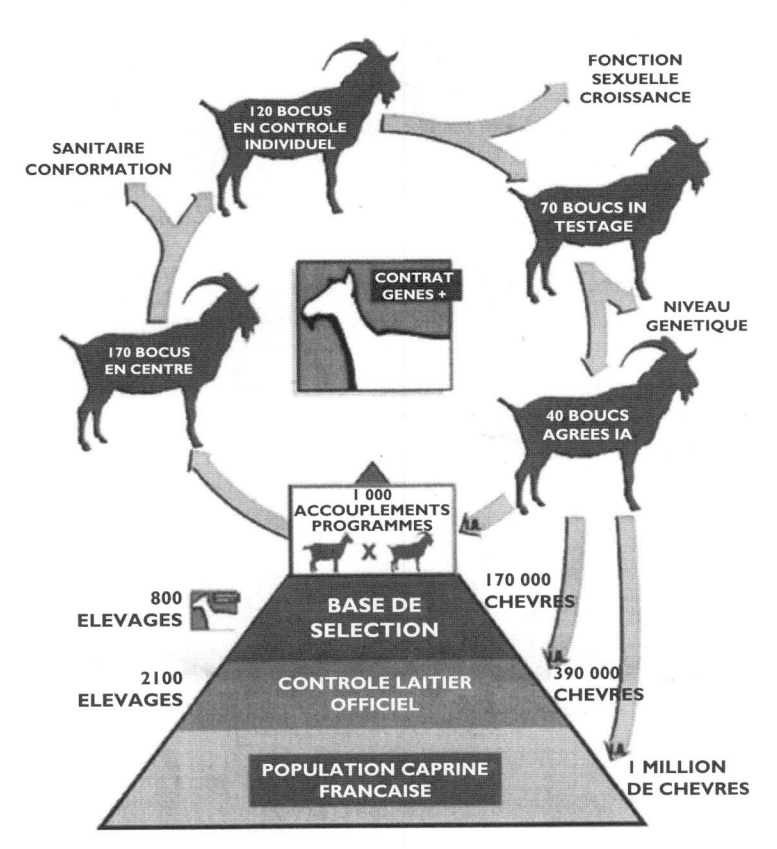

FIGURE 19.2 Example of a collective breeding process with collective ownership

As can be seen from this analysis, there are some major differences between the institutional characteristics of commons-based production with animal genetic resources and those using plant and microbial resources. The main differences involve the reliance on private actors (rather than public collections) whose resources are pooled in a collective breeding programme as well as the limited geographic scope of the commons-based improvement programmes. The geographical limitations of this sector remain the case despite the active international exchange of genetic material for the selective upgrading of domestic breeds, which creates a network of highly interrelated populations in various countries. The greatest institutional similarities are to be found between the animal and the microbial sector. In these two sectors, many exchanges still happen on an informal basis because the threat of possible misappropriation through patents or breeders' rights is relatively rare. However, the recent introduction of new business practices may cause this situation to change rapidly.

Comparative assessment of the design features

As shown in this chapter, an emerging body of research has identified new and emerging genetic resource commons, which share many features with other already well-researched fields of commons-based production such as the digital information commons and the natural resource commons. These commons share important features such as non-market motivations (which contribute to compliance with common rules without state intervention) and decentralized problem solving (in the form of modularity in the digital information commons, distributed infrastructures in the genetic resource commons and community-based governance in the natural resource commons).

The hypothesis of this chapter is that these general features are also relevant to an understanding of the practices for sharing in the global genetic resource commons. The main similarities shown by our comparative analysis of three relatively homogeneous problem situations are the reliance on non-market motivations and the adoption of modular organizational architectures for distributed collaboration. Although the balance between non-market and market motivations has shifted in the last decade due to increasing commercial pressures, it is fair to say that in all three cases, there is a mixed set of motivations for the commons-based exchange practices. Of these, the scientific research ethos, biodiversity conservation, animal health and food security, along with monetary recompense, are the most important. Moreover, in all three fields, distributed collaboration has been shown to be an effective solution for dealing with innovation processes where multiple inputs are needed to provide single outputs, which are, in turn, the inputs for further research and innovation and which need to accommodate both commercial and non-commercial uses of the outputs. We have represented a summary of these features in Table 19.1.

This analysis has also shown some important differences between the genetic resource and digital knowledge commons. One difference involves the cost of the collections. The costs of creating genetic resource collections are substantial since they involve the long-term conservation of the resources. By contrast, the physical capital required to participate in the digital information commons is mainly limited to an individual computer that is able to access the network (Benkler, 2006). In the field of microbial and plant genetic resources, part of this capacity problem is addressed since the distributed collections are already connected through global networking. Examples of such networks include the breeding programmes within the CGIAR network and the global network of culture collections coordinated by the WFCC. In the case of commons-based production in the field of animal genetic resources, the semen that is contributed to collective breeding improvement programmes is, to a certain extent, excess capacity, as local farm breeding would continue in the absence of the collective breeding programme. In this case, the analogy with the networking of the excess capacity of computing resources in peer-to-peer networks, as analysed by Y. Benkler (2006, 114), is much more straightforward.

Despite this difference, the importance of addressing non-market motivations in the design of a global genetic resource commons, once the initial investments in capacity have been made, should not be underestimated. As shown by the case studies, the sustainability of the various genetic resource commons always depends on a substantial investment in the strengthening of non-market values. Such an investment can be accomplished either through formal legal rules that modify the behavioural incentive structures for the participants or through informal means that act directly on the development of social and intrinsic motivations. An example of a formal means that applies to plant and microbial material is the use of standard viral licenses for preventing the misappropriation of the resource; an example from the animal resource field is the legislation that regulates the development of collective-breeding organizations in various jurisdictions. Examples of informal means are the guidelines for quality management schemes based on the broader social values developed by the collective animal-breeding organizations, the WFCC guidelines for the operation of culture collections and the CGIAR policy guidelines that pre-dated the ITPGRFA.

Conclusion

There has been a dramatic increase in interest in commons in the last 10 to 15 years, from traditional commons that manage the use of exhaustible natural resources by fixed numbers of people within natural borders to a global information commons that deals with non-rival, non-excludible goods by a potentially limitless number of unknown users. The emerging global genetic resource commons fits somewhere in between, shifting in the direction of information commons as digital information infrastructures allow physically distributed commons to be networked in virtual global pools.

The analysis of a selected set of cases in this chapter shows that networking common pools of genetic resources in a global commons is potentially a workable alternative to market-based solutions, which have been shown to be unable to generate sufficient investment in the vast quantities of genetic resources that are neglected because of their unknown and/or unlikely commercial value or which have been shown to under-represent the needs of poor countries. These neglected resources are the building blocks for future scientific research and have enormous value for sustaining biodiversity and local livelihoods in developing and industrialized countries. Research and breeding requires access to these multiple inputs, which can be combined into new compounds or screened to find organisms with new properties.

In the current legal environment, there are a range of obstacles that present a formidable challenge to fully realizing the new opportunities offered by global networks of genetic resources (Reichman et al., 2011). Such challenges demonstrate the need for appropriate organizational forms, legal arrangements and social practices that can help to better secure the global user community's need to address issues of common concern, such as global food security, global health, human development, biodiversity conservation and climate change. As discussed in this chapter, in response to this challenge, governments, non-profit organizations, global research communities and breeders have developed a range of initiatives for the exchange of materials and information, which have already delivered important outcomes. The key issue is how to build upon these initiatives and how to put the incipient global genetic resource commons on a solid institutional basis that will enable commons-based production to co-exist, whenever effective, with market-based and state-based contributions to collective goods.

Notes

The author gratefully acknowledges the support of the Belgian Science Policy Division of the Ministry of Science under Grant no. IUAPVI/06 DEMOGOV and the support of the sixth Framework Program in Research and Development of the European Commission under Grant no. RTD CIT3_513420 REFGOV. He also wishes to thank Jerome Reichman and Paul Uhlir for their invaluable insights in the discussion of this chapter; and Maria Iglesias, Unai Pascual, Per Stromberg and Sabine Weiland for their research assistance at many stages of this project. The editorial assistance of Valérie Hilson and the language assistance of Alisson Kelly and Stacy Belden are also thankfully acknowledged. All remaining errors are of course my own responsibility.

1 International Treaty on Plant Genetic Resources for Food and Agriculture, 29 June 2004, www.planttreaty.org/texts_en.htm (last accessed 11 May 2011).
2 Convention on Biological Diversity, 31 ILM 818 (1992).
3 Interview with Nina Chanisvili, 25 March 2010.
4 For more information on StrainInfo, see www.straininfo.net (last accessed 10 April 2011).
5 For more information on avain influenza, see the World Health Organization, www.who.int/csr/disease/avian_influenza/en/ (last accessed 10 April 2011).
6 See StrainInfo at www.straininfo.net (last accessed 10 April 2011).
7 Standard Material Transfer Agreement, 16 June 2006, ftp://ftp.fao.org/ag/agp/planttreaty/agreements/smta/SMTAe.pdf (last accessed 11 May 2011).
8 For more information on the International Treaty on Plant Genetic Resources for Food and Agriculture, see www.planttreaty.org/inclus_en.htm (last accessed 10 April 2011).
9 Capgènes, www.capgenes.com (last accessed 10 April 2011).

References

Agrawal, A. (2001) 'Common Property Institutions and Sustainable Governance of Resources', *World Development* (29): 1649–72.
Beattie, A.J., W. Barthlott, E. Elisabetsky, R. Farrel, C. Teck Kheng, I. Prance (2005) 'New Products and Industries from Biodiversity', in R. Hassan, R. Scholes and N. Ash (eds),

Ecosystems and Human Well-being: Current State and Trends, volume 1, Island Press, Washington, DC.
Beck, R. (2010) 'Farmer's Rights and Open Source Licensing', *Arizona Journal of Environmental Law and Policy* 1(2): 167–18.
Benkler, Y. (2006) *The Wealth of Networks: How Social Production Transforms Markets and Freedom*, Yale University Press, New Haven, CT.
Blakeney, M. (2001) 'Intellectual Property Aspects of Traditional Agricultural Knowledge', in P. Drahos and M. Blakeney (eds), *Intellectual Property in Biodiversity and Agriculture: Regulating the Biosphere*, Sweet and Maxwell, London.
Boyle, J. (2008) *The Public Domain: Enclosing the Commons of the Mind*, Yale University Press, New Haven, CT.
Braudel, F. (1992) [1979] *Civilization and Capitalism, 15th–18th Century*, Volume 1: *The Structure of Everyday Life*, University of California Press, Berkeley, CA.
Burhenne-Guilmin, F. (2008) 'Biodiversity and International Law: Historical Perspectives and Present Challenges: Where Do We Come from, Where Are We Going?' in M.I. Jeffery, J. Firestone and K. Bubna-Litic (eds), *Biodiversity Conservation, Law and Livelihoods. Bridging the North-South Divide*, Cambridge University Press, Cambridge.
Byerlee, D. (2010) 'Crop Improvement in the CGIAR as a Global Success Story of Open Access and International Collaboration', *International Journal of the Commons* 4(1): 452–80.
CGRFA (2009) 'The Use and Exchange of Animal Genetic Resources for Food and Agriculture', Background Study Paper of the CGRFA 43, 55 pp.
Consultative Group on International Agricultural Research (CGIAR) (2009) *CGIAR Joint Declaration*, 8 December, Washington, DC.
Chen, Y.-F., and C.-C. Liao (2004) 'Intellectual Property Rights for a Biological Resource Center as the Interface between Academia and Industry', in I. Kurtböke and J. Swings (eds), *Microbial Genetic Resources and Biodiscovery*, World Federation for Culture Collections and University of the Sunshine Coast, Queensland, Australia.
Dawyndt, P., T. Dedeurwaerdere and J. Swings (2006) 'Exploring and Exploiting Microbiological Commons: Contributions of Bioinformatics and Intellectual Property Rights in Sharing Biological Information', *International Social Science Journal* 188: 249–58.
Deci, E. (1976) 'The Hidden Costs of Rewards: The Entity from which ERIC Acquires the Content, including Journal, Organization, and Conference Names, or by Means of Online Submission from the Author', *Organizational Dynamics* 4(3): 61–72.
Dedeurwaerdere, T. (2005) 'From Bioprospecting to Reflexive Governance', *Ecological Economics* 53(4): 473–91.
Dedeurwaerdere, T., M. Iglesias, S. Weiland and M. Halewood (2009) *Use and Exchange of Microbial Genetic Resources Relevant for Food and Agriculture*, CGRFA Background Study Chapter no. 46, Commission on Genetic Resources for Food and Agriculture, Rome.
Dedeurwaerdere, T., V. Krishna and U. Pascual (2007) 'An Evolutionary Institutional Approach to the Economics of Bioprospecting', in A. Kontoleon and U. Pascual (eds), *Biodiversity Economics: Principles, Methods, and Applications*, Cambridge University Press, Cambridge.
Deek, F.P., and J.A. McHugh (2008) *Open Source: Technology and Policy*, Cambridge University Press, Cambridge.
Doyle, M., L-A. Jaykus and M. Metz (2005) *Research Opportunities in Food and Agriculture Microbiology*, Report from the American Academy of Microbiology, American Academy of Microbiology, Washington, DC.
Food and Agriculture Organization (FAO) (2007) *The State of the World's Animal Genetic Resources for Food and Agriculture – in Brief*, edited by Dafydd Pilling and Barbara Rischkowsky, FAO, Rome.
Fowler, C., M. Smale and S. Gaiji (2001) 'Unequal Exchange? Recent Transfers of Agricultural Resources and Their Implications for Developing Countries', *Development Policy Review* 19: 181–204.
Frey, B., and R. Jegen (2001) 'Motivation Crowding Theory', *Journal of Economic Surveys* 15(5): 589–611.

Genetic Resources Policy Committee (2007) Technical Issues Relating to Agricultural Microbial Genetic Resources (AMIGRs), including Their Characteristics, Utilization, Preservation and Distribution: A Draft Information Paper Prepared for the Genetic Resources Policy Committee (GRPC) of the CGIAR (based on the information document prepared by J.G. Howieson). Doc. CGRFA-11/07/Circ.3.

Goeschl, T., and T. Swanson (2002a) 'The Diffusion of Benefits from Biotechnological Developments: The Impact of Use Restrictions on the Distribution of Benefits', in T. Swanson (ed.), *The Economics of Managing Biotechnologies*, Kluwer Academic Publishers, Dordrecht.

——(2002b) 'The Social Value of Biodiversity for Research and Development', *Environmental and Resource Economics* 22: 477–504.

Halewood, M. (2010) 'Governing the Management and Use of Pooled Microbial Genetic Resources: Lessons from the Global Crop Commons', *International Journal of the Commons* 4(1): 404–36.

Halewood, M., and K. Nnadozie (2008) 'Giving Priority to the Commons: The International Treaty on Plant Genetic Resources', in G. Tansey and T. Rajotte (eds), *The Future Control of Food*, Earthscan, London.

Hammond, W.N.O., and P. Neuenschwander (1990) 'Sustained Biological Control of the Cassava Mealybug Phenacoccus Manihoti [Hom.: Pseudococcidae] by Epidinocarsis Lopezi [Hym.: Encyrtidae] in Nigeria', *BioControl* 35(4): 515–26.

Helfer, L. (2005) 'Agricultural Research and Intellectual Property Rights', in K.E. Maskus and J.H. Reichman (eds), *International Public Goods and Transfer of Technology*, Cambridge University Press, Cambridge.

Hess, C., and E. Ostrom (2007) *Understanding Knowledge as a Commons: From Theory to Practice*, MIT Press, Cambridge, MA.

Hope, J. (2008) *Biobazaar: The Open Source Revolution and Biotechnology*, Harvard University Press, Cambridge, MA.

Jinnah, S., and S. Jungcurt (2009) 'Global Biological Resources: Could Access Requirements Stifle Your Research?' *Science* 323(5913): 464–65.

Lessig, L. (2001) *The Future of Ideas: The Fate of the Commons in a Connected World*, Random House, New York.

——(2008) *Remix: Making Art and Commerce Thrive in the Hybrid Economy*, Penguin Press, New York.

Mäki-Tanila, A., M.W. Tvedt, H. Ekström and E. Fimland (2008) *Management and Exchange of Animal Genetic Resources*, Norden, Copenhagen.

Nguyen, T. (2007) 'Science Commons: Material Transfer Agreement Project', *Innovations: Technology, Governance, Globalization* 2(3): 137–43.

Organisation for Economic and Co-operative Development (OECD) (2007) *Best Practice Guidelines for Biological Resource Centres*, OECD Publications, Paris.

Ostrom, E. (1990) *Governing the Commons: The Evolution of Institutions for Collective Action*, Cambridge University Press, Cambridge.

——(2005) *Understanding Institutional Diversity*, Princeton University Press, Princeton, NJ.

Parry, B. (2004) *Trading the Genome*, Columbia University Press, New York.

Pilling, D. (2009) *The Use and Exchange of Animal Genetic Resources for Food and Agriculture*, Background Study Chapter no. 43, Commission on Genetic Resources for Food and Agriculture, Rome.

Rai, A.K. (1999) 'Regulating Scientific Research: Intellectual Property Rights and the Norms of Science', *Northwest University Law Review* 94: 77–152.

Reichman, J.H., T. Dedeurwaerdere and P.F. Uhlir (2011) *Designing the Microbial Research Commons: Strategies for Accessing, Managing and Using Essential Public Knowledge Assets* [under review; manuscript on file with the authors].

Roa-Rodríguez, C., and T. van Dooren (2008) 'Shifting Common Spaces of Plant Genetic Resources in the International Regulation of Property', *Journal of World Intellectual Property* 11(3): 176–202.

System-wide Genetic Resources Programme (SGRP) (2003) *Policy Instruments, Guidelines and Statements on Genetic Resources, Biotechnology and Intellectual Property Rights*, version 2, produced by the SGRP with the CGIAR Genetic Resources Policy Committee, Rome.

Simpson, R.D., R.A. Sedjo and J.W. Reid (1996) 'Valuing Biodiversity for Use in Pharmaceutical Research', *Journal of Political Economy* 104(1): 163–85.

Smith, D. (2007) 'The Implementation of OECD Best Practice in WFCC Member Culture Collections', in E. Stackebrandt, M. Wozniczka, V. Weihs and J. Sikorski (eds), *Connections between Collections*, World Federation for Culture Collections and German Resource Centre for Biological Material, Goslar.

Smith, J., J. Waage, J. Woodhall, S. Bishop and N. Spence (2008) 'The Challenge of Providing Plant Pest Diagnostic Services for Africa', *European Journal of Plant Pathology* 121(3): 365–75.

Staley, A. (2002) 'A Microbial Perspective of Biodiversity', in J. Staley and A.-L. Reysenbach (eds), *Biodiversity of Microbial Life*, Wiley-Liss, New York.

Stern, S. (2004) *Biological Resource Centers*, Brookings Institution Press, Washington, DC.

Stromberg, P., U. Pascual and T. Dedeurwaerdere (2006) *Information Sharing among Culture Collections*, survey report [unpublished; on file with the authors].

Stromberg, P., T. Dedeurwaerdere and U. Pascual (2007) 'An Empirical Analysis of Ex-Situ Conservation of Microbial Diversity', Presented at the Ninth International BIOECON Conference on Economics and Institutions for Biodiversity Conservation, King's College, Cambridge, 19–21 September 2007, Biodiversity and Economics for Conservation, online: www.bioecon.ucl.ac.uk/ (last accessed 11 May 2011).

Sulston, J., and G. Ferry (2003) *The Common Thread: Science, Politics, Ethics and the Human Genome*, Corgi Books, London.

Swanson, T., and T. Goeschl (1998) *The Management of Genetic Resources for Agriculture: Ecology and Information, Externalities and Policies*, Working Chapter no. GEC 98–12, Centre for Social and Economic Research on the Global Environment, Norwich.

——(2005) 'Diffusion and Distribution: The Impacts on Poor Countries of Technological Enforcement within the Biotechnology Sector', in K.E. Maskus and J.I. Reichman (eds), *International Public Goods and Transfer of Technology*, Cambridge University Press, Cambridge.

Tvedt, M.W., S.J. Hiemstra, A.G. Drucker, N. Louwaars, and K. Oldenbroek (2007) *Legal Aspects of Exchange, Use and Conservation of Farm Animal Genetic Resources*, Institute Report no. 1/2007, Fridtjof Nansen Institute, Lysaker, Norway.

Verslyppe, B., R. Kottman, W. De Smet, B. De Baets, P. De Vos and P. Dawyndt (2010) 'Microbiological Common Language (MCL): a standard for electronic information exchange in the Microbial Commons', Research in Microbiology 161(6): 439–445.

World Federation for Culture Collections (WFCC) (2010) *Guidelines for the Establishment and Operation of Collections of Cultures of Micro-organisms*, 3rd edition, WFCC Executive Board, Shizuoka.

INDEX

ABS law 129n, 199, 202–3
Africa: climate change and adaptive planting 83; nodal centres of excellence 334–35; plant genetic resources networks 237–38, 332–35, 337–38, 343n
African Group: Annex 1 crops, inclusion of 228, 237; PGRFA provision and recompense 166, 273
Africa Rice Center (WARDA) **122**
Agreement on Trade-Related Aspects of Intellectual Property Rights 5, 142, 254, 293, 307n
Annex 1 crops: access terms and conditions 273; AEGIS availability 333–34; deposit terms under Article 11.2 179–80, 203–4; IARCs holdings incorporated 275–76; ITPGRFA availability 206–7, *207*, 253; list compilation, attributing factors 268–70, 272–73; list compilation process 266–67, 276–79; list expansion, request for 236–37, 273; regional agendas 279–80; selection criteria 143, 274
Arab Organization for Agricultural Development (AOAD) 337
Association of Agricultural Research Institutes in the Near East and North Africa (AARINENA) 332–33
Association of Southeast Asia Nations (ASEAN) 64

benefit sharing fund: balancing member aims 213, 234–36, **349**; developing countries, priorities of 229, 233; operation and priorities 156–57, 207–8, 219, 221n; priority projects 336; voluntary donations 318
biofuel production 93
biopiracy 74, 214, 380
Bioversity International: Chinese accessions 71; collection gaps 118–19; EURISCO 297; gene bank acquisitions 103; GENESYS database 190, 299–300, 340, 360
Brazil 71, 212
breeders: dissatisfaction with SMTA 216–17; intellectual property rights 246, 247; interests' protection 268, 349, 382; support from GIPB 336

Canada 206
capacity building: global information, access to 287; Governing Body priority 157, 158, 219; Joint Capacity Building Programme 193–94, 200, 204, 238, 337; national incentives 212; plant breeding programme 336; regional initiatives 338
Centre for Agricultural Research for Development 45
Centre for Pacific Crops and Trees (CePaCT) 332, 335, 338–39
CGIAR centres: access and supply of PGRFA, study methodology 101–7, **102–3**, *103*, **115–16**, 124–26; accession criteria 103, 104, 107; accession holdings *205*; access policy 106–7; acquisition data (1980–2004) **102**, *103*, 105; acquisition data (2005–10) **114**;

Index **393**

acquisition obstacles, political and legal 107–10; acquisition priorities 118–20, 330; breeding programme 104; decline in deposit, cited reasons 100, 105; in-trust agreements with FAO 106–7, 110; legal uncertainty over supply 110–11, 120, 129n; sample distribution data (1980–2004) **102**, 105; *in situ* materials, collecting missions 114, **115–16**; SMTA and lifted restrictions 112–13, **127–28**; *see also* Consultative Group on International Agricultural Research (CGIAR); International Agricultural Research Centres (IARCs); CGN 209–10

China: agricultural technology, adoption of 62–63, **63**; crop genetic resources, collection development 63–66, 74–75; exchange of genetic resources 71–73, **72**, 75; food security 66, 73, 212; heavy dependency on IARC's 64, 66, 73–74; PGRFA, US major supplier 64, 66, 68, 69, 73; potato variety development 69–70; rice variety development 66–67; soybean breeding programme 69; sweet potato breeding programme 70–71; wheat variety development 67–68

Chinese Academy of Agricultural Sciences: exchange agreement 64, 71, 74; gene bank management 65–66

climate change: crop gene pools, protective measures 82–83, 240; crop interdependence, analysis of impact 84–90, **87**, **88**, **89**; crop interdependence, expected changes 90–93, 280; forages, adaptive planting 83; predicted impact on agriculture 81, 347

Cocoa Research Unit 51

Coconut Genetic Resources Network (COGENT) 179

'Columbian exchange' 3

commercialized products: ITPGRFA fund contribution 23, 33n, 207–8; monetary benefits, dissatisfaction with 24, 25–26; SMTA and benefit sharing 144, 156, 158, 158, 254–55

Commission on Genetic Resources for Food and Agriculture (CGRFA): access and benefit-sharing discussion 8; interim administrator for ITPGRFA 144; objectives of **350**; SMTA formulation 144–45; sovereign control exposed 33n; Working Group on Annex 1 crops 266–67, 273–74, 276

Commission on Plant Genetics Resources (CPGR): catalyst for PGRFA negotiations 13, 15, 110; *ex situ* collections, ownership agreement 247–48; forum for debate 138, 142, 245–46, 249; *see also* Commission on Genetic Resources for Food and Agriculture (CGRFA)

commons, definition of 9

community seed banks 2–3

Conference of the Parties to the CBD: access and benefit sharing negotiated 22–23, 161n; International Undertaking on PGRFA 13; Nagoya Protocol adopted 7, **348**, 363; public good research 295

Consultative Group on International Agricultural Research (CGIAR): China's use of 64; collection access, undermining Treaty rules 12–14, 19; FAO-GCIAR in-trust agreements 106–7, 108, 110, 253, 257; food security in China 66; germplasm sample distribution 40; governance features 4, 139, 369, 383; research partnerships and facilitation 321; SINGER database 64, 238–39, 297, 340; *see also* CGIAR centres; International Agricultural Research Centres (IARCs)

Contact Group: Annex 1 crops, access to 268, 274; third party beneficiary, discussion on 165–67, 173–74n

contracting parties: benefit sharing fund 156–57, 158; brokering role in implementation 313–14; funding strategy, management issues 235–36, 240–41, 259–60; Global Plan of Action on PGRFA 139; interaction with Secretariat 191–93; legal obligations 152, 154, 161n, 167, 232; limited contribution to SMTA 230; management and control of holdings 180–82, 256, 259, 311–13, 325n, 357–59; mandatory benefit sharing 270–72; non-compliance procedure 18, 159, 178, 233; PGRFA sources supplied 114, 159, 179–82; recruitment obligation 17, 153, 178, 204; role of GCDT 232; sharing of PGRFA information 294, 295, 306n; *in situ* sample provision 140, 180; third party beneficiary, role in 171, 172; treaty membership 11, 150

contract law: principles of 164; third party beneficiary rights 168

Convention on Biological Diversity (CBD): access and benefit-sharing negotiated 5, 22–23, 151, 161n, 244, 248–50; Agricultural Biodiversity programmes **348**, 362; approval of 141; farmers' rights 256; Nagoya Protocol 7, 152, **348**, 363; national sovereignty and access 249, 250–51, 267–68, 307–8n, 382; privatization of conservation 254–55, 258; sovereign rights, implication of 5, 54, 64, 110, 257
copyright 292, 301, 307n
crop resources: banana 53–56; cacao diversity 52–53; chocho 46–48; maize 41–44, 68, peanut 48–51; potato 69–70; rice 44–46, 66–67; soybeans 69, 71, 371, 377; sweet potato 66, 70–71; wheat 68
Crucible Group 141

Dalrymple, Dana 319
developed countries: Annex 1 crops, inclusion of 228; biotechnology, protected access 5–6; breeders, interests of 268, 349, 382; non-compliance issues 233
developing countries: facilitation through ITPGRFA 324, 340–41; farmers' role and needs unrecognized 51–53, 215–16; intellectual property rights 16, 32–33n, 228, 244, 247, 272; main recipient of germplasm 40, 92, 341; PGRFA provision and recompense 166, 172, 175n, 229, 232, 234–36, 273; SMTA training and implementation 193; widening the network 213
disease resistance, contribution to 40–41, 52–53

East African Plant genetic Resources Network 237, 238
Ecuador 46–48
EURISCO 72, 205, 297, 360, 377
European Commission 294
European Genebank Integrated System (AEGIS) 210, 238, 333–34, 339, 358
ex situ collections: access under ITPGRFA 180, 253, 319, 330–31; CGIAR endorsement 4; government owned 11, 349; holdings data inaccurate 12; ownership agreement 247–48; regional cooperation 71, 332–35; support of GCDT 232, 331–32; widening the network 54–55

FAO-CGIAR in-trust agreements: creation of 106–7; dispute resolution 257; dissatisfaction with 108, 110; influence on ITPGRFA 253
farmers' rights: commercial pressures 53–54; International Undertaking on PGRFA 139, 246, 247, 256; ITPGRFA inclusion 144, 229–30, 256; Nairobi Final Act, Resolution 3 141–42; *in situ* conservation 286, 306n; traditional knowledge, protection of 293–94, 295–96, 307–8n; training needs 215
Flores Palacios, Ximena 39
Food and Agriculture Organization (FAO): Commission on Genetic Resources for Food and Agriculture 8, 138, 142–43; information agencies **349–50**, 361; international focal point 345; International Undertaking, proposal, debate and approval 136–38, 244–48; International Undertaking, revision of 142; International Undertaking adopted 12; in-trust agreements with CGIAR centres 106–7, 110, 139; safeguarding diversity, early measures 3–4, 136; *Second Report on the State of the World's PGRFA* 117, 213, 229, 345, 357–58; *State of the World's PGRFA* 139, 247, 250, 259, 274; third party beneficiary 158, 163n
food production: adaption through PGFRA exchange 212, 324; Chinese Government priority 62–63, **63**; climate change, impact of 81; future challenges 78
food quality: cultivation and promotion 47; trait transference 43–44
food security: CGIAR, promotional role 13–14; China 66, 73; conditional germplasm distribution 237; diversity through exchange 56, 74, 79–80, 212; international debate 274; national responsibilities 323–24
forages: climate change impacts 83; pasture improvement 79–80
Fundación Hondureña de Investigación Agrícola (FHIA) 54–55
fungal diseases 54, 377

GENESYS database 190, 298–300, 340, **350**, 360
Genetic Resources Network for West and Central Africa (GRENEWECA) 237, 238

Index   395

Global Biodiversity Information Facility (GBIF) 297
Global Crop Diversity Trust (GCDT): climate change and adaptive breeding 337; contribution to ITPGRFA aims 120–21, 144, 230, 342n; funding strategy role 231–32, 251, 254, 358; GENESYS database 190, 299–300, 340, 360; information distribution 188–89; objectives of **350**; sample regeneration deposit scheme 115–16, 213, 357; wild species collection 119
global genetic resource commons: access terms and conditions 373; comparative analysis of institutions 375, **376**, 377, 386–87; digital networks 372; farm animal resources 383–86, *385*; microbial resources 377–81, *378*; non-market incentives 373–74, 379–80, 384–85; organizational structure 374–75, *385*, 385; survival threats addressed 371
global information system: cooperative projects 339–40; 'informational commons', advantages of 306–7; 'informational commons', operations of 296–302; multilateral system, role of 284–85; non-exclusive right and accessibility 292–93; PGRFA holder's rights 292, 307n; publicly funded information 294–95; standardized operations 287–88; technical challenges 289–91; traditional knowledge and farmer consent 293–94; universal tool for agricultural biodiversity 285–87; users and beneficiaries 288, **289**
Globally Important Agricultural Heritage System Initiative (GIAHS) 82–83, 358
Global Partnership Initiative Plant Breeding Capacity (GIPB) 336, **351**
Global Plan of Action on PGRFA 139, 236, 247, 331, 345, **352**, 364
global public goods: CGIAR interpretation 13, 315; definition of 11; impact of property rights 319; ITPGRFA key issue 250
Global Seed Vault (Svalbard) 239, 254, 332, **351**, 358
Global System for the Conservation and Sustainable Use of PGRFA: activity coordination 348–49, 366n; availability 359–60; conceptual framework **350–55**, 365; efficiency of conservation 357–59; environmental and human benefits 347–48, *348*; global functioning 356,
356–57, 362–63, 365; International Undertaking, basis for 244, 249–50; knowledge management 348, 361; limitations 363–65; operational elements 345, 349–57; operational objectives and user priorities 346–47; use 360–61
Governing Body: Ad Hoc Advisory Committee on Funding Strategy 158, 232–33; Ad Hoc Advisory Committee on SMTA 157–58, 192–93, 204; Ad Hoc Advisory Committee on the Third Party Beneficiary 171, 190; agreements with international organizations 231–32; agreement with CGIAR centres 230; balancing provision and benefit 234–36, 270–72; benefit sharing fund projects 336; Global Crop Diversity Trust 331–32; implementation delays 232–34, 322; information systems 238–39, 295; promotion of Treaty 239–40; regional cooperative initiatives 338–39; Treaty implementation, role in 144, 229–31
GRIN database 297–98
Group of 77 137

'informational commons': open access policy 301–2; operations of 296–300; voluntary deposit and shared access 300–301, 308–9n
innovation systems 320, 326n, 379–81
in situ conservation: access under ITPGRFA 180, 206; data transfer to ITPGRFA, lack of 117; financing of 251; global information, access to 286, 306n, 358; international networks **351**; wild species 82–83, 99–100, 269
Instituto Nacional Autónomo de Investigaciones Agropecuarias (INIAP) 46–48
intellectual property rights: biological and genetic resources 5–6; breeders and farmers rights 246, 247; clarification of material 272; developing countries' stance 228, 244; information, access and use 292, 303, 319, 369; ITPGRFA, voluntary inclusion 153; public domain 182–83, 251–52
International Agricultural Research Centres (IARCs): agreement under ITPGRFA 144, 178, 206, 253–54, 330–31, 358; Annex 1 crops list, influence on 275–76; China, germplasm recipient 64, 66, 73–74; developing countries, main recipients 40, 209, 341; funding by

GCDT 332; information distribution 189–90, 209; in-trust agreements with FAO 139, 248, 253, 274–75, 294; ITPGRFA intermediaries 312, 313, 314–16; research partnerships and facilitation 320–22, 324–25, 368; role of 99, 349, 360; Standard Material Transfer Agreement, adoption of 140, 231, 248, 331; supply to non-ITPGRFA members 210–11; *see also* CGIAR centres

International Board on Plant Genetic Resources (IBPGR): creation of 136, 244; criticism of 12; governance features 5; *see also* Bioversity International; International Plant Genetic Resources Institute (IPGRI)

International Center for Agricultural Research In Dry Areas (ICARDA) 103, **123–24**

International Center for Tropical Agriculture (CIAT): breeding programmes 40, 45; collaborative research 371; gene bank acquisitions 103, **121**; inter gene bank deposits 106

International Centre for Maize and Wheat Improvement: *see* International Maize and Wheat Improvement Center (CIMMYT)

International Cocoa Gene Bank *205*, 210

International Coconut Gene banks *205*

International Crop Research Institute for the Semi-Arid Tropics (ICRISAT): China, germplasm recipient 66; gene bank acquisitions **122–23**; inter gene bank deposits 106; peanut breeding 50; peanut germplasm collection 49

International Institute for the Unification of Private Law (UNIDROIT) 168, 174n

International Institute of Tropical Agriculture (IITA): China, germplasm recipient 65–66; gene bank acquisitions **122–23**; germplasm development programmes 42–43, 335

International Keystone Dialogue Series on Plant Genetic Resources 140–41

International Livestock Research Institute (ILRI): gene bank acquisitions 103, **122**; inter gene bank deposits 106; soybean accessions 71

International Maize and Wheat Improvement Center (CIMMYT): China, germplasm recipient 65–66, 68–69; collaborative research 383; gene bank acquisitions **121**; inter gene bank deposits 106, 345; Tuxpeño maize, shared accessions 42–43; Ug99 resistance 41

International Network for Genetic Evaluation of Rice (INGER): Chinese variety development 66–67; conditional distribution 108; future breeding challenges 45; SMTA and lifted restrictions 115, 210

International Network for the Improvement of Bananas and Plantains (INIBAP) 54–55

International Plant Genetic Resources Institute (IPGRI): gene bank acquisitions **122–23**; role in ITPGRFA 143, 248; *see also* Bioversity International; International Board on Plant Genetic Resources (IBPGR)

International Potato Center (CIP): China, germplasm recipient 66, 69–71; CIP-China Center for Asia and the Pacific (CCCAP) 69; gene bank acquisitions **121**; 'in trust' accessions 112; potato varieties parentage **70**

International Rice Research Institute (IRRI): acquisition priorities 118, 130n; China, germplasm recipient 64, 65, 67; gene bank acquisitions 103, **123–24**; germplasm development programmes 44–45; SMTA and lifted restrictions *113*, 115, 210; supply to non-ITPGRFA members 210–11

International Transit Centre (ITC) 55

International Treaty on Plant Genetic Resources for Food and Agriculture (ITPGRFA): access and benefit sharing 6–7, 145; adoption of 143–44, 368–69, 382; agreements leading to 136–42; Annex 1 crops, selection criteria 143; Annex 1 crops, selection process 276–79; benefit sharing fund 156–57, 207–8, 318; CGIAR centres, access abuse 12–14; CGIAR centres, access agreement 107, 140, 154; Chinese membership, benefits of 74; compliance/enforcement measures 21–22, 33n, 171–72, 302; crops omitted 50, 51; farmers' role and needs unrecognized 215–16; financial benefit sharing, failure of 23–25, 33n, 208, 217; financial benefit sharing, possible reforms 25–26; global challenges 322–25; member access only, viability of 19–21; multilateral system, acceptance of 120–21, 213;

national focal points and implementation 201, 214, 218, 220, 324; national intersectional consultation 202–3; national public organizations 11, 114–15, 153; negotiations, progress of 227–28; non-Annex 1 crops, necessary inclusion 94, 118; non-member exploitation 13–14; objectives of **350**, 361–62; operational challenges 363; private institutions, limited deposit 206; procedural and structural delays 111–12; provision and reciprocity, abuse of 8–9, 16–18, 214, 318; public domain, interpretation of 183–84, 204, 250–51; regional initiatives 338–39; reluctant deposit of diversity data 14, 55, 117, 217; resource networks, support of 237–38, 332–39; selective availability 16; training and promotion 193–94, 202, 382–83; voluntary deposit 153–54; weak negotiators 255; *see also* multilateral system

International Undertaking on PGRFA: benefit sharing, unrealistic system 259; proposal, debate and approval 138, 244–48; revision of 142; sovereign rights 15, 33n, 139; universal availability and benefit 12, 22, 139, 252

International Union for the Protection of New Varieties of Plants (UPOV) 244

Kenya 202

Leipzig Declaration, adoption of 139, 247

material transfer agreement (MTA) 107
memorandum of understanding, use of 332–33
Millennium Seed Bank 119
Morocco: localized seed exchange 216; treaty implementation issues 201
multilateral system: access terms and conditions 154–55, 167, 180, 216–17, 253, 261–62; agreements leading to 244–50; assessment of functionality, methodology of 200–201; benefit sharing, alternative approaches 259–61; benefit sharing rules and fund operation 155–57, 216, 323; contractual framework, reasons behind 257–58, 262; documentation, importance of 159, 163n, 207, 211, 236; educating decision makers 313–14; government and public domain PGRFA material 153, 259;

IARCs access agreement 107, 154, 231, 248, 252, 253–54; incentives for adoption 211–13; international legal support 10–11, 152–53, 213; maintenance by technical committees 157–58; member cooperation questioned 17–18; membership and shared rights 16, 151–52, 290–91, 310–13, 317–19, 359; non-food use, alternative approach 260–61; non-food use prohibited 154, 163n, 269; notification letter of deposit 188–89, 195–96, 205, 207; operational challenges 159–61; PGRFA, inclusion debate 268–69; PGRFA activities, funding of 158, 208, 251, 317, 321–22, 362; PGRFA distribution 341; PGRFA locations 205–7, *206*; private institutions, limited deposit 206; provision by contracting parties 179–82, 204, 230; SMTA and benefit sharing 158–59, 216–17, 252, 254–55; sources of ITPGRFA 178–79; third party beneficiary 167–68, 256–57; virtual pooling 6, 10, 31n, 150–51, 284, 360
Mutant Germplasm Repository 179, *205*

Nagoya Protocol 7, 152, 293, 307–8n, 363
Nairobi Final Act, Resolution 3 141, 244, 249
National Genetic Resources Services (US) 14
National Institute for Agricultural Research, Peru 201
National Programme for Andean Legumes and Grains 47
Netherlands 210
nodal centres of excellence (Africa) 334–35
non-Annex 1 crops: AEGIS availability 334; IARC and SMTA usage 107, 140, 154, 210, 253–54, 331; ITPGRFA inclusion vital 94, 118; pre-Treaty material 331
non-government organizations **352**, 364
novel climates 91, 92

on-farm conservation: direct funding, lack of 216; global information system, role of 286, 287, **289**; Governing Body priority 157, 158, 336; innovation systems 320; long-term sustainability 340–41
Organisation for Economic Co-operation and Development (OCED) 294–95

Ostrom, Elinor 9, 18
Ouadadougou Declaration 334, 335, 343n

patent pools 6, 31–32n
patent rights 292, 381
Peru 201, 212
pests: larger grain borer 43; mealybug 371; nematode 49
Philippines 201–2
plant genetic resources for food and agriculture (PGRFA): collective pooling, history of 3–5; cooperative agreements 64; crop interdependence and climate similarity, analysis of 84–90, **87**, *88*, *89*; crop interdependence and climatic trends 90–93; diversity and food security 56, 79; diversity and human usage 10, 371; future demand and facilitation 93–94, 347–49, 369; innovation systems 320, 321; property rights 181; regional cooperation 332–35; sample supply exceeds deposit 55; Standard Material Transfer Agreement, use of 14, 19–20, 23–24, 26; training programmes 239–40
plant genetic resources networks 237–38
plant induction stations 4
public domain: interpretation of 182–84, 204; publicly funded information 294–95

regional cooperation: Europe 333–34; Near East and North Africa 332–33, 337; objectives of **348–49**; Pacific Islands 335, 338–39, 343n; Southern Africa 332, 338; West and Central Africa 334–35
Ryan, Jim 320–21

SADC Plant Genetic Resources Centre (SPGRC) 332, 338, 339
Secretariat of ITPGRFA: formation of 230; GENESYS database 190, 299–300, 340, 360; policy guidance and development 191–93, 194, 219; role of 187–88; SMTA management 190–91, 194–95; SMTA training and promotion 193–94
Secretariat of the Pacific Community 335, 338–39
Seed networks 2, 216
Southern Africa Development Community Plant Genetic Resources Programme (SPGRC) 237, 238
sovereign rights: Chinese policy 74–75; conflicting approaches to 249–50; exercise of 33n, 55, 182, 312, 381–82; FAO recognition 15; implication of CBD 5, 54, 64, 110, 257; International Undertaking on PGRFA 139; national policy, CBD's influence 249, 250–51, 267–68
Standard Material Transfer Agreement (SMTA): adoption by CGIAR centres 66, 230–31; analysis of use 208–11; CGIAR centres, use of (2007–10) **115–16**; commercialized products 144, 156, 158, 158; compliance monitoring 168–69, 174–75n; dispute resolution 170, 171; FAO as third party beneficiary 158, 163n, 170–72; formulation by Expert Group 144–45, 165; ITPGRFA approval 144–45; ITPGRFA benefit sharing fund 156–57; legal instrument of supply 152, 154–55; non-food use prohibited 154, 163n; reporting obligations 158, 163n, 284, 290, 303–6n; Secretariat, operational management 190–93; third party beneficiary, concept discussions 165–67; third party beneficiary, operational role 168–70, 218; training and promotion 193–94
StrainInfo 297, 380
System-Wide Information Network for Genetic Resources (SINGER) 64, 71, 238–39, 297, 340, 360

Tesfaye, Bashah 320
third party beneficiary: FAO appointed 158, 163n, 170–72, 233–34; formulation within SMTA 165–67, 173–74n; rights in national law 168; SMTA, operational role 168–70, 191, 257
Tropical Agricultural Research and Higher Education Centre (CATIE): accession holdings *205*; agreement under ITPGRFA 144, 178; breeding programmes 349; cacao breeding programme 51–53; non-Annex 1 crops and SMTA 210; safeguarding diversity, early measures 136

UNESCO: Man and the Biosphere Programme 358–59
University of Reading 52
US Department of Agriculture: cooperative agreement with China 64, 68; Germplasm Resources Information Network (GRIN) 297–98, 340, 360; maize regeneration 106; National Plant Germplasm System 65; peanut germplasm collection 49; sample distribution 14–15

Vavilov, Nicolay 3

Wen, Jiabao 62
West and Central African Council for Agricultural Research and Development 334–35, 343n
wild species: climate change and adaption 337; disease resistance, contribution to 80; genetic erosion 82; habitat destruction threat 50–51; national collections 342n; *in situ* conservation 82–83, 92, 99–100; valuable gene pools 119
World Agroforestry Centre (ICRAF) **122**
World Federation for Culture Collections (WFCC) 378, 380
World Végétable Center (AVRDC) 69, 349